THE EXTRACELLULAR MATRIX *FactsBook*

Second Edition

Other books in the FactsBook Series:

A. Neil Barclay, Albertus D. Beyers, Marian L. Birkeland, Marion H. Brown, Simon J. Davis, Chamorro Somoza and Alan F. Williams
The Leucocyte Antigen FactsBook, 1st edn

Robin Callard and Andy Gearing
The Cytokine FactsBook

Steve Watson and Steve Arkinstall
The G-Protein Linked Receptor FactsBook

Rod Pigott and Christine Power
The Adhesion Molecule FactsBook

Grahame Hardie and Steven Hanks
The Protein Kinase FactsBook
The Protein Kinase FactsBook CD-Rom

Edward C. Conley
The Ion Channel FactsBook
I: Extracellular Ligand-Gated Channels

Edward C. Conley
The Ion Channel FactsBook
II: Intracellular Ligand-Gated Channels

Kris Vaddi, Margaret Keller and Robert Newton
The Chemokine FactsBook

Marion E. Reid and Christine Lomas-Francis
The Blood Group Antigen FactsBook

A. Neil Barclay, Marion H. Brown, S.K. Alex Law, Andrew J. McKnight, Michael G. Tomlinson and P. Anton van der Merwe
The Leucocyte Antigen FactsBook, 2nd edn

Jeff Griffiths and Clare Sansom
The Transporter Factsbook

Robin Hesketh
The Oncogene and Tumour Suppressor Gene FactsBook, 2nd edn

THE EXTRACELLULAR MATRIX *FactsBook*

Second Edition

Shirley Ayad
Ray Boot-Handford
Martin J. Humphries
Karl E. Kadler
Adrian Shuttleworth
School of Biological Sciences,
University of Manchester, Manchester, UK

Academic Press
Harcourt Brace & Company, Publishers
SAN DIEGO LONDON BOSTON NEW YORK
SYDNEY TOKYO TORONTO

This book is printed on acid-free paper

First edition published 1994

Academic Press
525 B Street, Suite 1900, San Diego, California 92101-4495, USA
http://www.apnet.com

Academic Press Limited
24–28 Oval Road, London NW1 7DX, UK
http://www.hbuk.co.uk/ap/

ISBN 0-12-068911-1

A catalogue record for this book is available from the British Library

Typeset in Great Britain by Alden, Oxford, Didcot and Northampton
Printed in Great Britain by WBC, Bridgend, Mid Glamorgan

98 99 00 01 02 03 WB 9 8 7 6 5 4 3 2 1

Contents

Section I INTRODUCTORY MATERIAL

Section II THE EXTRACELLULAR MATRIX PROTEINS

This book is dedicated to David S. Jackson and Michael E. Grant for establishing and fostering matrix research in Manchester.

Preface

The authors would like to thank those colleagues who provided sequence and other information that was invaluable in the preparation of this book. These included: Scott Argraves, Mon-Li Chu, David Eyre, Tim Hardingham, Richard Mayne, Taini Pihlajaniemi, Francesco Ramirez, Jaro Sodek, Michel van der Rest, Marvin Tanzer and Arthur Veis. The authors would also like to acknowledge the funding of their own work by The Wellcome Trust, The Medical Research Council, and The Arthritis and Rheumatism Council. The authors would be grateful if they could be notified of any errors or omissions; the nature of the book means that these are inevitable but they can be corrected in future editions. Please send such correspondence to the Editor, Extracellular Matrix FactsBook, Academic Press, 24–28 Oval Road, London, NW1 7DX.

From top left: Martin Humphries, Adrian Shuttleworth, Ray Boot-Handford; bottom left: Karl Kadler, Shirley Ayad.

Abbreviations

BEHAB	Brain enriched hyaluronan binding protein
BMP-1	Bone morphogenetic protein-1
CCP	Complement control protein
COL	Collagenous
COMP	Cartilage oligomeric matrix protein
COOH	Carboxyl
CS	Chondroitin sulphate
CUB	Three regions of internal sequence homology found in the complement components C1r/C1s, the sea urchin Uegf, and BMP-1
DEAE	Diethylaminoethyl
DS	Dermatan sulphate
EGF	Epidermal growth factor
EM	Electron microscopy
FACIT	Fibril-associated collagen with interrupted triple helix
FN	Fibronectin
FPLC	Fast protein liquid chromatography
GAG	Glycosaminoglycan
HA	Hyaluronan/hyaluronic acid
HS	Heparan sulphate
(k)Da	(Kilo)Dalton
KS	Keratan sulphate
LDL	Low density lipoprotein
MMP	Matrix metalloproteinase
NC	Non-collagenous
NH_2	Amino
N-linked	Asparagine-linked
NMR	Nuclear magnetic resonance
O-linked	Hydroxyl-linked
PARP	Proline- and arginine-rich protein
PG	Proteoglycan
PRELP	Proline arginine-rich end leucine-rich repeat protein
SDS–PAGE	Polyacrylamide gel electrophoresis in sodium dodecyl sulphate
SLS	Segment-long spacing
TGF	Transforming growth factor
TIMP	Tissue inhibitor of metalloproteinases

Section I

INTRODUCTORY MATERIAL

Introduction

The purpose of this book is to collate and present key facts relating to the diverse group of macromolecules that assemble to form the extracellular matrix. An in-depth introduction to the structure and function of extracellular matrix is not within the remit of this book, but several excellent reviews and monographs have been published recently on this topic and these are listed below. However, it is necessary to define the authors' perception of the extracellular matrix and our criteria for the inclusion and exclusion of various classes of macromolecules within this book.

The extracellular matrix is the substance which underlies all epithelia and endothelia, and surrounds all connective tissue cells providing mechanical support and physical strength to tissues, organs and the organism as a whole. In simple, multicellular animals such as the hydra, the physical support for the whole organism may consist solely of a basement membrane. In more complex animals such as man, different types of extracellular matrix have evolved in connective tissues such as skin, bone, tendon, ligament and cartilage in addition to the ubiquitous basement membrane. The extracellular matrix should not be viewed as merely providing strength and physical support for tissues and organisms. It is now quite clear that this matrix exerts profound influences on both the behaviour (e.g. adherence, spreading and migration) and the pattern of gene expression of the cells in contact with it and, where identified, peptide sequences responsible for these effects are listed.

In deciding which groups of molecules should be either included or excluded from the first edition of this book, we took the view that to be included, a molecule must perform a structural role within the matrix and must be secreted, in its entirety, into the extracellular compartment (the exception to this is the inclusion of a number of collagens that may be membrane associated). Accordingly, the following groups of molecules were excluded: (1) membrane intercalated proteoglycans, since, although the extracellular domain of the molecule may well be performing a structural role within the extracellular matrix, a key function of this class of molecules may be to act as signal transducers across the plasma membrane and thus be more appropriately included in a volume dedicated to cell surface receptors; and (2) extracellular molecules that become either intimately associated with the matrix or simply trapped but whose function is clearly not structural. Such molecules include growth factors and plasma proteins like albumin. In the second edition we have modified this view slightly and have included, where known, enzymes involved in the processing of matrix macromolecules, and that group of enzymes and inhibitors most closely associated with matrix turnover – the matrix metalloproteinases and tissue inhibitors of metalloproteinases (TIMPs). The field of proteinases and matrix turnover is large and complex and a complete review of proteinases in the extracellular matrix would have been beyond the scope of this book.

General format

Each entry in this book has been prepared using a common format. Factual information is allocated under a number of subheadings; an introductory paragraph, and sections entitled Molecular structure, Isolation, Primary structure, Structural and functional sites, Gene structure and References. The criterion used to determine where particular information should be included in the entry reflects the level at which the information is defined. Where a particular structural or functional property has been shown to reside in a specific amino acid or series of amino acids, it is included under Structural and functional sites; where the property has been narrowed down to a specific region of the molecule but cannot be defined as a particular amino acid sequence, it is included in a cartoon and is described under Molecular structure; and where the property cannot be assigned to a specific region of the molecule, it is described in the introductory paragraph. Accession number(s) are given under Primary structure to allow the reader to access sequences from databases. The particular numbers quoted usually reflect the first deposited sequence together with those for subsequent variant forms of the molecule. As a result, the codes vary in their origin and may refer to EMBL/GenBank, SwissProt or PIR databases. A complete listing of codes is rarely necessary and, if needed, these can usually be found by searching with the code that is given. References are given at the end of each entry and are given over primarily to review articles and major cloning and sequencing reports since there is insufficient space for supporting each factual statement with a reference. Titles are provided so that the reader can identify references of interest.

In general, estimates of the relative mobilities of proteins in polyacrylamide gel electrophoresis in sodium dodecyl sulphate (SDS–PAGE) are not presented (except for collagen α chains in Table 1) because of the anomalous migration of most extracellular matrix molecules. Primary sequence information and database accession numbers are given for human molecules where possible, but only if the complete sequence is available. In addition, there are a number of known matrix molecules that were not completely sequenced at the time the book went to press, and these have therefore been left for future editions except where they are part of a defined family of molecules, e.g. laminin α5. All potential N-linked oligosaccharide attachment sequences are listed under Structural and functional sites. O-Linked oligosaccharide attachment sites are listed only if established chemically.

Cartoon

It is now well established that many protein molecules are built up from a series of independently folded polypeptide building blocks or repeats. Several of these repeats are found in a large number of proteins, suggesting that they might have arisen originally as autonomous polypeptides and that they were subsequently duplicated at the genomic level and incorporated into larger assemblies. Extracellular matrix macromolecules are good examples of proteins constructed in this way, since they contain many examples of protein repeats and in a number of cases, these repeats are catenated into mixed arrays.

The presence of independent repeats has implications not only for protein evolution, but also for function. It is wrong to assume that a repeat always serves the same function, and instead they should be viewed as structural units specialized for different functions in different proteins. Long arrays of repeats might provide a spacer function, particularly relevant for extracellular matrix molecules like fibrillin or laminin that form interconnected networks, or alternatively they may function as templates onto which specific recognition signals, or motifs, are grafted. One specialized example of this kind of function would be the evolution of tripeptide cell adhesive sequences within a small minority of fibronectin type III repeats.

In most cases, the repeat structure of extracellular matrix proteins has been determined by matrix-based database searching. The level of sequence identity is often in the range of 20–30% and the level of sequence similarity is slightly higher at 40–50%. Frequently, however, there are particular residues in a repeat that are crucial for structural integrity, and these are therefore more highly conserved. The presence of these residues also provides strong supportive evidence when assigning a particular repeat type to a polypeptide sequence.

For the purpose of this book, which is primarily concerned with facts about the extracellular matrix, it is important to present a perception of the repeat structure of extracellular matrix molecules. Since it is possible to represent this structure simply and visually in the form of a cartoon that complements detailed sequence data, almost every entry in this book contains a view of the molecule concerned. The cartoons contain some symbols that represent well-characterized protein repeats and others that portray regions of polypeptide that are less well defined in terms of structure. The latter are included simply to give as much information as possible about the substructure of the molecules. In addition, thick horizontal bars on some cartoons indicate regions of the molecule which may undergo alternative splicing of pre-mRNA.

Thirty-nine different symbols are used in the cartoons in this book (including that for alternatively spliced sequences; see Figure 1). The repeats that are represented are as follows:

- Acidic. Short stretches of polypeptide containing a high proportion of aspartic acid and/or glutamic acid residues. Structure unknown.
- Anaphylatoxin. Contain 30–40 amino acids including six cysteine residues. Found in complement components C3a, C4a and C5a and albumin. Structure unknown.
- Basic. Short stretches of polypeptide containing a high proportion of lysine, arginine and/or histidine residues. Structure unknown.
- Collagen triple helix. Variable size. Composed of repeating glycine-X-Y triplets (see Introduction for structural description).
- CUB. Contain 100 amino acids. Adopt a nine-stranded anti-parallel β-barrel topology similar to immunoglobulin modules.
- Elastin cross-link. Contain 10–30 amino acids including two lysine residues three or four amino acids apart that take part in inter-chain cross-linking. Structure unknown.
- Elastin hydrophobic. Contain 10–80 amino acids including a high content of hydrophobic residues (particularly alanine and valine). Structure unknown.
- Epidermal growth factor (EGF). Contain 30–80 amino acids. Six- or eight-cysteine versions are possible. Found in numerous proteins. The structure of EGF itself and the EGF repeat in factor IX have been determined[1,2].
- Fibrinogen β, γ. Contain approximately 200 amino acids. Structure unknown.
- Fibronectin type I. Contain 40–50 amino acids. Contain two intra-repeat disulphide bonds. Found in tissue plasminogen activator and factor XII. The structure of fibronectin type I-7 has been determined[3].
- Fibronectin type II/Kringle. Contain approximately 60 amino acids. Contain two intra-repeat disulphide bonds. Found in tissue plasminogen activator and plasminogen. The structure of the fibronectin type II repeat in a bovine seminal fluid protein has been determined[4].
- Fibronectin type III. Contain 90–100 amino acids. No intra-repeat disulphides. Found in numerous proteins. Several structures have been determined, e.g. RGD-containing repeats in fibronectin and tenascin[5,6]. These have a similar folding pattern to immunoglobulin domains, but have limited sequence homology.

- Frizzled. A module found originally in the *Drosophila* frizzled protein, and also known to exist in two rat proteins. Contain 100–110 amino acids, including 10 cysteine residues. Structure unknown.
- Gla. Region of polypeptide containing a number of γ-carboxyglutamic acid residues.
- Hemopexin. Contain 140–190 amino acids including three cysteine residues. Structure unknown.
- Heptad coiled coil. A chain association domain that mediates the formation of double- or triple-helical coiled coils. Heptad repeats contain non-polar residues at positions 1 and 4 and polar residues at 5 and 7 [7].
- Immunoglobulin (Ig). Contain 90–100 amino acids. Some Ig folds have an inter-repeat disulphide bond. Found in numerous proteins. Many structures have been determined[8]. All consist of a sandwich of two opposing sheets made up of β strands. Ig folds can be divided into V- and C-domains and further subdivisions are possible. C-domains contain three β strands on one face and four on the other; V-domains have extra sequence which adds two strands to the three-strand face (making five and four in total). All strands are connected by variable loops.
- Kazal. Contains 70–80 amino acids. A non-globular structure stabilized by a small hydrophobic core and five disulphide bonds.
- Kunitz proteinase inhibitor. Contains approximately 50 amino acids including six cysteines. Structure unknown.
- Laminin G. Contain 160–230 amino acids including four cysteine residues. Structure unknown.
- Low density lipoprotein (LDL) receptor. Contain 40–50 amino acids with no inter-repeat disulphide bonds. Structure unknown.
- Lectin. Contain 120–130 amino acids with two intra-repeat disulphide bonds. Also called C-type lectin because some proteins containing this repeat bind carbohydrate in a calcium-dependent manner. The structure of a C-type lectin domain from rat mannose binding protein has been solved by X-ray crystallography and shown to contain two regions, one with irregular structure, the other containing both α-helix and β-sheet[9].
- Leucine-rich domain. Short stretches of polypeptide containing closely spaced repeats of leucine residues. Structure unknown.
- Link protein. Contain 70–80 amino acids including four cysteine residues. Found in CD44. Structure unknown.
- Lysine/proline-rich. Region of polypeptide containing a high content of lysine and proline residues. Structure unknown.
- Matrix metalloproteinase (MMP). The catalytic domain of matrix metalloproteinases.
- Ovomucoid. Contains approximately 55 amino acids including six cysteines. Found in several serine proteinase inhibitors. Structure unknown.
- Proline- and arginine-rich protein (PARP). Contain approximately 210 amino acids. Structure unknown.
- Properdin. Contain approximately 50 amino acids including six cysteines. Found in complement components C6 and C9. Structure unknown.
- Transforming growth factor (TGF) $\beta 1$ receptor repeat. Contain 70–80 amino acids including eight cysteine residues. Structure unknown.
- Thrombospondin type 3. Contain 20–40 amino acids including two cysteines. These repeats also contain one or two copies of a 12-residue EF-hand-type cation-binding loop (coordination from residues 1, 3, 5, 7, 9 and 12). Structure otherwise unknown.

- Thyroglobulin. Contain 30–40 amino acids including several cysteine residues. Structure unknown.
- Tyrosine sulphate-rich. Region of polypeptide containing several sulphated tyrosine residues.
- von Willebrand factor A. Contain 190–230 amino acids with no inter-repeat disulphide bonds. Found in many proteins. Structure unknown.
- von Willebrand factor B. Contain 25–35 amino acids including several cysteine residues. Structure unknown.
- von Willebrand factor C/procollagen. Contain approximately 70 amino acids including several cysteine residues. Structure unknown.
- von Willebrand factor D. Contain 270–290 amino acids including many cysteine residues. Structure unknown.

Figure 1 *Key to the artwork symbols used.*

3D Structures

In recent years, several 3D structures for fragments of extracellular matrix molecules have been solved, either through the use of nuclear magnetic resonance (NMR) spectroscopy or X-ray crystallography. The number of such structures is increasing rapidly, but as yet, relatively few are deposited in databases such as Brookhaven. For this reason, we have decided not to include 3D structures under the individual molecular entries. However, to give the reader an idea of the topology of those specific modules and fragments that have been solved, a composite figure is included as Figure 2. Since overall module structures are very highly conserved between proteins, it is likely that an extrapolation of the images presented in Figure 2 can be made from the parent protein. What is largely unknown at present is the

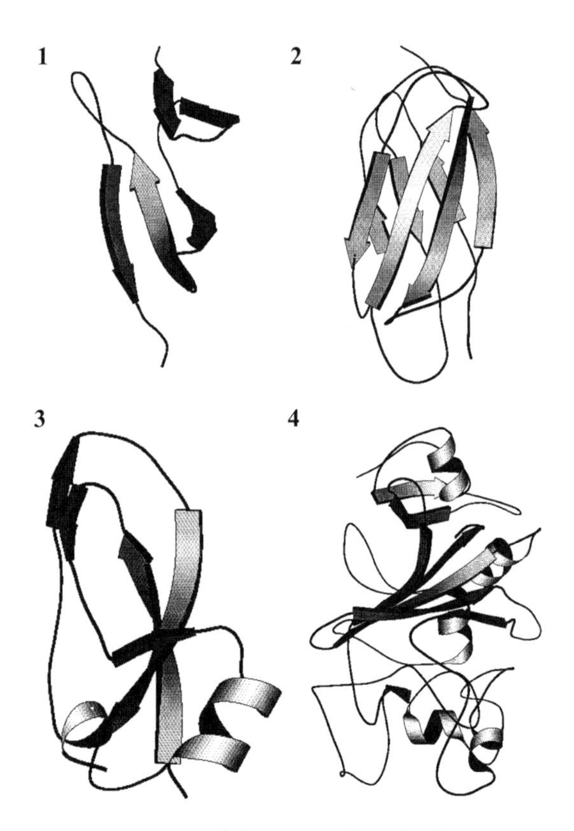

Figure 2 *Schematic representations of the currently solved 3D structures of fragments of extracellular matrix molecules displayed as MOLSCRIPT images. Readers are referred to the original references for a full description of the structures: (1) Fibronectin type I module pair* [10]*; (2) Fibronectin type III module* [5]*; (3) Collagen type VI Kunitz domain* [11]*; (4) Fibrinogen carboxy-terminal domain* [12]*; (5) Fibrillin calcium-binding EGF module pair* [13]*; (6) Osteonectin carboxy-terminal domain* [14]*; (7) MMP1* [15]*. In addition, the web site http://www.bork.embl-heidelberg.de/ modules/ contains an excellent review of module structures.*

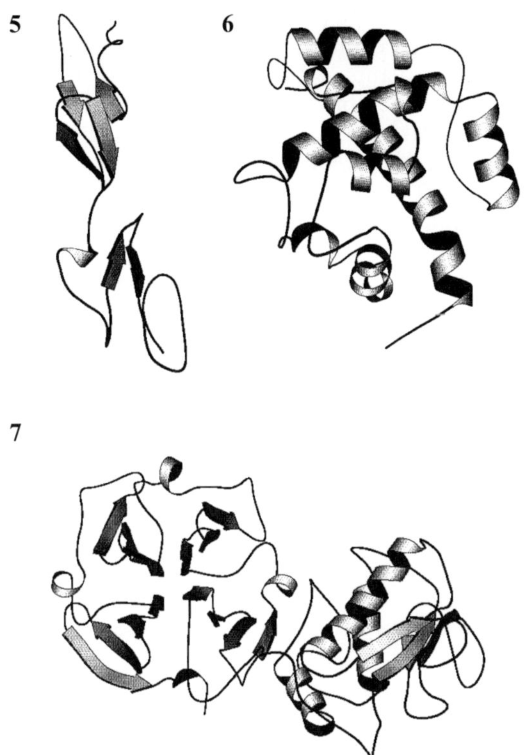

Figure 2 *Continued.*

way in which arrays of modules are folded together, and the extent to which flexibility exists at module–module boundaries. This will only become apparent as more structures are determined.

References

[1] Cooke, R.M. et al. (1987) The solution structure of human epidermal growth factor. Nature 327: 339–341.

[2] Handford, P.A. et al. (1990) The first EGF-like domain from human factor IX contains high-affinity calcium binding site. EMBO J. 9: 475–480.

[3] Baron, M. et al. (1990) Structure of the fibronectin type I module. Nature 345: 642–646.

[4] Constantine, K.L. et al. (1992) Refined solution structure and ligand-binding properties of PDC-109 domain b. A collagen-binding type II domain. J. Mol. Biol. 223: 281–298.

[5] Leahy, D.J. et al. (1992) Structure of a fibronectin type III domain from tenascin phased by MAD analysis of the selenomethionyl protein. Science 258: 987–991.

[6] Main, A.L. et al. (1992) The three-dimensional structure of the tenth type III module of fibronectin: an insight into RGD-mediated interactions. Cell 71: 671–678.

[7] Cohen, C. and Parry, D. (1990) α-Helical coiled coils and bundles: how to design an α-helical protein. Proteins: Struct. Func. Genet. 7: 1–15.

[8] Williams, A.F. and Barclay, A.N. (1988) The immunoglobulin superfamily – domains for cell surface recognition. Annu. Rev. Immunol. 6: 381–405.
[9] Weis, W.I. et al. (1992) Structure of the calcium-dependent lectin domain from a rat mannose-binding protein determined by MAD phasing. Science 254: 1608–1615.
[10] Williams, M.J. et al. (1994) Solution structure of a pair of fibronectin type-I modules with fibrin binding-activity. J. Mol. Biol. 235: 1302–1311.
[11] Arnoux, B. et al. (1995) The 1.6-Angstrom structure of Kunitz-type domain from the alpha-3 chain of human type VI collagen. J. Mol. Biol. 246: 609–617.
[12] Yee, V.C. et al. (1997) Crystal structure of a 30 kDa C-terminal fragment from the gamma chain of human fibrinogen. Structure 5: 125–138.
[13] Downing, A.K. et al. (1996) Solution structure of a pair of calcium-binding epidermal growth factor-like domains – Implications for the Marfan Syndrome and other genetic disorders. Cell 85: 597–605.
[14] Hohenester, E. et al. (1996) Structure of a novel extracellular Ca^{2+}-binding module in BM-40. Nature Structural Biol. 3: 67–73.
[15] Li, J. et al. (1995) Structure of full-length porcine synovial collagenase reveals a C-terminal domain containing a calcium-linked, 4-bladed beta propeller. Structure 3: 541–549.

Further reading

Bork, P. (1992) The modular architecture of vertebrate collagens. FEBS Lett. 307: 49–54.
Bork, P. et al. (1996) Structure and distribution of modules in extracellular proteins. Quart. Rev. Biophys. 29: 119–167.
Neame, P.J. et al. (1990) Isolation and primary structure of PARP, a 24-kDa proline- and arginine-rich protein from bovine cartilage closely related to the amino-terminal domain in collagen α1 (XI). J. Biol. Chem. 265: 20401–20408.
See also P. Bork on http://www.bork.embl-heidelberg.de/modules/.

Extracellular matrices

Connective tissues join the other tissues of the body together. They take the stress of movement, maintain shape, and can be considered as a composite of insoluble fibres and soluble polymers. The principal fibres are collagen and elastin, while the soluble molecules include proteoglycans and glycoproteins. The structure (and hence function) of any connective tissue depends on the relative proportions of these constituent molecules. Those tissues that have to withstand large tensional force (such as tendon) tend to be particularly rich in fibrillar collagens, while a tissue that has to withstand compressive forces (such as cartilage) contains high levels of proteoglycans. The key features of the main classes of extracellular matrix molecules are presented below.

Collagens

The collagens constitute a highly specialized family of glycoproteins of which there are now at least 19 genetically distinct types encoded by at least 34 genes. Many have very complex structures and it is becoming increasingly difficult to define what is a collagen and what is not. At least three collagens comprise the protein cores of proteoglycans. Several proteins (e.g. the C1q component of complement and the enzyme acetylcholinesterase) contain the basic unit of a collagen – the triple helix – but for a protein to be classified as a collagen it must be an integral component of the extracellular matrix.

The triple helix

The triple helix comprises three polypeptide (α) chains, each with a left-handed helical configuration, wound round each other to form a right-handed super helix. Glycine occupies every third residue and this is an absolute requirement as glycine is the only residue with a small enough side-chain to fit into the centre of the triple helix without distorting it. Approximately 20–22% of the remaining X and Y residues in the repeating triplet $[\text{Gly-X-Y}]_n$ are, respectively, the imino acids proline and hydroxyproline. The hydroxyl group of hydroxyproline is essential for the formation of hydrogen bonds that stabilize the triple helix. Lysine and hydroxylysine residues in specific regions of both helical and non-helical regions are important in the formation of stable covalent cross-links within and between the collagen molecules in many of their supramolecular forms. Hydroxylysine residues are also potential sites for glycosylation with either galactose or glucosylgalactose. The hydroxylysine/lysine ratio as well as the degree of hydroxylysine glycosylation vary for the different collagen types and for the same collagen type in different tissues and with age. The triple helical domains also vary in length for the different collagen types and can either be continuous or interrupted with non-helical domains.

The genetically distinct collagen types

The molecular configuration, supramolecular structure, approximate α chain molecular weight (from the matrix form, estimated by SDS–PAGE) and tissue distribution for types I–XIX collagens are listed in Table 1.

Collagen biosynthesis

The synthesis of collagen α chains follows the established pathway of other secretory proteins but involves several co- and post-translational modifications. These include

Table 1 *The collagen family*

Type	Molecular configuration	Supramolecular structure	$M_r \times 10^{-3}$ α chain	Examples of tissue location
I	$[\alpha1(I)]_2\alpha2(I)$	Large-diameter 67 nm banded fibrils	95	Bone, cornea, skin, tendon
	$[\alpha1(I)]_3$ trimer	67 nm banded fibrils		Tumours, skin
II	$[\alpha1(II)]_3$	67 nm banded fibrils	95	Cartilage, vitreous
III	$[\alpha1(III)]_3$	Small-diameter 67 nm banded fibrils	95	Skin, aorta, uterus, gut
IV	$[\alpha1(IV)]_2\alpha2(IV)$, plus α3(IV), α4(IV), α5(IV), α6(IV) chains	Non-fibrillar meshwork	170–180	Basement membranes
V*	$[\alpha1(V)]_2\alpha2(V)$ $[\alpha1(V)\alpha2(V)\alpha3(V)]$ $[\alpha1(V)]_3$	9 nm diameter non-banded fibrils	120–145	Placental tissue, bone, skin
VI	$[\alpha1(VI)\alpha2(VI)\alpha3(VI)]$ and other forms?	5–10 nm diameter beaded microfibrils, 100 nm periodicity	140, 340	Uterus, skin, cornea, cartilage
VII	$[\alpha1(VII)]_3$	Anchoring fibrils	170	Amniotic membrane, skin, oesophagus
VIII	$[\alpha1(VIII)]_2\alpha2(VIII)$	Non-fibrillar hexagonal lattice	61	Descemet's membrane endothelial cells
IX	$[\alpha1(IX)\alpha2(IX)\alpha3(IX)]$	FACIT, non-fibrillar	68–115	Cartilage, vitreous
X	$[\alpha1(X)]_3$	Non-fibrillar, hexagonal lattice?	59	Calcifying cartilage
XI*	$[\alpha1(XI)\alpha2(XI)\alpha3(XI)]$	Fine fibrils similar to type V collagen	110–145	Cartilage, intervertebral disc
XII	$[\alpha1(XII)]_3$	FACIT, non-fibrillar	220, 340	Skin, tendon, cartilage
XIII	α1(XIII) chain	Membrane-intercalated?	62–67	Endothelial cells, epidermis
XIV	$[\alpha1(XIV)]_3$	FACIT, non-fibrillar	220	Skin, tendon, cartilage
XV	α1(XV) chain	MULTIPLEXIN, non-fibrillar	125	Placenta, kidney, heart, ovary, testis
XVI	$[\alpha1(XVI)]_3$	FACIT, non-fibrillar	150–160	Heart, kidney, smooth muscle
XVII	$[\alpha1(XVII)]_3$	Membrane-intercalated	180	Hemidesmosomes of specialized epithelia
XVIII	α1(XVIII) chain	MULTIPLEXIN, non-fibrillar	200	Kidney, liver, lung
XIX	α1(XIX) chain	FACIT, non-fibrillar	Unknown	Fibroblast cell lines

* Molecules comprising both type V and XI α chains are known.

the hydroxylation of proline and lysine, glycosylation of hydroxylysine, sulphation of tyrosine, disulphide bond formation and chain association, and the addition of complex carbohydrates and glycosaminoglycans (generally in non-collagenous domains) prior to secretion and deposition in the extracellular matrix. Additional modifications can occur in the extracellular matrix and depend on the collagen type and specific supramolecular structure formed.

The 'classical' fibrillar collagens

Collagen types I, II and III constitute the classical fibril-forming collagens and account for 80–90% of all the collagen in the body. They are synthesized intracellularly as large precursor procollagens comprising a continuous triple helix at each end of which are non-helical domains (propeptides). The propeptides are cleaved extracellularly by specific N- and C-proteinases during fibrillogenesis giving rise to the collagen monomers consisting of the triple helix (length 285 nm, diameter 1.4 nm) flanked by short non-helical regions. The monomers spontaneously assemble to form the fibril, each monomer staggered by 234 amino acid residues (a D unit 67 nm long) which results in maximal electrostatic and hydrophobic interactions between adjacent monomers and allows specific lysine/hydroxylysine residues in the helix and non-helical regions within the 0.4 D overlap region to form stable covalent cross-links. The cross-links are formed by the oxidation of the non-helical ε-NH_2 group of lysine/hydroxylysine by the enzyme lysyl oxidase and the subsequent condensation of the resulting aldehyde with the ε-NH_2 group of the helical lysine/hydroxylysine residue to form a Schiff's base. These bifunctional cross-links condense with additional lysine/hydroxylysine or histidine residues on adjacent molecules to form the more mature stable trifunctional cross-links.

Types V and XI collagens are also classified as fibrillar collagens on the basis of their homology with types I–III collagens. All other collagens deviate considerably from this class of collagen in both structure and assembly within the extracellular matrix.

Recent studies indicate that there may be no such thing as a 'type I, II or III collagen fibril'. Research suggests that these major collagens can exist as heterotypic fibrils with I and III forming copolymers, and with type V and XI collagens copolymerized largely on the inside of type I and II collagen fibrils, respectively. Other collagens known as FACIT collagens (*F*ibril *A*ssociated *C*ollagens with *I*nterrupted *T*riple *H*elix) such as types IX, XII and XIV associate with the surface of fibrils and modify their interactive properties.

Other collagen types

The interaction of fibrillar collagens with other collagens present either on the surface or in the interior of the fibril may alter their interactive capacity. This situation is likely to vary from tissue to tissue, and in the same tissue during development. In addition, there are a variety of collagenous molecules that can form three-dimensional networks (type IV), beaded filaments (type VI), anti-parallel dimers which form anchoring fibrils (type VII) and a hexagonal lattice (type VIII).

Nomenclature

All triple helical collagenous (COL) domains and the non-helical, non-collagenous (NC) domains are numbered from the carboxy-terminal domain. In the case of the fibrillar collagens which are synthesized as larger precursor procollagen (pro-) forms, only the COL domains are numbered, the NC domains being the propeptides that are completely or partially cleaved on processing. The term preprocollagen, previously denoting the procollagen form with attached signal sequence, will not be used for the complete primary structure: only the individual chain will be specified, e.g. $\alpha 1(I)$ chain.

Proteoglycans

Proteoglycans are a diverse family of molecules characterized by a core protein to which is attached one or more glycosaminoglycan (GAG) side-chains. These molecules appear to be distributed ubiquitously among animal cells, and participate in a variety of biological processes. Proteoglycans are found outside cells, intercalated in cell membranes and intracellularly in storage granules. This book is only concerned with those proteoglycans that are found in the extracellular matrix.

The heterogeneity of proteoglycan structure is a reflection not only of a variation in protein core, but also variation in the type and size of the GAG chains. Indeed, cells at different stages of development express different variants of the same proteoglycan. In recent times, proteoglycan nomenclature reflects the protein core and/or function of the molecule, whereas previously classification has been according to the chemical composition of the GAG chains. The glycosaminoglycan group of complex carbohydrates includes chondroitin sulphate (CS), dermatan sulphate (DS), heparan sulphate (HS), keratan sulphate (KS) and hyaluronan (HA). The characteristic feature of these molecules is that they are composed of a disaccharide repeat sequence of two different sugars; one of these is normally a hexuronate while the other is a hexosamine. Configurational variation in the disaccharide bonds and the position of sulphation leads to increased diversity in the chemical and physical properties of the chains.

Chondroitin sulphate

The basic repeating unit is formed from a glucuronic acid joined to *N*-acetylgalactosamine by β1–3 linkage. Sulphate residues can be added to the 4 and 6 position of the amine residues, and are referred to as chondroitin-4-sulphate and chondroitin-6-sulphate. More commonly, a given chain exhibits stretches of 4-sulphation followed by 6-sulphation, and there are frequent occasions where neither residue in the disaccharide is sulphated or where both positions in an individual residue are sulphated.

Dermatan sulphate

This GAG has a similar structure to chondroitin sulphate except that some glucuronic acid residues are epimerized and converted to iduronic acid. Once formed, the iduronic acid residues may be sulphated at the 2 position. The glycosidic bond is also changed to α1–3. A CS chain with one or more iduronic acid residues is called dermatan sulphate.

Keratan sulphate

In the case of KS, the uronic acid residue is replaced by galactose. The repeat sequence is galactose joined to *N*-acetylglucosamine by a β1–4 linkage. Sulphation can occur on position 6 of each residue.

Heparan sulphate

The repeating sequence is glucuronic acid joined to *N*-acetylglucosamine by a β1–4 linkage. Both residues can be extensively sulphated at O- and N-positions, and the uronic acid can undergo epimerization.

Hyaluronan (hyaluronic acid)

Hyaluronan consists of an alternating polymer of glucuronic acid and *N*-acetylglucosamine joined by β1–3 linkage.

The high content of sulphate and the presence of uronic acid confer a large negative charge on glycosaminoglycans and permit them to associate with a large number of ligands by electrostatic interactions. Only a limited amount of information is available on sequences that confer specific binding.

Apart from hyaluronan, all the other GAG chains are found in tissues covalently attached to protein and function as proteoglycans. In general, the GAG chains are attached via a xylose residue linkage to specific serine residues in the protein core, the exception being keratan sulphate chains which are attached to protein via N- and O-linked glycosidic linkages to asparagine or serine/threonine, respectively.

Hyaluronan is a unique member of the GAG family since it functions *in vivo* as a free carbohydrate. It is polyanionic and large; a single molecule can have a molecular weight up to 10 million. HA assumes a randomly kinked, coil structure which occupies a large solution volume and endows solutions with high viscosity. The individual molecules can self-associate and form networks. A number of proteoglycans can associate with HA and form supramolecular aggregates in tissues. In developing tissues, HA can be the major structural macromolecule and supports both cell proliferation and migration. Our exclusion of HA from this book is in no way a reflection of the contribution of the molecule to matrix formation and function. Despite considerable developments in the structure, cellular interactions and nature of receptors, there is still very little functional data available relating to specific sites on the molecule.

Proteoglycans are diverse in both structural and functional terms. The highly acidic and hydrophilic GAG chains will have a major influence on tissue hydration and elasticity, and can show high affinity binding to a variety of ligands. Protein cores have also been shown to have specific interactions in the matrix. Developments in both protein and carbohydrate chemistry will help define the molecular basis for the diverse biological functions in which these molecules participate.

Glycoproteins

The simple picture of connective tissues as frameworks of insoluble fibrils and soluble polymers serves to highlight the importance of collagen fibres in resisting tensile stress and proteoglycans in associating with large amounts of water, swelling and resisting compressive forces. While these functions of the major structural elements of matrices are critical to many aspects of the physical properties of connective tissues, they take no account of the role of the large number of glycoproteins that are present in the extracellular matrix. A variety of extracellular glycoproteins have been isolated and the key feature of many of these molecules is their ability to interact not only with cells, but also with other macromolecules in the matrix.

Many of the matrix glycoproteins contain distinct and functionally active peptide domains that prescribe interactions with cell surface receptors as well as other matrix molecules. In the case of molecules found in mineralized connective tissue, this may also involve interaction with the inorganic phase of bone or dentine. Some of the molecules form oligomers by disulphide bond formation or by non-covalent association. In addition, alternative splicing is a frequent finding producing families of closely related proteins. This heterogeneous group of molecules collectively contains carbohydrate covalently attached to the protein core which in some cases is O-linked, in others N-linked and some contain both O- and N-linked carbohydrate. Phosphorylation of serine and threonine, and sulphation of tyrosine residues, are other post-translational modifications that might be found. A number of these molecules have been described as adhesive glycoproteins, the best studied of which is fibronectin.

The term adhesive glycoproteins includes such molecules as fibronectin, vitronectin, laminin, thrombospondin and von Willebrand factor, which have several features in common. Distinct structural and functional sites are found in these molecules and some repetitive structural motifs (fibronectin type III and EGF repeats) are common to a number of adhesive molecules. Many possess short aspartate-containing sequences such as Arg-Gly-Asp (RGD), which mediate cellular adhesion through the integrin family of membrane receptors. This ability to interact with cells is clearly a key function, and we have chosen to include fibrinogen in this book because it also functions as a major adhesive molecule in blood, and where there has been damage to and leakage from the endothelium, it appears in the extracellular matrix and can influence fibroblast function.

In addition to the adhesive molecules, a number of other matrix glycoproteins have been characterized, particularly from skeletal tissue (cartilage, bone and dentine). One characteristic feature of many of the bone-associated molecules is their anionic nature. This fixed negative charge is a consequence of a variety of features; many are rich in the acidic amino acids (aspartic and glutamic), some contain stretches of consecutive aspartic acid residues (osteopontin), others contain stretches of consecutive glutamic acid residues (bone sialoprotein). We have chosen to present these as acidic-rich motifs. Other molecules contain γ-carboxyglutamic acid residues (osteocalcin and matrix Gla protein), and these have been shown as Gla-containing protein motifs. No Gla-recognition sequences have been shown. Sometimes the negative charge is provided by sialic acid residues on the oligosaccharide side-chains (osteopontin, bone sialoprotein). A number of post-translational modifications also contribute to the anionic nature of the molecules, notably phosphorylation of serine (phosphoryns) and threonine, and sulphation of tyrosine residues (bone sialoprotein). This produces a variety of molecules that have been implicated in the process of calcification and calcium ion binding, although no specific functions have been demonstrated. In general, these molecules are of low molecular weight (4000–40 000), and tend to exist and function in monomeric form.

Apart from the adhesive glycoproteins and those isolated from skeletal tissues, another class of glycoproteins have been described which are associated with elastin and elastin deposition, notably fibrillin and MAGP. The best characterized of these is fibrillin, which despite its incomplete sequence has been included because some aspects of its protein structure are well defined, and it has been clearly shown to be related to the incidence of Marfan's syndrome. A number of other molecules which have been implicated with elastin fibre formation have not been included because full identification and structures are not available.

Glycoproteins of the extracellular matrix clearly have a number of important functions and are implicated in a variety of developmental and pathological changes in the extracellular matrix. The ability to influence cell behaviour by allowing attachment and migration of cells is clearly a major function of the adhesive glycoproteins. The ability to influence ion concentrations is a probable mode of action of a number of the skeletal glycoproteins, although some appear able to influence bone cell metabolism directly. Protein–protein interactions are also important in the functioning of some molecules, as evidenced by the ability of fibrillin to influence the deposition of elastin. The diversity and versatility of these molecules serves to highlight the importance of this class of molecule in the extracellular matrix.

Matrix metalloproteinases

A major change in the second edition of this book has been the inclusion of enzymes and proteinases involved in the deposition and degradation of the extracellular matrix. These enzymes include the procollagen N- and C-proteinases that cleave procollagen to collagen and the matrix metalloproteinases (MMPs) that degrade collagen and other structural macromolecules in the matrix. A dynamic equilibrium exists between the synthesis and the degradation of the extracellular matrix in order to generate the elaborate architectures seen in tissues. Current evidence suggests that MMPs play a pivotal role in maintaining this equilibrium.

The MMPs are synthesized as inactive precursors that contain a prodomain. This domain resides within the active site of the proteinases and blocks access of the substrate. Dislocation of the prodomain from the active site by mercurial compounds or by proteolysis generates active MMP molecules. Dislocation of the prodomain occurs by a specific cysteine switch mechanism.

Much is known about the MMPs that catalyse matrix degradation; however, little is known about the proteinases that catalyse matrix synthesis. Three functionally distinct procollagen metalloproteinases (PMPs) convert soluble procollagens to the insoluble collagens that self-assemble into the fibres that are the major source of mechanical strength of connective tissue. The critical importance of the PMPs and MMPs in extracellular matrix biology, and their intimate association in the assembly and biology of other extracellular matrix molecules, are reasons for their inclusion in this updated volume.

Nomenclature of the matrix metalloproteinases

Two systems are currently used to describe members of the MMP superfamily of proteinases. The first is a numerical system in which the MMP number generally coincides with the chronological discovery of the enzymes. On some occasions MMPs have been assigned a number and have later been shown to be identical to a previously identified MMP.

Consequently, gaps have occurred in the numerical numbering system such that MMP4–6 have disappeared from the literature. The other system used to describe the MMPs has been to add the suffix 'ase' to the most frequently cleaved substrate for a particular MMP. Consequently names such as gelatinase and collagenase have appeared in the literature. However, the promiscuous activities of many of the MMPs and overlapping substrate specificities have resulted in confusion within the field. In this FactsBook the proteinases are described using both the numerical and substrate cleavage nomenclature systems.

Bibliography

General

Hay, E.D., ed. (1991) Cell Biology of the Extracellular Matrix, 2nd edition, Plenum Press, New York.

Royce, P.M. and Steinmann, B., eds (1993) Connective Tissue and its Heritable Disorders, Molecular, Genetic, and Medical Aspects, Wiley-Liss Inc., New York.

Comper, W.D., ed. (1996) Extracellular Matrix, Vols 1 and 2, Harwood Academic, Australia.

Haralson, M.A. and Hassell, J.R., eds (1995) Extracellular Matrix: A Practical Approach, IRL Press, Oxford.

Mecham, R., ed. Biology of the Extracellular Matrix: A Series, Academic Press, New York. This series includes the following volumes;

(1) Mecham, R.P., ed. (1986) Regulation of Matrix Accumulation.
(2) Wight, T.N. and Mecham, R.P., eds. (1987) Biology of Proteoglycans.
(3) Mayne, R. and Burgeson, R.E., eds. (1987) Structure and Function of Collagen Types.
(4) Mosher, D.F., ed. (1989) Fibronectin.
(5) Adair, W.S. and Mecham, R.P., eds. (1990) Organisation and Assembly of Plant and Animal Extracellular Matrix.
(6) Sandell, L.J. and Boyd, C.D., eds. (1990) Extracellular Matrix Genes.
(7) Yurchenko, P.D., Birk, D.E. and Mecham, R.P., eds. (1994) Extracellular Matrix Assembly and Structure.

Collagens

Burgeson, R.E. and Nimni, M.E. (1992) Collagen types. Molecular structure and tissue distribution. Clin. Orthop. 282: 250–272.

Eyre, D.R. (1987) Collagen cross-linking amino acids. Methods Enzymol. 144: 115–139.

Kadler, K. (1995) Extracellular matrix 1: fibril forming collagens. In: Protein Profile, Vol. 2, issue 5, ed. Sheterline, P., Academic Press, London.

Kagan, H.M. (1986) Characterization and regulation of lysyl oxidase. In: Regulation of Matrix Accumulation: Biology of the Extracellular Matrix, Vol. 1, ed. Mecham, R.P., Academic Press, Orlando, pp. 321–398.

Kielty, C.M. et al. (1993) The collagen family: Structure, assembly and organization in the extracellular matrix. In: Connective Tissue and its Heritable Disorders. Molecular Genetics and Medical Aspects, ed. Royce, P.M. and Steinmann, B., Wiley-Liss, Inc., New York, pp. 103–147.

Kivirikko, K. (1993) Collagens and their abnormalities in a wide spectrum of diseases. Annals of Med. 25: 113–126.

Kivirikko, K. and Myllyla, R. (1985) Post-translational processing of procollagens. Ann. N.Y. Acad. Sci. 460: 187–201.

Kuhn, K. (1987) The classical collagens: types I, II and III. In: Structure and Function of Collagen Types, ed. Mayne, R. and Burgeson, R.E., Academic Press, London, pp. 1–42.

Olsen, B.R. (1995) New insights into the function of collagens from genetic analysis. Curr. Opin. Cell Biol. 7: 720–727.

Ramachandran, G.N. and Ramakrishnan, C. (1976) Molecular structure. In: Biochemistry of Collagen, ed. Ramachandran, G.N. and Reddi, A.H., Plenum Press, New York, pp. 45–84.

van der Rest, M. and Garrone, R. (1991) Collagen family of proteins. FASEB J. 5: 2814–2823.

Proteoglycans

Bourdon, M.A. et al. (1987) Identification and synthesis of a recognition signal for the attachment of glycosaminoglycans to proteins. Proc. Natl Acad. Sci. USA 84: 3194–3198.

Esko, J.D. (1991) Genetic analysis of proteoglycan structure, function and metabolism. Curr. Opin. Cell Biol. 3: 805–816.

Gallagher, J.T. (1989) The extended family of proteoglycans: social residents of the pericellular zone. Curr. Opin. Cell Biol. 1: 1201–1218.

Heinegard, D. and Oldberg, A. (1993) Glycosylated matrix proteins. In: Connective Tissue and its Heritable Disorders, ed. Royce, P.M. and Steinman, B., Wiley-Liss, New York, pp. 189–209.

Iozzo, R.V. and Murdoch, A.D. (1996) Proteoglycans of the extracellular environment: clues from the gene and protein side offer novel perspectives in molecular diversity and function. FASEB J. 10: 598–614.

Jackson, R.L. et al. (1991) Glycosaminoglycans: Molecular properties, protein interactions, and role in physiological processes. Physiol. Rev. 71: 481–539.

Kjellen, L. and Lindahl, U. (1991) Proteoglycans: Structure and interactions. Annu. Rev. Biochem. 60: 443–475.

Scott, J.E. (1992) Supramolecular organization of extracellular matrix glycosaminoglycans in vitro and in the tissues. FASEB J. 6: 2639–2645.

Wight, T.N. et al. (1992) The role of proteoglycans in cell adhesion, migration and proliferation. Curr. Opin. Cell Biol. 4: 793–801.

Glycoproteins

Bork, P. (1992) Mobile modules and motifs. Curr. Opin. Struct. Biol. 2: 413–421.

Heinegard, D. and Oldberg, A. (1989) Structure and biology of cartilage and bone matrix noncollagenous macromolecules. FASEB J. 3: 2042–2051.

Edelman, G.M. et al., eds (1990) Morphoregulatory Molecules. Wiley & Sons, New York.

Mosher, D.F. et al. (1992) Assembly of the extracellular matrix. Curr. Opin. Cell Biol. 4: 810–818.

Paulsson, M. (1992) Basement membrane proteins: Structure, assembly and cellular interactions. Crit. Rev. Biochem. Mol. Biol. 27: 93–127.

Schwarzbauer, J.E. (1991) Fibronectin: From gene to protein. Curr. Opin. Cell Biol. 3: 786–791.

Timpl, R. (1989) Structure and biological activity of basement membrane proteins. Eur. J. Biochem. 180: 487–502.

Timpl, R. and Brown, J.C. (1995) Supramolecular assembly of basement membranes. BioEssays 18: 123–132.

von der Mark, K. and Goodman, S. (1993) Adhesive glycoproteins. In: Connective Tissue and its Heritable Disorders, ed. Royce, P.M. and Steinmann, B., Wiley-Liss, New York, pp. 211–236.

Yurchenco, P.D. and Schittny, J.C. (1990) Molecular architecture of basement membranes. FASEB J. 4: 1577–1590.

Yurchenco, P.D. and O'Rear, J.J. (1994) Basal lamina assembly. Curr. Opin. Cell Biol. 6: 674–681.

Proteinases

Birkedal-Hansen, H. (1995) Proteolytic remodeling of extracellular matrix. Curr. Opin. Cell Biol. 7: 728–735.

Coussens, L.M. and Werb, Z. (1996) Matrix metalloproteinases and the development of cancer. Chem. Biol. 3: 895–904.
Murphy, G. (1997) Matrix metalloproteinases. Where next? A sample of opinions. Matrix Biol. 15: 511–518.
Nagase, H. (1997) Activation mechanisms of matrix metalloproteinases. Biol. Chem. 378: 151–160.
Parsons, S.L. et al. (1997) Matrix metalloproteinases. Br. J. Surg. 84: 160–168.

Section II

THE EXTRACELLULAR MATRIX PROTEINS

Aggrecan is a large chondroitin sulphate proteoglycan (CS-PG) that accounts for about 10% of the dry weight of cartilage. Aggrecan interacts with hyaluronan (HA) via an HA-binding domain. The interaction has a dissociation constant of about 10^{-8} M and is additionally strengthened by the binding of link protein. Aggrecan is usually found as part of a large aggregate containing approximately 100 proteoglycan molecules per HA molecule. Aggrecan carries a large number of fixed negatively charged side-groups that result in high osmotic pressure in the tissue. It is generally accepted that the primary role of aggrecan is to swell and hydrate the framework of collagen fibrils in cartilage.

Molecular structure

Aggrecan comprises about 87% chondroitin sulphate (CS), 6% keratan sulphate (KS) and 7% protein and has a molecular mass of 2.6×10^6 Da with a core protein of molecular mass approximately 220 kDa. By rotary shadowing electron microscopy and comparison of cDNAs of aggrecans from several species, the core protein is seen to have seven domains. The first three domains at the NH_2-terminus comprise two globular domains called G1 and G2 (separated by an inter-globule domain, IGD) which have structural similarities to the link protein of cartilage. G1 is a copy of the complete link protein sequence that can be divided into A (immunoglobulin repeat), B and B′ subdomains (link repeats) and contains the hyaluronan binding domain (subdomain A), G2 is a copy of the COOH-terminal two-thirds of link protein and contains the B and B′ subdomains (both link repeats). The next three domains contain sites for attachment of glycosaminoglycan (GAG). Keratan sulphate is the principal GAG in the fourth domain and CS is the principal GAG in the fifth and sixth domains. The seventh domain at the COOH-terminus contains a globule called G3 that is composed of a lectin repeat. There are at least three forms of aggrecan transcripts generated by alternative exon usage.

Alternatively spliced sequences are a 38 amino acid sequence inserted before the seventh domain that is highly homologous to an EGF repeat and a complement control protein repeat that can be spliced out after the seventh domain. Aggrecan is cleaved by recombinant stromelysin between residues N341 and F342 that separate G1 from the remainder of the molecule. Evidence exists for a similar cleavage *in vivo* producing a non-aggregating form of aggrecan[1–3].

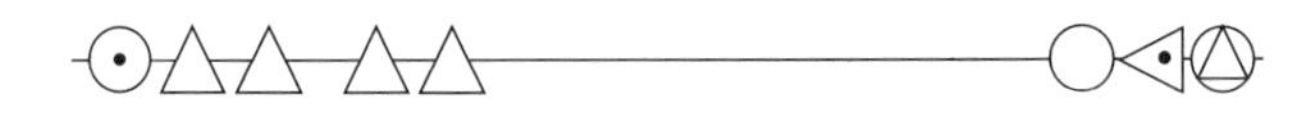

Isolation

Aggrecan is readily isolated from cartilage by extraction with 4 M guanidine–HCl/0.05 M sodium acetate pH 5.8 at 4°C, and is purified by sequential density gradient centrifugation and ion exchange chromatography[4]. A simple one-step purification procedure using 1.32–10% polyacrylamide gels has been described[5].

Accession number

J05062; A39086

Primary structure

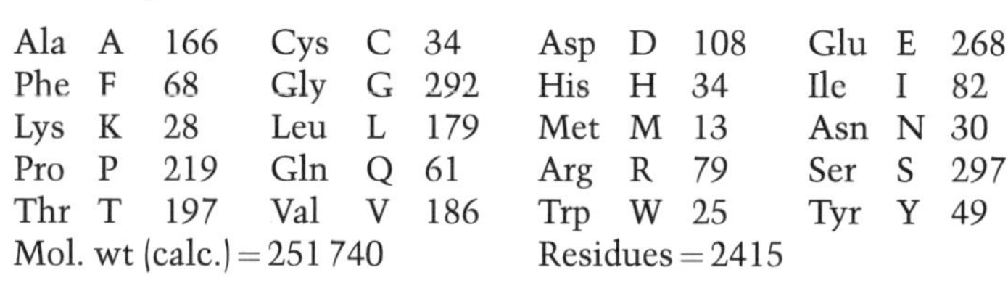

Ala	A	166	Cys	C	34	Asp	D	108	Glu	E	268
Phe	F	68	Gly	G	292	His	H	34	Ile	I	82
Lys	K	28	Leu	L	179	Met	M	13	Asn	N	30
Pro	P	219	Gln	Q	61	Arg	R	79	Ser	S	297
Thr	T	197	Val	V	186	Trp	W	25	Tyr	Y	49

Mol. wt (calc.) = 251 740 Residues = 2415

```
1     MTTLLWVFVT LRVITAAVTV ETSDHDNSLS VSIPQPSPLR VLLGTSLTIP
51    CYFIDPMHPV TTAPSTAPLA PRIKWSRVSK EKEVVLLVAT EGRVRVNSAY
101   QDKVSLPNYP AIPSDATLEV QSLRSNDSGV YRCEVMHGIE DSEATLEVVV
151   KGIVFHYRAI STRYTLDFDR AQRACLQNSA IIATPEQLQA AYEDGFHQCD
201   AGWLADQTVR YPIHTPREGC YGDKDEFPGV RTYGIRDTNE TYDVYCFAEE
251   MEGEVFYATS PEKFTFQEAA NECRRLGARL ATTGHVYLAW QAGMDMCSAG
301   WLADRSVRYP ISKARPNCGG NLLGVRTVYV HANQTGYPDP SSRYDAICYT
351   GEDFVDIPEN FFGVGGEEDI TVQTVTWPDM ELPLPRNITE GEARGSVILT
401   VKPIFEVSPS PLEPEEPFTF APEIGATAFA EVENETGEAT RPWGFPTPGL
451   GPATAFTSED LVVQVTAVPG QPHLPGGVVF HYRPGPTRYS LTFEEAQQAC
501   PGTGAVIASP EQLQAAYEAG YEQCDAGWLR DQTVRYPIVS PRTPCVGDKD
551   SSPGVRTYGV RPSTETYDVY CFVDRLEGEV FFATRLEQFT FQEALEFCES
601   HNATATTGQL YAAWSRGLDK CYAGWLADGS LRYPIVTPRP ACGGDKPGVR
651   TVYLYPNQTG LPDPLSRHHA FCFRGISAVP SPGEEEGGTP TSPSGVEEWI
701   VTQVVPGVAA VPVEEETTAV PSGETTAILE FTTEPENQTE WEPAYTPVGT
751   SPLPGILPTW PPTGAETEES TEGPSATEVP SASEEPSPSE VPFPSEEPSP
801   SEEPFPSVRP FPSVELFPSE EPFPSKEPSP SEEPSASEEP YTPSPPEPSW
851   TELPSSGEES GAPDVSGDFT GSGDVSGHLD FSGQLSGDRA SGLPSGDLDS
901   SGLTSTVGSG LTVESGLPSG DEERIEWPST PTVGELPSGA EILEGSASGV
951   GDLSGLPSGE VLETSASGVG DLSGLPSGEV LETTAPGVED ISGLPSGEVL
1001  ETTAPGVEDI SGLPSGEVLE TTAPGVEDIS GLPSGEVLET TAPGVEDISG
1051  LPSGEVLETT APGVEDISGL PSGEVLETAA PGVEDISGLP SGEVLETAAP
1101  GVEDISGLPS GEVLEIAAPG VEDISGLPSG EVLETAAPGV EDISGLPSGE
1151  VLETAAPGVE DISGLPSGEV LETAAPGVED ISGLPSGEVL ETAAPGVEDI
1201  SGLPSGEVLE TAAPGVEDIS GLPSGEVLET AAPGVEDISG LPSGEVLETA
1251  APGVEDISGL PSGEVLETAA PGVEDISGLP SGEVLETTAP GVEEISGLPS
1301  GEVLETTAPG VDEISGLPSG EVLETTAPGV EEISGLPSGE VLETSTSAVG
1351  DLSGLPSGGE VLEISVSGVE DISGLPSGEV VETSASGIED VSELPSGEGL
1401  ETSASGVEDL SRLPSGEEVL EISASGFGDL SGVPSGGEGL ETSASEVGTD
1451  LSGLPSGREG LETSASGAED LSGLPSGKED LVGSASGDLD LGKLPSGTLG
1501  SGQAPETSGL PSGFSGEYSG VDLGSGPPSG LPDFSGLPSG FPTVSLVDST
1551  LVEVVTASTA SELEGRGTIG ISGAGEISGL PSSELDISGR ASGLPSGTEL
1601  SGQASGSPDV SGEIPGLFGV SGQPSGFPDT SGETSGVTEL SGLSSGQPGV
1651  SGEASGVLYG TSQPFGITDL SGETSGVPDL SGQPSGLPGF SGATSGVPDL
1701  VSGTTSGSGE SSGITFVDTS LVEVAPTTFK EEEGLGSVEL SGLPSGEADL
1751  SGKSGMVDVS GQFSGTVDSS GFTSQTPEFS GLPSGIAEVS GESSRAEIGS
```

```
1801 SLPSGAYYGS GTPSSFPTVS LVDRTLVESV TQAPTAQEAG EGPSGILELS
1851 GAHSGAPDMS GEHSGFLDLS GLQSGLIEPS GEPPGTPYFS GDFASTTNVS
1901 GESSVAMGTS GEASGLPEVT LITSEFVEGV TEPTISQELG QRPPVTHTPQ
1951 LFESSGKVST AGDISGATPV LPGSGVEVSS VPESSSETSA YPEAGFGASA
2001 APEASREDSG SPDLSETTSA FHEANLERSS GLGVSGSTLT FQEGEASAAP
2051 EVSGESTTTS DVGTEAPGLP SATPTASGDR TEISGDLSGH TSQLGVVIST
2101 SIPESEWTQQ TQRPAETHLE IESSSLLYSG EETHTVETAT SPTDASIPAS
2151 PEWKRESEST AAAPARSCAE EPCGAGTCKE TEGHVICLCP PGYTGEHCNI
2201 DQEVCEEGWN KYQGHCYRHF PDRETWVDAE RRCREQQSHL SSIVTPEEQE
2251 FVNNNAQDYQ WIGLNDRTIE GDFRWSDGHP MQFENWRPNQ PDNFFAAGED
2301 CVVMIWHEKG EWNDVPCNYH LPFTCKKGTV ACGEPPVVEH ARTFGQKKDR
2351 YEINSLVRYQ CTEGFVQRHM PTIRCQPSGH WEEPRITCTD PTTYKRRLQK
2401 RSSRHPRRSR PSTAH
```

Structural and functional sites

Signal peptide: 1–19
G1: 48–350
  Immunoglobulin repeat: 48–141
  Link repeats: 152–247, 253–349
IGD: 351–477
G2: 478–673
  Link repeats: 478–572, 579–673
KS attachment domain: 677–861
CS attachment domain 1: 864–1510
CS attachment domain 2: 1511–2162
G3: 2163–2391
  EGF (6C) repeat: 2163–2200
  Lectin repeat: 2201–2329
  CCP repeat: 2330–2391
Potential N-linked glycosylation sites: 126, 239, 333, 387, 434, 602, 657, 737, 1898

Gene structure

The human aggrecan gene is located on chromosome 15q26[6].

References

[1] Baldwin, C.T. et al. (1989) A new epidermal growth factor-like repeat in the human core protein for the large cartilage-specific proteoglycan. J. Biol. Chem. 264: 15747–15750.

[2] Doege, K. et al. (1991) Complete coding sequence and deduced primary structure of the human cartilage large aggregating proteoglycan, aggrecan. J. Biol. Chem. 262: 894–902.

[3] Flannery, C.R. et al. (1992) Identification of a stromelysin cleavage site within the interglobular repeat of human aggrecan. Evidence for proteolysis at this site in vivo in human articular cartilage. J. Biol. Chem. 267: 1008–1014.

[4] Heinegård, D. (1977) Polydispersity of cartilage proteoglycans. Structural variations with size and buoyant density of the molecules. J. Biol. Chem. 252: 1980–1989.

[5] Vilim, V. and Krajickova, J. (1991) Electrophoretic separation of large proteoglycans in large-pore polyacrylamide gradient gels (1.32–10.0% T) and a one-step procedure for simultaneous staining of proteins and proteoglycans. Anal. Biochem. 197: 34–39.

[6] Korenberg, J.R. et al. (1993) Assignment of the human aggrecan gene (AGC1) to 15q26 using fluorescent in situ hybridization analysis. Genomics 16: 546–551.

Agrin

Agrin was identified as a major heparan sulphate proteoglycan in embryonic chick brain and is a component of muscle fibre basal lamina where it is concentrated at the neuromuscular junction. During synapse formation, agrin is believed to promote differentiation of the post-synaptic muscle fibres and the pre-synaptic motor neuron. Extraction of agrin-like proteins from brain homogenates are active in inducing acetylcholine receptor clusters on isolated myotubes, suggesting it might be active in inducing synaptic specializations at inter-neuronal connections. This property appears to be associated with particular splice variants, and null mutants for these variants die shortly before or after birth. These animals never breathe and there is no receptor clustering associated with the motor nerves.

Molecular structure

As deduced from cDNA sequences, the protein has a molecular weight of 225 kDa and contains several domains. The N-terminal part contains nine follistatin-like (kazal-like) domains, implicated in proteinase resistance. The C-terminal portion contains laminin G-like domains interspersed with epidermal growth factor (EGF)-like domains. This region is responsible for receptor clustering. In addition there are two laminin EGF-like domains, and a serine/threonine region interspersed by an unusual module first found in sea urchin sperm protein. A number of splice variants are found that differ in their acetylcholine receptor clustering activity. Chick agrin contains five potential sites for N-linked glycosylation and three sites for glycosaminoglycan attachment [1–6].

Isolation

Agrin has been isolated from the electric organ of *Torpedo californica* by isotonic saline and detergent extraction, the insoluble material was suspended in bicarbonate buffer and the supernatant applied to a Cibacron Blue 3GA column and eluted with a linear salt gradient 0–3 M. Fractions were then applied to a Bio-Gel A 1.5 m column and finally purified on diethylaminoethyl cellulose [1]

Accession number

P25304

Primary structure (rat)

Ala	A	145	Cys	C	141	Asp	D	93	Glu	E	113
Phe	F	61	Gly	G	203	His	H	55	Ile	I	42
Lys	K	50	Leu	L	168	Met	M	22	Asn	N	47
Pro	P	158	Gln	Q	92	Arg	R	115	Ser	S	150
Thr	T	127	Val	V	129	Trp	W	15	Tyr	Y	33

Mol. wt (calc.) = 208 645   Residues = 1959

```
1     MPPLPLEHRP  RQEPGASMLV  RYFMIPCNIC  LILLATSTLG  FAVLLFLSNY
51    KPGIHFTPAP  PTPPDVCRGM  LCGFGAVCEP  SVEDPGRASC  VCKKNACPAT
101   VAPVCGSDAS  TYSNECELQR  AQCNQQRRIR  LLRQGPCGSR  DPCANVTCSF
151   GSTCVPSADG  QTASCLCPTT  CFGAPDGTVC  GSDGVDYPSE  CQLLSHACAS
201   QEHIFKKFNG  PCDPCQGSMS  DLNHICRVNP  RTRHPEMLLR  PENCPAQHTP
251   ICGDDGVTYE  NDCVMSRIGA  TRGLLLQKVR  SGQCQTRDQC  PETCQFNSVC
301   LSRRGRPHCS  CDRVTCDGSY  RPVCAQDGHT  YNNDCWRQQA  ECRQQRAIPP
351   KHQGPCDQTP  SPCHGVQCAF  GAVCTVKNGK  AECECQRVCS  GIYDPVCGSD
401   GVTYGSVCEL  ESMACTLGRE  IQVARRGPCD  PCGQCRFGSL  CEVETGRCVC
451   PSECVESAQP  VCGSDGHTYA  SECELHVHAC  THQISLYVAS  AGHCQTCGEK
501   VCTFGAVCSA  GQCVCPRCEH  PPPGPVCGSD  GVTYLSACEL  REAACQQQVQ
551   IEEAHAGPCE  PAECGSGGSG  SGEDDECEQE  LCRQRGGIWD  EDSEDGPCVC
601   DFSCQSVPRS  PVCGSDGVTY  GTECDLKKAR  CESQQELYVA  AQGACRGPTL
651   APLLPVAFPH  CAQTPYGCCQ  DNFTAAQGVG  LAGCPSTCHC  NPHGSYSGTC
701   DPATGQCSCR  PGVGGLRCDR  CEPGFWNFRG  IVTDGHSGCT  PCSCDPRGAV
751   RDDCEQMTGL  CSCRPGVAGP  KCGQCPDGQV  LGHLGCEADP  MTPVTCVEIH
801   CEFGASCVEK  AGFAQCICPT  LTCPEANSTK  VCGSDGVTYG  NECQLKAIAC
851   RQRLDISTQS  LGPCQESVTP  GASPTSASMT  TPRHILSKTL  PFPHNSLPLS
901   PGSTTHDWPT  PLPISPHTTV  SIPRSTAWPV  LTVPPTAAAS  DVTSLATSIF
951   SESGSANGSG  DEELSGDEEA  SGGGSGGLEP  PVGSIVVTHG  PPIERASCYN
1001  SPLGCCSDGK  TPSLDSEGSN  CPATKAFQGV  LELEGVEGQE  LFYTPEMADP
1051  KSELFGETAR  SIESTLDDLF  RNSDVKKDFW  SVRLRELGPG  KLVRAIVDVH
1101  FDPTTAFQAS  DVGQALLRQI  QVSRPWALAV  RRPLQEHVRF  LDFDWFPTFF
1151  TGAATGTTAA  MATARATTVS  RLPASSVTPR  VYPSHTSRPV  GRTTAPPTTR
1201  RPPTTATNMD  RPRTPGHQQP  SKSCDSQPCL  HGGTCQDQDS  GKGFTCSCTA
1251  GRGGSVCEKV  QPPSMPAFKG  HSFLAFPTLR  AYHTLRLALE  FRALETEGLL
1301  LYNGNARGKD  FLALALLDGR  VQFRFDTGSG  PAVLTSLVPV  EPGRWHRLEL
1351  SRHWRQGTLS  VDGETPVVGE  SPSGTDGLNL  DTNLYVGGIP  EEQVAMVLDR
1401  TSVGVGLKGC  IRMLDINNQQ  LELSDWQRAA  VQSSGVGECG  DHPCLPNPCH
1451  GGALCQALEA  GMFLCQCPPG  RFGPTCADEK  SPCQPNPCHG  AAPCRVLSSG
1501  GAKCECPLGR  SGTFCQTVLE  TAGSRPFLAD  FNGFSYLELK  GLHTFERDLG
1551  EKMALEMVFL  ARGPSGLLLY  NGQKTDGKGD  FVSLALHNRH  LEFCYDLGKG
1601  AAVIRSKEPI  ALGTWVRVFL  ERNGRKGALQ  VGDGPRVLGE  SPKSRKVPHT
1651  MLNLKEPLYI  GGAPDFSKLA  RGAAVSSGFS  GVIQLVSLRG  HQLLTQEHVL
1701  RAVDVSPFAD  HPCTQALGNP  CLNGGSCVPR  EATYECLCPG  GFSGLHCEKG
1751  LVEKSVGDLE  TLAFDGRTYI  EYLNAVIESE  LTNEIPAPET  LDSRALFSEK
1801  ALQSNHFELS  LRTEATQGLV  LWIGKAAERA  DYMALAIVDG  HLQLSYDLGS
1851  QPVVLRSTVK  VNTNRWLRIR  AHREHREGSL  QVGNEAPVTG  SSPLGATQLD
1901  TDGALWLGGL  QKLPVGQALP  KAYGTGFVGC  LRDVVVGHRQ  LHLLEDAVTK
1951  PELRPCPTP
```

Structural and functional sites

Signal peptide: 1–29
Kazal-like repeats: 65–137, 141–212, 213–284, 287–356, 361–429, 430–494, 495–559, 563–645, 794–864
EGF (8C) repeats: 661–718, 719–772
EGF (6C) repeats: 1220–1258, 1440–1477, 1479–1516, 1709–1748
Ser/Thr-rich region: 869–992, 1147–1215
G repeats: 1259–1439, 1517–1708, 1799–1959
Potential N-linked glycosylation sites: 145, 672, 827, 957
Alternatively spliced sites: 1643–1646, 1779–1787, 1788–1798 (N.B. An additional splice variant with the sequence EHRKLLA has been observed in chick agrin starting at approximately residue 49 in the rat sequence)

Gene structure

Theoretically, eight different isoforms are possible, but only five are expressed *in vivo*[6].

References

1. Nitkin, R.M. et al. (1987) Identification of agrin, a synaptic organizing protein from Torpedo electric organ. J. Cell Biol. 105: 2471–2478.
2. Ruegg, M.A. (1996) Agrin, laminin β2 (s-laminin) and ARIA: their role in neuromuscular development. Curr. Opin. Neurobiol. 6: 97–103.
3. Denzer, A.J. et al. (1996) Diverse functions of the extracellular matrix molecule agrin. The Neurosciences 8: 357–366.
4. Tsen, G. et al. (1995) Agrin is a heparan sulphate proteoglycan. J. Biol. Chem. 270: 3392–3399.
5. Gautam, M. et al. (1996) Defective neuromuscular synaptogenesis in agrin-deficient mutant mice. Cell 85: 525–535.
6. O'Connor, L.T. et al. (1994) Localisation and alternative splicing of agrin mRNA in adult rat brain. J. Neurosci. 14: 1141–1152.

Bamacan

BM-CSPG

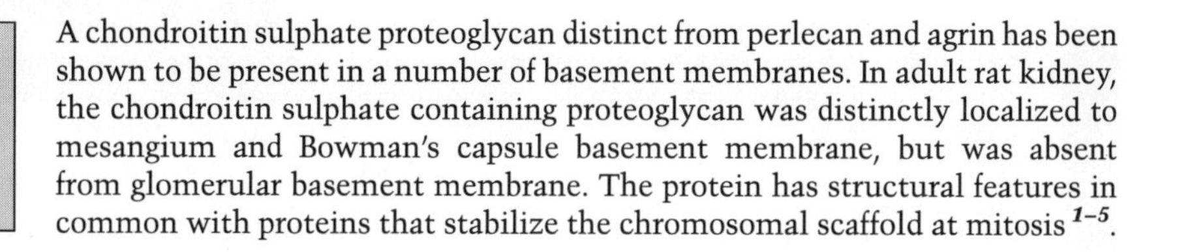

A chondroitin sulphate proteoglycan distinct from perlecan and agrin has been shown to be present in a number of basement membranes. In adult rat kidney, the chondroitin sulphate containing proteoglycan was distinctly localized to mesangium and Bowman's capsule basement membrane, but was absent from glomerular basement membrane. The protein has structural features in common with proteins that stabilize the chromosomal scaffold at mitosis [1–5].

Molecular structure

As deduced from cDNA sequences, the protein has a predicted molecular weight of 138 000. The predicted protein has been divided into five domains. The first 165 amino acids lack cysteine and are hydrophilic. Domain II and IV contain 335 and 364 amino acids, respectively, and have a high potential for coiled-coil structure. Domain III lies between two coiled-coil regions and comprises 165 amino acids; it contains four cysteines and a VTXG potential cell adhesion site. The C-terminal domain V contains 162 amino acids in which there are four Ser-Gly, potential glycosaminoglycan attachment sites [5].

Isolation

Proteoglycans have been extracted from L2 yolk sac and Engelbreth–Holm–Swarm (EHS) tumours with 4 M guanidine–HCl chromatographed on Sephadex G-50 and on DEAE-Sephacel in 8 M urea. Final purification was achieved by isopycnic density gradient centrifugation and size fractionation on Sepharose CL-4B [1].

Accession number

U82626

Primary structure

Ala	A	70	Cys	C	7	Asp	D	66	Glu	E	141
Phe	F	38	Gly	G	52	His	H	17	Ile	I	59
Lys	K	125	Leu	L	122	Met	M	34	Asn	N	56
Pro	P	21	Gln	Q	85	Arg	R	84	Ser	S	74
Thr	T	55	Val	V	52	Trp	W	4	Tyr	Y	29

Mol. wt (calc.) = 138 448 Residues = 1191

```
1     MYIKQVIIQG  FRSYRDQTIV  DPFSSKHNVI  VGRNGSGKSN  FFYAIQFVLS
51    DEFSHLRPEQ  RLALLHEGTG  PRVISAFVEI  IFDNSDNRLP  IDKEEVSLRR
101   VIGAKKDQYF  LDKKMVTKND  VMNLLESAGF  SRSNPYYIVK  QGKINQMATA
151   PDSQRLKLLR  EVAGTRVYDE  RKEESISLMK  ETEGKREKIN  ELLKYIEERL
201   HTLEEEKEEL  AQYQKWDKMR  RALEYTIYNQ  ELNETRAKLD  ELSAKRETSG
251   EKSRQLRDAQ  QDARDKMEDI  ERQVRELKTK  ISAMKEEKEQ  LSAERQEQIK
301   QRTKLELKAK  DLQDELAGNS  EQRKRLLKER  QKLLEKIEEK  QKELAETEPK
351   FNSVKEKEER  GIARLAQATQ  ERTDLYAKQG  RGSQFTSKEE  RDKWIKKELK
401   SLDQAINDKK  RQIAAIHKDL  EDTEANKEKN  LEQYNKLDQD  LNEVKARVEE
451   LDRKYYEVKN  KKDELQSERN  YLWREENAEQ  QALAAKREDL  EKKQQLLRAA
```

```
501   TGKAILNGID SINKVLDHFR RKGINQHVQN GYHGIVMNNF ECEPAFYTCV
551   EVTAGNRLFY HIVDSDEVST KILMEFNKMN LPGEVTFLPL NKLDVRDTAY
601   PETNDAIPMI SKLRYNPRFD KAFKHVFGKT LICRSMEVST QLARAFTMDC
651   ITLEGDQVSH RGALTGGYYD TRKSRLELQK DVRKAEEELG ELEAKLNENL
701   RRNIERINNE IDQLMNQMQQ IETQQRKFKA SRDSTLSEMK MLKEKRQQSE
751   KTFMPKQRSL QSLEASLHAM ESTRESLKAE LGTDLPSQLS LEDQKRVDAL
801   NDEIRQLQQK NRQLLNERIK LEGIITRVET YLNENLRKRL DQVEQELNEL
851   RETEGGTVLT ATTSQLEAIN KRVKDTMARS EDLDNSIDKT EAGIKELQKS
901   MERWKNMEKE HMDAINHDTK ELEKMTNRQG MLLKKKEECM KKIRELGSLP
951   QEAFEKYQTL SLKQLFRKLE QCNTELKKYS HVNKKALDQF VNFSEQKEKL
1001  IKRQEELDRG YKSIMELMNV LELRKYEAIQ LTFKQVSKNF SEVFQKLVPG
1051  AKATLVMKKG DVEGSQSQDE GEGSGESERG SGSQSSVPSV DQFTGVGIRV
1101  SFTGKQGEMR EMQQLSGGQK SLVALALIFA IQKCDPAPFY LFDEIDQALD
1151  AQHRKAVSDM IMELAVHAQF ITTTFRPELL ESADKSSGKS E
```

Structural and functional sites

Signal peptide: Not defined
Potential N-linked glycosylation sites: 233, 992, 1039
O-linked glycosylation sites: 36–37, 249–250, 1073–1074, 1081–1082, 1116–1117, 1187–1188

Gene structure

No details known

References

[1] Iozzo, R.V. and Clark, C.C. (1986) Biosynthesis of proteoglycans by rat embryo parietal yolk sacs in organ culture. J. Biol. Chem. 261: 6658–6669.

[2] Couchman, J.R. et al. (1984) Mapping by monoclonal antibody detection of glycosaminoglycans in connective tissues. Nature 307: 650–652.

[3] Danielson, K.G. et al. (1992) Establishment of a cell line from EHS tumor: biosynthesis of basement membrane constituents and characterisation of a hybrid proteoglycan containing heparan and chondroitin sulphate chains. Matrix 11: 22–35.

[4] Couchman, J.R. et al. (1996) Perlecan and BM-CSPG (Bamacan) are two basement membrane chondroitin/dermatan sulphate proteoglycans in the Engelbreth–Holm–Swarm tumor matrix. J. Biol. Chem. 271: 9595–9602.

[5] Wu, R-R. and Couchman, J.R. (1997) cDNA cloning of the basement membrane chondroitin sulphate proteoglycan core protein, bamacan: a five domain structure including coiled-coil motifs. J. Cell Biol. 136: 433–444.

BEHAB

BEHAB (brain enriched hyaluronan binding protein) is a recently described hyaluronan-binding protein that is restricted to brain. Its expression is developmentally regulated, being detected in late embryonic development and peaking two weeks postnatally. Size and sequence suggest that it functions like link protein in stabilizing interactions between hyaluronan and proteoglycans in brain [1–4]. BEHAB is identical to the N-terminal half of brevican [5].

Molecular structure

As deduced from cDNA sequences, the protein comprises 371 amino acids, which contains a putative signal peptide cleavage site at alanine 22. The resulting mature protein has a predicted molecular weight of 38 447 000, with two potential *N*-glycosylation sites. BEHAB shows sequence homology to the proteoglycan tandem repeat (PTR) family of hyaluronan-binding proteins and the first 350 amino acids contains ten conserved cysteine residues which generate three disulphide-bonded loops. The first loop has homology to the immunoglobulin superfamily followed by two PTR folds (PTR1 and PTR2) [1–3].

Isolation

At present no protein has been isolated.

Accession number

Z28366

Primary structure

Ala	A	41	Cys	C	10	Asp	D	23	Glu	E	19
Phe	F	14	Gly	G	35	His	H	5	Ile	I	10
Lys	K	8	Leu	L	37	Met	M	3	Asn	N	9
Pro	P	28	Gln	Q	14	Arg	R	29	Ser	S	22
Thr	T	10	Val	V	29	Trp	W	6	Tyr	Y	19

Mol. wt (calc.) = 40 633 Residues = 371

```
1     MIPLLLSLLA  ALVLTQAPAA  LADDLKEDSS  EDRAFRVRIG  AAQLRGVLGG
51    WVAIPCHVHH  LRPPPSRRAA  PGFPRVKWTF  LSGDREVEVL  VARGLRVKVN
101   EAYRFRVALP  AYPASLTDVS  LVLSELRPND  SGVYRCEVQH  GIDDSSDAVE
151   VKVKGVVFLY  REGSARYAFS  FAGAQEACAR  IGARIATPEQ  LYAAYLGGYE
201   QCDAGWLSDQ  TVRYPIQNPR  EACYGDMDGY  PGVRNYGVVG  PDDLYDVYCY
251   AEDLNGELFL  GAPPGKLTWE  EARDYCLERG  AQIASTGQLY  AAWNGGLDRC
301   SPGWLADGSV  RYPIITPSQR  CGGGLPGVKT  LFLFPNQTGF  PSKQNRFNVY
351   CFRDSAHPSA  FSEPPAQPLM  D
```

Structural and functional sites

Signal peptide: 1–22
Immunoglobulin repeat: 20–157
PTR repeat: 158–244, 245–338
Potential N-linked glycosylation sites: 129, 336

Gene structure

A single 3.9 kb transcript has been detected.

References

[1] Perkins, S.J. et al. (1989) Immunoglobulin fold and tandem repeat structures in proteoglycan NH_2 terminal domains and link protein. J. Mol. Biol. 206: 737–753.

[2] Doege, K. et al. (1990) Molecular biology of cartilage proteoglycan (aggrecan) and link protein. In: Extracellular Matrix Genes. ed. Sandell, L. and Boyd, C.D. Academic Press, New York. 137–156.

[3] Jaworski, D.M. et al. (1994) BEHAB, a new member of the proteoglycan tandem repeat family of hyaluronan-binding proteins that is restricted to the brain. J. Cell Biol. 125: 495–509.

[4] Yamada, H. et al. (1995) cDNA cloning and the identification of an aggrecanase-like cleavage site in rat brevican. Biochem. Biophys. Res. Comm. 216: 957–963.

[5] Seidenbecher, C.I. et al. (1995) Brevican, a chondroitin sulphate proteoglycan of rat brain, occurs as secreted and cell surface glycosylphosphatidylinositol-anchored isoforms. J. Biol. Chem. 270: 27206–27212.

Biglycan

PG-S1, PGI, byglycan

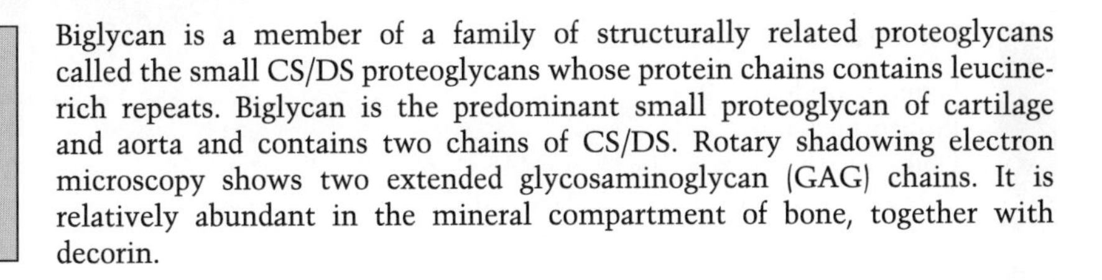

Biglycan is a member of a family of structurally related proteoglycans called the small CS/DS proteoglycans whose protein chains contains leucine-rich repeats. Biglycan is the predominant small proteoglycan of cartilage and aorta and contains two chains of CS/DS. Rotary shadowing electron microscopy shows two extended glycosaminoglycan (GAG) chains. It is relatively abundant in the mineral compartment of bone, together with decorin.

Molecular structure

Biglycan contains two CS/DS side-chains, but most of the protein consists of 12 repeats of 24 residues. These leucine-rich sequences share marked homology (80% identity) with sequences in other proteoglycans, such as decorin and fibromodulin, and in other proteins including the serum protein LRG, platelet surface protein GPIb, ribonuclease/angiotensin inhibitor (RAI), chaoptin, toll protein and adenylate cyclase [1,2].

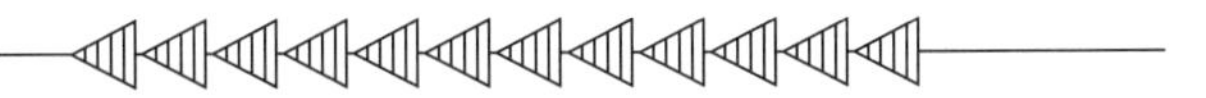

Isolation

Biglycan can be isolated from a variety of tissues including human bone and articular cartilage by guanidine HCl, density gradient centrifugation, gel filtration, ion exchange chromatography and octyl-Sepharose chromatography [3,4].

Accession number

P21810; P13247

Primary structure

Ala	A	14	Cys	C	7	Asp	D	23	Glu	E	18
Phe	F	16	Gly	G	24	His	H	11	Ile	I	18
Lys	K	22	Leu	L	53	Met	M	8	Asn	N	27
Pro	P	24	Gln	Q	11	Arg	R	18	Ser	S	24
Thr	T	12	Val	V	23	Trp	W	4	Tyr	Y	11

Mol. wt (calc.) = 41 550 Residues = 368

```
1    MWPLWRLVSL LALSQALPFE QRGFWDFTLD DGPFMMNDEE ASGADTSGVL
51   DPDSVTPTYS AMCPFGCHCH LRVVQCSDLG LKSVPKEISP DTTLLDLQNN
101  DISELRKDDF KGLQHLYALV LVNNKISKIH EKAFSPLRNV QKLYISKNHL
151  VEIPPNLPSS LVDVRIHDNR IRKVPKGVFS GLRNMNCIEM GGNPLENSGF
201  EPGAFDGLKL NYLRISEAKL TGIPKDLPET LNELHLDHNK IQAIELEDLL
251  RYSKLYRLGL GHNQIRMIEN GSLSFLPTLR ELHLDNNKLA RVPSGLPDLK
301  LLQVVYLHSN NITKVGVNDF CPMGFGVKRA YYNGISLFNN PVPYWEVQPA
351  TFRCVTDRLA IQFGNYKK
```

Structural and functional sites

Signal peptide: 1–19
Propeptide: 20–37
Leucine-rich repeats: 71–94, 95–115, 116–139, 140–163, 164–184, 185–210, 211–230, 231–254, 255–275, 276–301, 302–320, 335–342
Potential GAG attachment sites: 270, 311
Determined GAG attachment sites: 42, 47

Gene structure

The human biglycan gene is a single copy located on chromosome Xq27-qter. It is approximately 8 kb in size and contains eight exons [5].

References

1. Fisher, L.W. et al. (1989) Deduced protein sequence of bone small proteoglycan (biglycan) shows homology with proteoglycan II (decorin) and several non-connective tissue proteins in a variety of species. J. Biol. Chem. 264: 4571–4576.
2. Roughley, P.J. and White, R.J. (1989) Dermatan sulphate proteoglycans of human articular cartilage. The properties of dermatan sulphate proteoglycans I and II. Biochem. J. 262: 823–827.
3. Fisher, L.W. et al. (1987) Purification and partial characterization of small proteoglycans I and II, bone sialoproteins I and II, and osteonectin from the mineral compartment of developing human bone. J. Biol. Chem. 262: 9702–9708.
4. Choi, H.U. et al. (1989) Characterization of the dermatan sulfate proteoglycans, DS-PGI and DS-PGII, from bovine articular cartilage and skin isolated by octyl-Sepharose chromatography. J. Biol. Chem. 264: 2876–2884.
5. Fisher, L.W. et al. (1991) Human biglycan gene – putative promoter, intron–exon junctions, and chromosomal localization. J. Biol. Chem. 266: 14371–14377.

Bone sialoprotein

BSP, bone sialoprotein II, BSPII

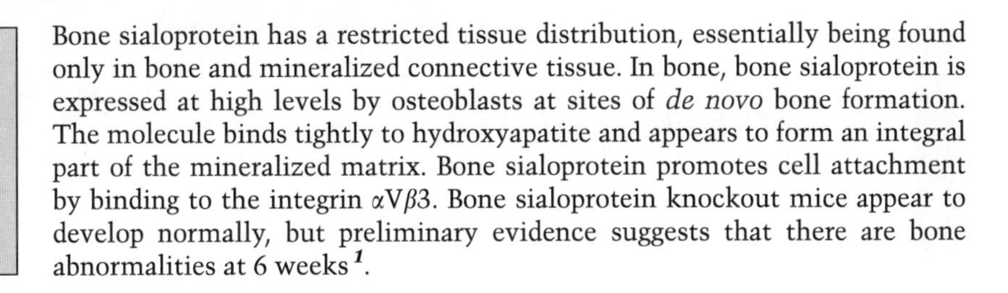

Bone sialoprotein has a restricted tissue distribution, essentially being found only in bone and mineralized connective tissue. In bone, bone sialoprotein is expressed at high levels by osteoblasts at sites of *de novo* bone formation. The molecule binds tightly to hydroxyapatite and appears to form an integral part of the mineralized matrix. Bone sialoprotein promotes cell attachment by binding to the integrin $\alpha V\beta 3$. Bone sialoprotein knockout mice appear to develop normally, but preliminary evidence suggests that there are bone abnormalities at 6 weeks [1].

Molecular structure

Bone sialoprotein is composed of a single polypeptide chain. Its primary sequence reveals a paucity of hydrophobic residues which predicts an open, flexible structure. Bone sialoprotein is isolated from bone matrix as a component of approximate molecular weight 60 000, containing both O- and N-linked oligosaccharides that are rich in sialic acid. The molecule is extensively glycosylated, about half of its serine residues are phosphorylated, and it contains extended sequences of acidic amino acid residues. Glutamic acid, glycine and aspartic acid account for about 30% of all amino acids. The glutamates are clustered, in two cases as eight consecutive residues. Both ends of the molecule, but particularly the COOH-terminus, contain high contents of tyrosine residues, many of which are sulphated. A single RGD cell adhesion site is located near the COOH-terminus in the middle of clusters of sulphated tyrosine residues. The rabbit molecule contains a keratan sulphate chain and is therefore a proteoglycan [2–7].

Isolation

Bone sialoprotein can be isolated from bone by extraction with demineralizing solutions (e.g. 0.5 M EDTA), and purified by molecular sieve and DEAE ion exchange chromatography [8].

Accession number

P21815

Primary structure

Ala	A	16	Cys	C	1	Asp	D	16	Glu	E	60
Phe	F	7	Gly	G	38	His	H	5	Ile	I	7
Lys	K	15	Leu	L	11	Met	M	3	Asn	N	24
Pro	P	12	Gln	Q	8	Arg	R	8	Ser	S	25
Thr	T	30	Val	V	8	Trp	W	0	Tyr	Y	23

Mol. wt (calc.) = 34 982 Residues = 317

```
1     MKTALILLSI  LGMACAFSMK  NLHRRVKIED  SEENGVFKYR  PRYYLYKHAY
51    FYPHLKRFPV  QGSSDSSEEN  GDDSSEEEEE  EEETSNEGEN  NEESNEDEDS
101   EAENTTLSAT  TLGYGEDATP  GTGYTGLAAI  QLPKKAGDIT  NKATKEKESD
151   EEEEEEEEGN  ENEESEAEVD  ENEQGINGTS  TNSTEAENGN  GSSGGDNGEE
201   GEEESVTGAN  AEGTTETGGQ  GKGTSKTTTS  PNGGFEPTTP  PQVYRTTSPP
251   FGKTTTVEYE  GEYEYTGVNE  YDNGYEIYES  ENGEPRGDNY  RAYEDEYSYF
301   KGQGYDGYDG  QNYYHHQ
```

Structural and functional sites

Signal peptide: 1–16
Poly-Glu region: 76–83, 151–158
Potential N-linked glycosylation sites: 104, 177, 182, 190
RGD cell adhesion site: 286–288
Phosphorylation sites: 67, 74, 149, 184
Sulphation sites: 271, 275, 290, 293, 297, 299

Gene structure

The gene appears to be a single copy located on human chromosome 4q28–q31. The human gene is approximately 15 kb in length and contains seven exons, the first of which is non-coding[5,9]. Two messages have been detected, a major 2.0 kb and a minor 3.0 kb, presumably from alternative poly-A sites.

References

1. Aubin, J.E. et al. (1996) Knockout mice lacking bone sialoprotein expression have bone abnormalities. J. Bone Mineral Res. 11: 30A.
2. Kinne, R.W. and Fisher, L.W. (1987) Keratan sulfate proteoglycan in rabbit compact bone is bone sialoprotein II. J. Biol. Chem. 262: 10206–10211.
3. Oldberg, A. et al. (1988) The primary structure of a cell-binding bone sialoprotein. J. Biol. Chem. 263: 19430–19432.
4. Oldberg, A. et al. (1988) Identification of a bone sialoprotein receptor in osteosarcoma cells. J. Biol. Chem. 263: 19433–19436.
5. Fisher, L.W. et al. (1990) Human bone sialoprotein. Deduced protein sequence and chromosomal localization. J. Biol. Chem. 265: 2347–2351.
6. Midura, R.J. et al. (1990) A rat osteogenic cell line (UMR.106.01) synthesizes a highly sulfated form of bone sialoprotein. J. Biol. Chem. 265: 5285–5291.
7. Sodek, J. et al. (1992) Elucidating the functions of bone sialoprotein and osteopontin in bone formation. In: Chemistry and Biology of Mineralized Tissues, ed. Slavkin, H. and Price, P., Elsevier, Amsterdam, pp. 297–306.
8. Franzen, A. and Heinegard, D. (1985) Isolation and characterisation of two sialoproteins present only in bone calcified matrix. Biochem. J. 232: 715–724.
9. Kerr, J.M. et al. (1993) The human bone sialoprotein gene (IBSP): Genomic localization and characterization. Genomics 17: 408–415.

Brevican

Brevican is a chondroitin sulphate proteoglycan found in brain tissue, which is a member of the hyaluronan-binding family of proteoglycans, aggrecan, versican, neurocan. While the actual glycosaminoglycan (GAG) attachment sites have not been delineated, the potential number is three. Like a number of other molecules, brevican is sometimes found without attached GAG chains in brain. In the rat there is evidence of a truncated version that contains a glypiation site, and may be membrane bound. Role of brevican is unclear, but it has been suggested that the spatial and temporal expression of chondroitin sulphates in brain play a critical role in the growth and guidance of axons. BEHAB, a hyaluronan binding protein found in brain, is identical to the N-terminal half of brevican [1–4].

Molecular structure

cDNA predicts a protein of 883 amino acids with a calculated mass of 95 916 Da. The hyaluronate-binding domain at the N-terminus consists of an Ig-like loop and two link protein-like tandem repeats, there is a central non-homologous domain which is much shorter than that found in other members of this family. The COOH-terminal domain of brevican contains one epidermal growth factor (EGF) repeat, a lectin-like domain and a complement regulatory protein domain [1,2].

Isolation

Total soluble proteoglycans have been extracted from brain tissue with 0.3 M sucrose/HEPES buffer and separated from other proteins by DEAE Sepharose chromatography, after chondroitinase digestion an 80 kDa fragment of the protein core could be purified by reverse-phase HPLC on a Vydac C4 column [1].

Accession number

S57563

Primary structure (mouse)

Ala	A	78	Cys	C	27	Asp	D	44	Glu	E	77
Phe	F	23	Gly	G	82	His	H	14	Ile	I	22
Lys	K	16	Leu	L	86	Met	M	6	Asn	N	18
Pro	P	80	Gln	Q	35	Arg	R	63	Ser	S	79
Thr	T	35	Val	V	52	Trp	W	17	Tyr	Y	29

Mol. wt (calc.) = 96 013 Residues = 883

```
1     MIPLLLSLLA  ALVLTQAPAA  LADDLKEDSS  EDRAFRVRIG  AAQLRGVLGG
51    ALAIPCHVHH  LRPPRSRRAA  PGFPRVKWTF  LSGDREVEVL  VARGLRVKVN
101   EAYRFRVALP  AYPASLTDVS  LVLSELRPND  SGVYRCEVQH  GIDDSSDAVE
151   VKVKGVVFLY  REGSARYAFS  FAGAQEACAR  IGARIATPEQ  LYAAYLGGYE
201   QCDAGWLSDQ  TVRYPIQNPR  EACSGDMDGY  PGVRNYGVVG  PDDLYDVYCY
```

```
251  AEDLNGELFL GAPPSKLTWE EARDYCLERG AQIASTGQLY AAWNGGLDRC
301  SPGWLADGSV RYPIITPSQR CGGGLPGVKT LFLFPNQTGF PSKQNRFNVY
351  CFRDSAHPSA SSEASSPASD GLEAIVTVTE KLEELQLPQE AMESESRGAI
401  YSIPISEDGG GGSSTPEDPA EAPRTPLESE TQSIAPPTES SEEEGVALEE
451  EERFKDLEAL EEEKEQEDLW VWPRELSSPL PTGSETEHSL SQVSPPAQAV
501  LQLDASPSPG PPRFRGPPAE TLLPPREWSA TSTPGGAREV GGETGSPELS
551  GVPRESEEAG SSSLEDGPSL LPATWAPVGP RELETPSEEK SGRTVLAGTS
601  VQAQPVLPTD SASHGGVAVA PSSGDCIPSP CHNGGTCLEE KEGFRCLCLP
651  GYGGDLCDVG LHFCSPGWEA FQGACYKHFS TRRSWEEAES QCRALGAHLT
701  SICTPEEQDF VNDRYREYQW IGLNDRTIEG DFLWSDGAPL LYENWNPGQP
751  DSYFLSGENC VVMVWHDQGQ WSDVPCNYHL SYTCKMGLVS CGPPPQLPLA
801  QIFGRPRLRY AVDTVLRYRC RDGLAQRNLP LIRCQENGLW EAPQISCVPR
851  RPGRALRSMD APEGPRGQLS RHRKAPLTPP SSL
```

Structural and functional sites

Signal peptide: 1–22
Hyaluronic acid binding domain: 35–352
  Immunoglobulin repeat: 35–158
  Link protein repeats: 159–353
Potential N-linked glycosylation sites: 129, 336
EGF (6C) repeat: 621–660
Lectin repeat: 661–814
CCP repeat: 815–883
Potential N-linked glycosylation sites: 130, 337

Gene structure

Northern blot analysis to total brain RNA revealed a single band of 3.3 kb. In rat brain tissue two transcripts of 3.3 and 3.6 kb are found, and suggest that the secreted and GPI-anchored forms are synthesized from alternatively processed transcripts of the same gene [1,2].

References

[1] Yamada, H. et al. (1994) Molecular cloning of brevican, a novel brain proteoglycan of the aggrecan/versican family. J. Biol. Chem. 269: 10119–10126.

[2] Seidenbecher, C.I. et al. (1995) Brevican, a chondroitin sulfate proteoglycan of rat brain, occurs as secreted and cell surface glycosylphosphatidylinositol-anchored isoforms. J. Biol. Chem. 270: 27206–27212.

[3] Yamada, H. et al. (1995) cDNA cloning and the identification of an aggrecanase-like cleavage site in rat brevican. Biochem. Biophys. Res. Comm. 216: 957–963.

[4] Jaworski, D.M. et al. (1994) BEHAB, a new member of the proteoglycan tandem repeat family of hyaluronan-binding proteins that is restricted to the brain. J. Cell Biol. 125: 495–509.

Cartilage matrix protein

Cartilage matrix protein is a major component of the extracellular matrix of non-articular cartilage. It is found in tracheal, nasal septal, xiphisternal, auricular and epiphyseal cartilage, but not in articular cartilage or extracts of the intervertebral disc. While cartilage matrix protein has been shown to account for 5% of the wet weight of aged tracheal cartilage, its function is not known, although it has been suggested that it can bind to and bridge type II collagen fibrils. Recent work suggests that cartilage matrix protein may act as a marker for post-mitotic chondrocytes [1].

Molecular structure

Cartilage matrix protein is a 54 kDa protein which occurs in cartilage as a disulphide-bonded multimer. The protein is normally isolated as a homotrimer (molecular weight approximately 148 000). Cartilage matrix protein contains two repeating sequences of 190 amino acids that are homologous to the A-type repeats of von Willebrand factor, separated by a six-cysteine epidermal growth factor repeat. The CMP1 and CMP2 repeats contain a cysteine at each end which may facilitate intra-domain disulphide bonding [2–4].

Isolation

Cartilage matrix protein can be isolated from cartilage by extraction with 4 M guanidine–HCl followed by fractionation on a caesium chloride density gradient under associative conditions. The top of the gradient is then fractionated by size exclusion chromatography on Sephadex G-200 and/or Sepharose 4B [5].

Accession number

P21941

Primary structure

Ala	A	42	Cys	C	14	Asp	D	29	Glu	E	28
Phe	F	22	Gly	G	36	His	H	7	Ile	I	22
Lys	K	33	Leu	L	43	Met	M	8	Asn	N	11
Pro	P	15	Gln	Q	23	Arg	R	26	Ser	S	48
Thr	T	26	Val	V	53	Trp	W	0	Tyr	Y	10

Mol. wt (calc.) = 53 639 Residues = 496

```
1    MRVLSGTSLM LCSLLLLLQA LCSPGLAPQS RGHLCRTRPT DLVFVVDSSR
51   SVRPVEFEKV KVFLSQVIES LDVGPNATRV GMVNYASTVK QEFSLRAHVS
101  KAALLQAVRR IQPLSTGTMT GLAIQFAITK AFGDAEGGRS RSPDISKVVI
151  VVTDGRPQDS VQDVSARARA SGVELFAIGV GSVDKATLRQ IASEPQDEHV
201  DYVESYSVIE KLSRKFQEAF CVVSDLCATG DHDCEQVCIS SPGSYTCACH
251  EGFTLNSDGK TCNVCSGGGG SSATDLVFLI DGSKSVRPEN FELVKKFISQ
301  IVDTLDVSDK LAQVGLVQYS SSVRQEFPLG RFHTKKDIKA AVRNMSYMEK
351  GTMTGAALKY LIDNSFTVSS GARPGAQKVG IVFTDGRSQD YINDAAKKAK
401  DLGFKMFAVG VGNAVEDELR EIASEPVAEH YFYTADFKTI NQIGKKLQKK
451  ICVEEDPCAC ESLVKFQAKV EGLLQALTRK LEAVSKRLAI LENTVV
```

Structural and functional sites

Signal peptide: 1–22
von Willebrand factor A repeats: 23–222, 264–453
EGF (6C) repeat: 223–263
Potential N-linked glycosylation site: 76
Potential intra-molecular disulphide bonds: 35–221, 265–452

Gene structure

The human cartilage matrix protein gene spans 12 kb and has eight exons. It appears to be a single copy gene located on chromosome 1 (locus p35)[3,4].

References

[1] Chen, Q. et al. (1995) Progression and recapitulation of the chondrocyte differentiation program: Cartilage matrix protein is a marker for cartilage maturation. Develop. Biol. 172: 293–306.
[2] Heinegard, D. and Paulsson, M. (1987) Cartilage. Methods Enzymol. 145: 336–363.
[3] Kiss, I. et al. (1989) Structure of the gene for cartilage matrix protein, a modular protein of the extracellular matrix. J. Biol. Chem. 264: 8126–8134.
[4] Jenkins, R.N. et al. (1990) Structure and chromosomal location of the human gene encoding cartilage matrix protein. J. Biol. Chem. 265: 19624–19631.
[5] Winterbottom, N. et al. (1992) Cartilage matrix protein is a component of the collagen fibril of cartilage. Dev. Dynamics 193: 266–276.

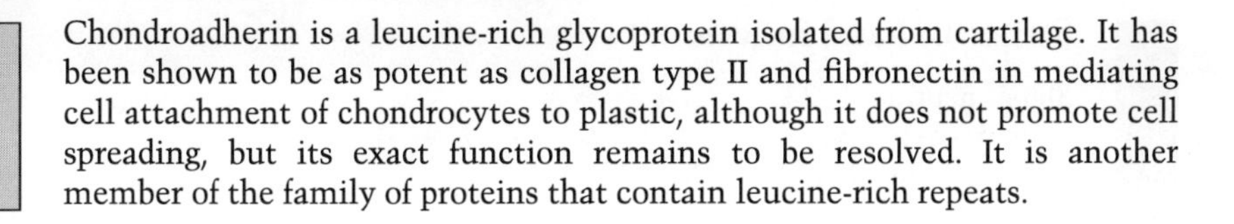

Chondroadherin

38 kDa leucine protein, 36 kDa

Chondroadherin is a leucine-rich glycoprotein isolated from cartilage. It has been shown to be as potent as collagen type II and fibronectin in mediating cell attachment of chondrocytes to plastic, although it does not promote cell spreading, but its exact function remains to be resolved. It is another member of the family of proteins that contain leucine-rich repeats.

Molecular structure

Chondroadherin exists in two major isoforms, one with a calculated molecular weight of 38 353 and a pI of 9.76, the other with a molecular weight of 37 304 and a pI of 9.5. The principal feature of the protein chain is a series of ten leucine-rich repeats, the most N-terminal of which contains a free cysteine. The majority of chondroadherin is monomeric, although it can form oligomers, particularly dimers under conditions which promote disulphide rearrangement [1–4].

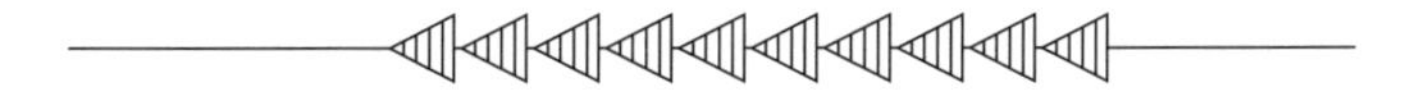

Isolation

Chondroadherin can be extracted from articular cartilage in 4 M guanidine hydrochloride, the extract allowed to reassociate and fractionated on caesium chloride density gradient centrifugation. The low-density fraction was applied to a Sephacryl S300 column and finally separated by reverse-phase chromatography [1].

Accession number

A53860

Primary structure

Ala	A	26	Cys	C	9	Asp	D	17	Glu	E	14
Phe	F	16	Gly	G	16	His	H	13	Ile	I	9
Lys	K	22	Leu	L	65	Met	M	3	Asn	N	24
Pro	P	18	Gln	Q	15	Arg	R	24	Ser	S	27
Thr	T	17	Val	V	14	Trp	W	4	Tyr	Y	8

Mol. wt (calc.) = 40 861 Residues = 361

```
1     MARPMLLLSL  SLGLLASLLP  ALAACPQNCH  CHSDLQHVIC  DKVGLQKIPK
51    VSEKTKLLNL  QRNNFPVLAT  NSFRAMPNLV  SLHLQHCQIR  EVAAGAFRGL
101   KQLIYLYLSH  NDIRVLRAGA  FDDLTELTYL  YLDHNKVTEL  PRGLLSPLVN
151   LFILQLNNNK  IRELRSGAFQ  GAKDLRWLYL  SENSLSSLQP  GALDDVENLA
201   KFYLDRNQLS  SYPSAALSKL  RVVEELKLSH  NPLKSIPDNA  FQSFGRYLET
251   LWLDNTNLEK  FSDGAFLGVT  TLKHVHLENN  RLHQLPSNFP  FDSLETLTLT
301   NNPWKCTCQL  RGLRRWLEAK  TSRPDATCAS  PAKFRGQHIR  DTDAFRGCKF
351   PTKRSKKAGR  H
```

Structural and functional sites

Signal peptide: 1–24
Leucine-rich repeats: 21–44, 45–68, 69–92, 93–116, 117–140, 141–164, 165–188, 189–212, 213–237, 238–261

Gene structure

The mRNA is 1.6 kb of which 511 base pairs is 3′ untranslated. There is evidence of heterogeneity at the 5′ end of the message[4].

References

[1] Larsson, T. et al. (1991) Cartilage matrix protein. A basic 36 kDa protein with restricted distribution to cartilage and bone. J. Biol. Chem. 266, 20428–20433.

[2] Sommarin, Y. et al. (1989) Chondrocyte–matrix interactions. Attachment to proteins isolated from cartilage. Exp. Cell Res. 184, 181–192.

[3] Paulsson, M. et al. (1983) Metabolism of cartilage proteins in cultured tissue sections. Biochem. J. 212, 659–667.

[4] Neame, P.J. et al. (1994) The structure of a 38-kDa leucine-rich protein (chondroadherin) isolated from bovine cartilage. J. Biol. Chem. 269, 21547–21554.

Collagen type I

Type I collagen is the major fibrillar collagen of bone, tendon and skin; it provides these and many other tissues with tensile strength. Type I collagen forms rope-like structures in tendon, sheet-like structures in skin, and in bone is reinforced with calcium hydroxyapatite.

More than one hundred mutations have been observed in the COL1A1 and COL1A2 genes, resulting in various forms of osteogenesis imperfecta (brittle bone disease) and in the Ehlers–Danlos syndrome type VII which is characterized by joint hypermobility and skin fragility[1–2]. A mouse strain (Mov13) which is homozygous for a viral insertion in the first intron of the COL1A1 gene is essentially a COL1A1 knockout and the mutation is lethal[1].

Molecular structure

Type I collagen is synthesized primarily as a heterotrimeric procollagen comprising two proα1(I) chains and one proα2(I) chain. The procollagen is processed extracellularly by N- and C-proteinases to give a triple-helical molecule that can assemble with a stagger of 234 amino acids into cross-banded fibrils with a 67 nm (D) periodicity. These fibrils are stabilized by inter-molecular cross-links derived from specific lysine/hydroxylysine residues in both the non-helical (telopeptide) and helical domains. The fibrillar form of type I collagen interacts with the protein cores of the proteoglycans decorin and fibromodulin. The triple-helical domain interacts with cells via integrin receptors, principally $\alpha 2\beta 1$, and a tetrapeptide Asp-Gly-Glu-Ala (DGEA) has been reported to account for this binding. A homopolymer comprising three identical α1(I) chains (designated collagen type I trimer) has also been observed in fetal and diseased tissues but it is not a significant component of normal adult tissues[1,3–7].

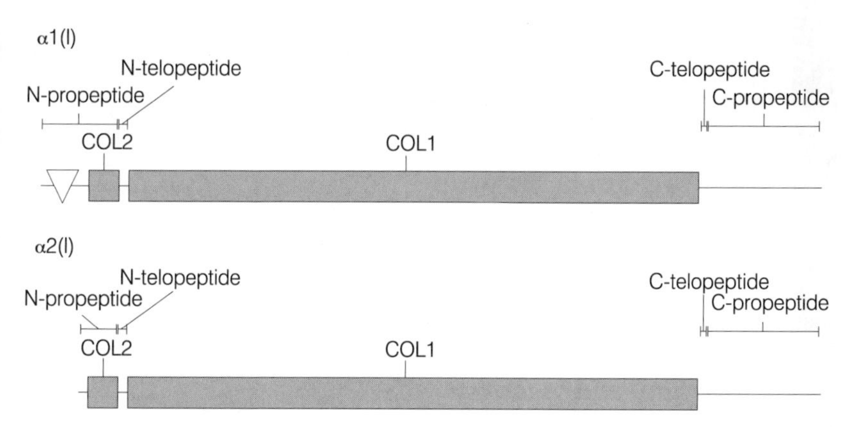

Isolation

Type I procollagen can be isolated from cultured fibroblasts[1]. Type I collagen can be prepared from fetal skin in its intact, processed form by extraction with 0.5 M acetic acid or in a slightly shorter form by pepsin digestion[8].

Accession number

P02452

Primary structure: α1(I) chain

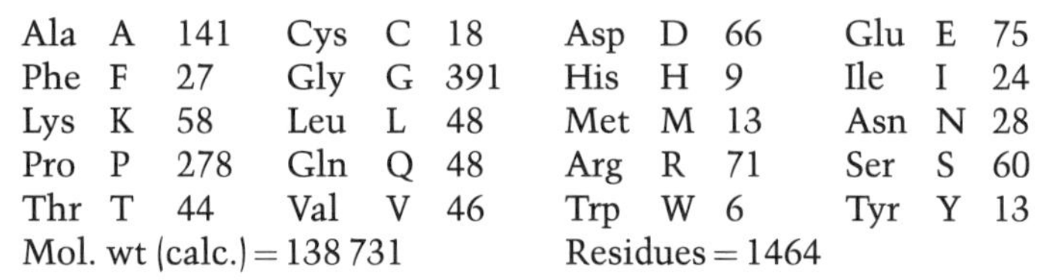

Ala	A	141	Cys	C	18	Asp	D	66	Glu	E	75
Phe	F	27	Gly	G	391	His	H	9	Ile	I	24
Lys	K	58	Leu	L	48	Met	M	13	Asn	N	28
Pro	P	278	Gln	Q	48	Arg	R	71	Ser	S	60
Thr	T	44	Val	V	46	Trp	W	6	Tyr	Y	13

Mol. wt (calc.) = 138 731 Residues = 1464

```
1     MFSFVDLRLL LLLAATALLT HGQEEGQVEG QDEDIPPITC VQNGLRYHDR
51    DVWKPEPCRI CVCDNGKVLC DDVICDETKN CPGAEVPEGE CCPVCPDGSE
101   SPTDQETTGV EGPKGDTGPR GPRGPAGPPG RDGIPGQPGL PGPPGPPGPP
151   GPPGLGGNFA PQLSYGYDEK STGGISVPGP MGPSGPRGLP GPPGAPGPQG
201   FQGPPGEPGE PGASGPMGPR GPPGPPGKNG DDGEAGKPGR PGERGPPGPQ
251   GARGLPGTAG LPGMKGHRGF SGLDGAKGDA GPAGPKGEPG SPGENGAPGQ
301   MGPRGLPGER GRPGAPGPAG ARGNDGATGA AGPPGPTGPA GPPGFPGAVG
351   AKGEAGPQGP RGSEGPQGVR GEPGPPGPAG AAGPAGNPGA DGQPGAKGAN
401   GAPGIAGAPG FPGARGPSGP QGPGGPPGPK GNSGEPGAPG SKGDTGAKGE
451   PGPVGVQGPP GPAGEEGKRG ARGEPGPTGL PGPPGERGGP GSRGFPGADG
501   VAGPKGPAGE RGSPGPAGPK GSPGEAGRPG EAGLPGAKGL TGSPGSPGPD
551   GKTGPPGPAG QDGRPGPPGP PGARGQAGVM GFPGPKGAAG EPGKAGERGV
601   PGPPGAVGPA GKDGEAGAQG PPGPAGPAGE RGEQGPAGSP GFQGLPGPAG
651   PPGEAGKPGE QGVPGDLGAP GPSGARGERG FPGERGVQGP PGPAGPRGAN
701   GAPGNDGAKG DAGAPGAPGS QGAPGLQGMP GERGAAGLPG PKGDRGDAGP
751   KGADGSPGKD GVRGLTGPIG PPGPAGAPGD KGESGPSGPA GPTGARGAPG
801   DRGEPGPPGP AGFAGPPGAD GQPGAKGEPG DAGAKGDAGP PGPAGPAGPP
851   GPIGNVGAPG AKGARGSAGP PGATGFPGAA GRVGPPGPSG NAGPPGPPGP
901   AGKEGGKGPR GETGPAGRPG EVGPPGPPGP AGEKGSPGAD GPAGAPGTPG
951   PQGIAGQRGV VGLPGQRGER GFPGLPGPSG EPGKQGPSGA SGERGPPGPM
1001  GPPGLAGPPG ESGREGAPGA EGSPGRDGSP GAKGDRGETG PAGPPGAPGA
1051  PGAPGPVGPA GKSGDRGETG PAGPAGPVGP AGARGPAGPQ GPRGDKGETG
1101  EQGDRGIKGH RGFSGLQGPP GPPGSPGEQG PSGASGPAGP RGPPGSAGAP
1151  GKDGLNGLPG PIGPPGPRGR TGDAGPVGPP GPPGPPGPPG PPSAGFDFSF
1201  LPQPPQEKAH DGGRYYRADD ANVVRDRDLE VDTTLKSLSQ QIENIRSPEG
1251  SRKNPARTCR DLKMCHSDWK SGEYWIDPNQ GCNLDAIKVF CNMETGETCV
1301  YPTQPSVAQK NWYISKNPKD KRHVWFGESM TDGFQFEYGG QGSDPADVAI
1351  QLTFLRLMST EASQNITYHC KNSVAYMDQQ TGNLKKALLL KGSNEIEIRA
1401  EGNSRFTYSV TVDGCTSHTG AWGKTVIEYK TTKTSRLPII DVAPLDVGAP
1451  DQEFGFDVGP VCFL
```

Structural and functional sites

Signal peptide: 1–22
N-Propeptide: 23–161
von Willebrand factor C repeat: 35–103
  COL2 domain: 109–159
N-Telopeptide: 162–178
Helical domain: 179–1192

C-Telopeptide: 1193–1218
C-Propeptide: 1219–1464
Lysine/hydroxylysine cross-linking sites: 170, 265, 1108, 1208
Potential N-linked glycosylation site: 1365
N-Proteinase cleavage site: 161–162
C-Proteinase cleavage site: 1218–1219
DGEA cell adhesion site: 613–616
Mammalian collagenase cleavage site: 953–954

Accession number

P02464, P08123

Primary structure: α2(I) chain

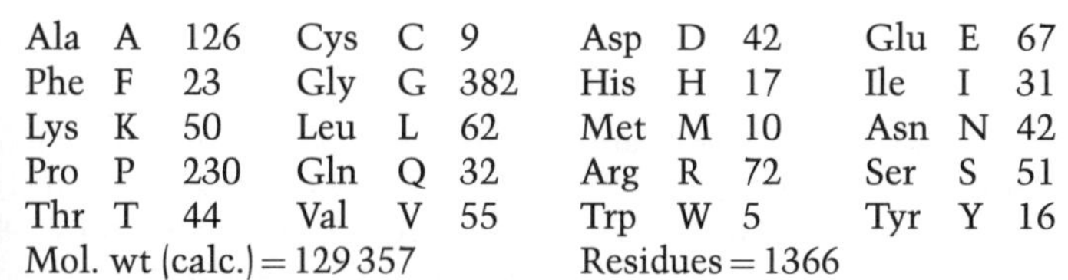

Ala	A	126	Cys	C	9	Asp	D	42	Glu	E	67
Phe	F	23	Gly	G	382	His	H	17	Ile	I	31
Lys	K	50	Leu	L	62	Met	M	10	Asn	N	42
Pro	P	230	Gln	Q	32	Arg	R	72	Ser	S	51
Thr	T	44	Val	V	55	Trp	W	5	Tyr	Y	16

Mol. wt (calc.) = 129 357 Residues = 1366

```
1     MLSFVDTRTL LLLAVTLCLA TCQSLQEETV RKGPAGDRGP RGERGPPGPP
51    GRDGEDGPTG PPGPPGPPGP PGLGGNFAAQ YDGKGVGLGP GPMGLMGPRG
101   PPGAAGAPGP QGFQGPAGEP GEPGQTGPAG ARGPAGPPGK AGEDGHPGKP
151   GRPGERGVVG PQGARGFPGT PGLPGFKGIR GHNGLDGLKG QPGAPGVKGE
201   PGAPGENGTP GQTGARGLPG ERGRVGAPGP AGARGSDGSV GPVGPAGPNG
251   SAGPPGFPGA PGPKGEIGAV GNAGPTGPAG PRGEVGLPGL SGPVGPPGNP
301   GANGLTGAKG AAGLPGVAGA PGLPGPRGIP GPPGAAGTTG ARGLVGEPGP
351   AGSKGESGNK GEPGSAGPQG PPGPSGEEGK RGPNGEAGSA GPPGPPGLRG
401   SPGSRGLPGA DGRAGVMGPP GSRGASGPAG VRGPNGDAGR PGEPGLMGPR
451   GLPGSPGNIG PAGKEGPVGL PGIDGRPGPI GPVGARGEPG NIGFPGPKGP
501   TGDPGKNGDK GHAGLAGARG APGPDGNNGA QGPPGPQGVQ GGKGEQGPAG
551   PPGFQGLPGP SGPAGEVGKP GERGLHGEFG LPGPAGPRGE RGPPGESGAA
601   GPTGPIGSRG PSGPPGPDGN KGEPGVVGAV GTAGPSGPSG LPGERGAAGI
651   PGGKGEKGEP GLRGEIGNPG RDGARGAHGA VGAPGPAGAT GDRGEAGAAG
701   PAGPAGPRGS PGERGEVGPA GPNGFAGPAG AAGQPGAKGE RGGKGPKGEN
751   GVVGPTGPVG AAGPAGPNGP PGPAGSRGDG GPPGMTGFPG AAGRTGPPGP
801   SGISGPPGPP GPAGKEGLRG PRGDQGPVGR TGEVGAVGPP GFAGEKGPSG
851   EAGTAGPPGT PGPQGLLGAP GILGLPGSRG ERGLPGVAGA VGEPGPLGIA
901   GPPGARGPPG AVGSPGVNGA PGEAGRDGNP GNDGPPGRDG QPGHKGERGY
951   PGNIGPVGAA GAPGPHGPVG PAGKHGNRGE TGPSGPVGPA GAVGPRGPSG
1001  PQGIRGDKGE PGEKGPRGLP GFKGHNGLQG LPGIAGHHGD QGAPGSVGPA
1051  GPRGPAGPSG PAGKDGRTGH PGTVGPAGIR GPQGHQGPAG PPGPPGPLGP
1101  LGVSGGGYDF GYDGDFYRAD QPRSAPSLRP KDYEVDATLK SLNNQIETLL
1151  TPEGSRKNPA RTCRDLRLSH PEWSSGYYWI DPNQGCTMEA IKVYCDFPTG
1201  ETCIRAQPEN IPAKNWYRSS KDKKHVWLGE TINAGSQFEY NVEGVTSKEM
1251  ATQLAFMRLL ANYASQNITY HCKNSIAYMD EETGNLKKAV ILQGSNDVEL
1301  VAEGNSRFTY TVLVDGCSKK TNEWGKTIIE YKTNKPSRLP FLDIAPLDIG
1351  GADHEFFVDI GPVCFK
```

Structural and functional sites

Signal peptide: 1–22
N-Propeptide: 23–79
 COL2 domain: 33–77
N-Telopeptide: 80–90
Helical domain: 91–1102
C-Telopeptide: 1103–1119
C-Propeptide: 1120–1366
Lysine/hydroxylysine cross-linking sites: 84, 177, 1023
Histidine cross-linking site: 182
Hydroxylysine glycosylation sites: 177, 264
Potential N-linked glycosylation site: 1267
N-Proteinase cleavage site: 79–80
C-Proteinase cleavage site: 1119–1120
Mammalian collagenase cleavage site: 865–866

Gene structure

The proα1(I) and proα2(I) chains are encoded by single genes located on human chromosomes 17 (locus q21.3–22) and 7 (locus q21.3–22), respectively. The proα1(I) gene contains 51 exons and the proα2(I) gene 52 exons. All the exons encoding the triple-helical domain are multiples of 9 bp corresponding to Gly-X-Y triplets (commonly 54 bp). The exon arrangement within the uninterrupted triple-helical domain [exons 7–47 for α1(I), exons 7–48 for α2(I)] are almost identical to each other and to that for exons 9–50 in the α1(II) gene. In cartilage, the use of a different transcription site within the first intron of the chick proα2(I) gene results in exons 1 and 2 being replaced by a new exon of 96 bp. The resulting transcript contains several open reading frames that are out of frame with the α2(I) coding sequence and therefore encode non-collagenous proteins, one of which appears to be a DNA-binding protein. Whether the α1(I) chain of collagen type I trimer is the same or a distinct gene product from that in the heterotrimer is still controversial[9–11].

References

[1] Kadler, K.E. (1995) Extracellular matrix 1: fibril-forming collagens. In: Protein Profile, Vol. 2, ed. Sheterline, P., Academic Press, London, pp. 491–619.
[2] Kivirikko, K.I. (1993) Collagens and their abnormalities in a wide spectrum of diseases. Ann. Med. 25: 113–126.
[3] Fietzek, P.P. and Kuhn, K. (1976) The primary structure of collagen. Int. Rev. Connective Tissue Res. 7: 1–60.
[4] Schupp-Byrne, D.E. and Church, R.L. (1982) Embryonic collagen (type I-trimer) α1-chains are genetically distinct from type I collagen α1-chains. Collagen Rel. Res. 2: 481–494.
[5] Kuhn, K. (1987) The classical collagens: types I, II and III. In: Structure and Function of Collagen Types, ed. Mayne, R. and Burgeson, R.E., Academic Press, London, pp. 1–42.

[6] Vuorio, E. and de Crombrugghe, B. (1990) The family of collagen genes. Annu. Rev. Biochem. 59: 837–872.

[7] Staatz, W.D. et al. (1991) Identification of a tetrapeptide recognition sequence for the $\alpha2\beta1$ integrin in collagen. J. Biol. Chem. 266: 7363–7367.

[8] Miller, E.J. and Rhodes, R.K. (1982) Preparation and characterization of the different types of collagen. Methods Enzymol. 82: 33–64.

[9] Bennett, V. and Adams, S.L. (1990) Identification of a cartilage-specific promoter within intron 2 of the chick $\alpha2$(I) collagen gene. J. Biol. Chem. 265: 2223–2230.

[10] Sandell, L.J. and Boyd, C.D. (1990) Conserved and divergent sequence and functional elements within collagen genes. In: Extracellular Matrix Genes, ed. Sandell, L.J. and Boyd, C.D., Academic Press, New York, pp. 1–56.

[11] Chu, M.L. and Prockop, D.J. (1993) Collagen: Gene structure. In: Connective Tissue and Its Heritable Disorders, ed. Steinmann, B. and Royce, P., Wiley-Liss, New York, pp. 149–165.

Collagen type II

Type II collagen is the principal collagenous component of cartilage, intervertebral disc and vitreous humour, but is also found in other tissues during development. Its function is to provide tensile strength and confers on cartilage the ability to resist shearing forces. Type II collagen supports chondrocyte adhesion and may influence the differentiated phenotype of these cells. More than 50 mutations in the COL2A1 gene have been observed which result in a number of clinical phenotypes ranging from mild to lethal chondrodysplasias and include the Stickler syndrome, Kniest dysplasia, hypochondrogenesis, spondyloepiphyseal dysplasia and achondrogenesis II. Some mutations are also associated with early onset of osteoarthritis [1–3]. Mice that are homozygous for an inactivated COL2A1 gene die just prior to or shortly after birth. The chondrocytes within the cartilage are highly disorganized and there are no discernible fibrils in the extracellular matrix. The long bones are devoid of endochondral bone and growth plates. Intramembranous ossification, however, is normal [4].

Molecular structure

Type II collagen is synthesized as a homotrimeric procollagen comprising three identical proα1(II) chains. Different molecular forms of type II procollagen arise by alternative splicing of the cysteine-rich domain within the N-propeptide. The unspliced and spliced variants are expressed in non-chondrogenic and chondrogenic tissues, respectively. The procollagen is processed extracellularly by N- and C-proteinases to produce a triple-helical molecule that can assemble with a stagger of 234 amino acids into cross-banded fibrils with 67 nm (D) periodicity. These fibrils are stabilized by intermolecular cross-links derived from specific lysine/hydroxylysine residues in both the non-helical (telopeptide) and helical domains. The COL1 domain of type II collagen interacts with integrin receptors, principally $\alpha 2\beta 1$, and the non-integrin binding protein anchorin CII. The fibrillar form of type II collagen is cross-linked to type IX collagen via lysine/hydroxylysine residues in the N- and C-telopeptides of type II collagen and in the helical COL2 domain of type IX collagen. The fibrillar form of type II collagen also interacts with the protein cores of the proteoglycans fibromodulin and decorin [1,5–7].

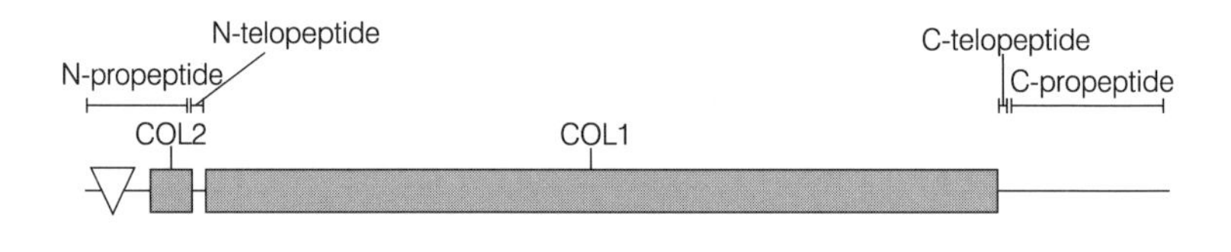

Isolation

Type II procollagen can be prepared from cultured chondrocytes [1]. Type II collagen is prepared from young cartilage by guanidine–HCl treatment (to remove proteoglycans) and pepsin digestion [8].

Accession number

P02458

Primary structure: α1(II) chain

Ala	A	132	Cys	C	19	Asp	D	63	Glu	E	79
Phe	F	24	Gly	G	405	His	H	8	Ile	I	35
Lys	K	67	Leu	L	57	Met	M	16	Asn	N	32
Pro	P	268	Gln	Q	61	Arg	R	72	Ser	S	50
Thr	T	44	Val	V	38	Trp	W	7	Tyr	Y	10

Mol. wt (calc.) = 141 788 Residues = 1487

```
1     MIRLGAPQSL VLLTLLVAAV LRCQGQDVQE AGSCVQDGQR YNDKDVWKPE
51    PCRICVCDTG TVLCDDIICE DVKDCLSPEI PFGECCPICP TDLATASGQP
101   GPKGQKGEPG DIKDIVGPKG PPGPQGPAGE QGPRGDRGDK GEKGAPGPRG
151   RDGEPGTLGN PGPPGPPGPP GPPGLGGNFA AQMAGGFDEK AGGAQLGVMQ
201   GPMGPMGPRG PPGPAGAPGP QGFQGNPGEP GEPGVSGPMG PRGPPGPPGK
251   PGDDGEAGKP GKAGERGPPG PQGARGFPGT PGLPGVKGHR GYPGLDGAKG
301   EAGAPGVKGE SGSPGENGSP GPMGPRGLPG ERGRTGPAGA AGARGNDGQP
351   GPAGPPGPVG PAGGPGFPGA PGAKGEAGPT GARGPEGAQG PRGEPGTPGS
401   PGPAGASGNP GTDGIPGAKG SAGAPGIAGA PGFPGPRGPP DPQGATGPLG
451   PKGQTGKPGI AGFKGEQGPK GEPGPAGPQG APGPAGEEGK RGARGEPGGV
501   GPIGPPGERG APGNRGFPGQ DGLAGPKGAP GERGPSGLAG PKGANGDPGR
551   PGEPGLPGAR GLTGRPGDAG PQGKVGPSGA PGEDGRPGPP GPQGARGQPG
601   VMGFPGPKGA NGEPGKAGEK GLPGAPGLRG LPGKDGETGA EGPPGPAGPA
651   GERGEQGAPG PSGFQGLPGP PGPPGEGGKP GDQGVPGEAG APGLVGPRGE
701   RGFPGERGSP GAQGLQGPRG LPGTPGTDGP KGASGPAGPP GAQGPPGLQG
751   MPGERGAAGI AGPKGDRGDV GEKGPEGAPG KDGGRGLTGP IGPPGPAGAN
801   GEKGEVGPPG PAGSAGARGA PGERGETGPP GTSGIAGPPG ADGQPGAKGE
851   QGEAGQKGDA GAPGPQGPSG APGPQGPTGV TGPKGARGAQ GPPGATGFPG
901   AAGRVGPPGS NGNPGPPGPP GPSGKDGPKG ARGDSGPPGR AGEPGLQGPA
951   GPPGEKGEPG DDGPSGAEGP PGPQGLAGQR GIVGLPGQRG ERGFPGLPGP
1001  SGEPGQQGAP GASGDRGPPG PVGPPGLTGP AGEPGREGSP GADGPPGRDG
1051  AAGVKGDRGE TGAVGAPGAP GPPGSPGPAG PTGKQGDRGE AGAQGPMGPS
1101  GPAGARGIQG PQGPRGDKGE AGEPGERGLK GHRGFTGLQG LPGPPGPSGD
1151  QGASGPAGPS GPRGPPGPVG PSGKDGANGI PGPIGPPGPR GRSGETGPAG
1201  PPGNPGPPGP PGPPGPGIDM SAFAGLGPRE KGPDPLQYMR ADQAAGGLRQ
1251  HDAEVDATLK SLNNQIESIR SPEGSRKNPA RTCRDLKLCH PEWKSGDYWI
1301  DPNQGCTLDA MKVFCNMETG ETCVYPNPAN VPKKNWWSSK SKEKKHIWFG
1351  ETINGGFHFS YGDDNLAPNT ANVQMTFLRL LSTEGSQNIT YHCKNSIAYL
1401  DEAAGNLKKA LLIQGSNDVE IRAEGNSRFT YTALKDGCTK HTGKWGKTVI
1451  EYRSQKTSRL PIIDIAPMDI GGPEQEFGVD IGPVCFL
```

Structural and functional sites

Signal peptide: 1–25
N-Propeptide: 26–181 (amino acid 29 is Q if unspliced, R if spliced)
von Willebrand factor C repeats (alternatively spliced domain): 29–97
  COL2 domain: 98–179
N-Telopeptide: 182–200
Helical domain: 201–1214
C-Telopeptide: 1215–1241

C-Propeptide (chondrocalcin): 1242–1487
Lysine/hydroxylysine cross-linking sites: 190, 287, 1130, 1231
Potential N-linked glycosylation sites: 317, 1388
Interchain disulphide bond residues: 1283, 1289
N-Proteinase cleavage site: 181–182
C-Proteinase cleavage site: 1241–1242
Mammalian collagenase cleavage site: 975–976
Stromelysin cleavage sites: 194–195, 198–199

Gene structure

Type II collagen is encoded by a single gene found on human chromosome 12 at locus q13.11–12. The gene spans approximately 30 kb and contains 54 exons, a number of which are multiples of 9 bp corresponding to Gly-X-Y triplets (commonly 54 bp). Exon 2, the sequence of which is conserved in types I and III collagen and which codes for part of the N-propeptide, can be alternatively spliced[6,9–11].

References

1. Kadler, K.E. (1995) Extracellular Matrix 1: fibril-forming collagens. In: Protein Profile, Vol. 2, ed. Sheterline, P., Academic Press, London, pp. 491–619.
2. Vikkula, M. et al. (1994) Type II collagen mutations in rare and common cartilage diseases. Ann Med. 26: 107–114.
3. Prockop, D.J. and Kivirikko, K.I. (1995) Collagens: molecular biology, diseases, and potential for therapy. Annu. Rev. Biochem. 64: 403–434.
4. Li, S.W. et al. (1995) Transgenic mice with targeted inactivation of the COL2A1 gene for collagen II develop a skeleton with membranous and periostal bone but no endochondral bone. Genes Develop. 9: 2821–2830.
5. Kuhn, K. (1987) The classical collagens: Types I, II and III. In: Structure and Function of Collagen Types, ed. Mayne, R. and Burgeson, R.E., Academic Press, London, pp. 1–42.
6. Sandell, L.J. et al. (1994) Alternatively splice form of type II procollagen (IIA) is predominant in skeletal precursors and non-cartilaginous tissues during early mouse development. Develop. Dynamics 199: 129–140.
7. Wu, J.-J. et al. (1991) Sites of stromelysin cleavage in collagen types II, IX, X and XI of cartilage. J. Biol. Chem. 266: 5625–5628.
8. Miller, E.J. and Rhodes, R.K. (1982) Preparation and characterization of the different types of collagen. Methods Enzymol. 82: 33–64.
9. Ala-Kokko, L. and Prockop, D.J. (1990) Completion of the intron–exon structure of the gene for human type II procollagen (COL2A1): variations in the nucleotide sequences of the alleles from three chromosomes. Genomics 8: 454–460.
10. Sandell, L.J. and Boyd, C.D. (1990) Conserved and divergent sequence and functional elements within collagen genes. In: Extracellular Matrix Genes, ed. Sandell, L.J. and Boyd, C.D., Academic Press, New York, pp. 1–56.
11. Chu, M.L. and Prockop, D.J. (1993) Collagen: Gene structure. In: Connective Tissue and Its Heritable Disorders, ed. Steinmann, B. and Royce, P., Wiley-Liss, New York, pp. 149–165.

Collagen type III

Type III collagen is a major fibrillar collagen in skin and vascular tissues. These tissues also contain collagen type I. Type III collagen fibres are thin and result in a more compliant tissue. Mutations in the COL3A1 gene result in Ehlers–Danlos syndrome type IV and in aortic aneurysms[1,2]. Mice that are homozygous for the inactivated COL3A1 gene have a short life span and die from the rupture of major blood vessels. Ultrastructural analysis of the tissues indicates that type III collagen is essential for type I collagen fibrillogenesis during normal development of the cardiovascular system and other organs[3].

Molecular structure

Type III collagen is synthesized as a homotrimeric procollagen comprising three identical proα1(III) chains. The procollagen is processed extracellularly by N- and C-proteinases to produce a triple-helical molecule in which the three α chains are linked by disulphide bonds at the COOH-terminal end of the triple helix. The molecules assemble with a stagger of 234 amino acids into cross-banded fibrils with a 67 nm (D) periodicity. N-Proteinase cleavage is often incomplete giving rise to a partially processed collagen (pN-collagen) which affects fibril formation. The fibrils are stabilized by inter-molecular cross-links derived from specific lysine/hydroxylysine residues in both the non-helical (telopeptide) and helical domains[4–6].

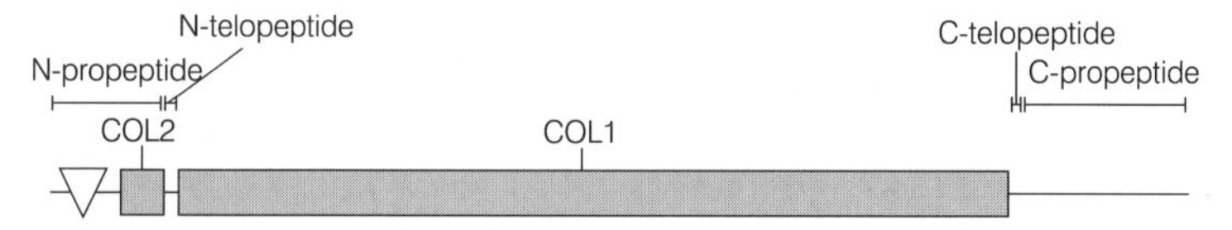

Isolation

Type III collagen can be prepared from fetal skin as pN- and processed forms by NaCl extraction[7] or as the shortened triple-helical α1(III) chain by pepsin digestion[8].

Accession number

P02461

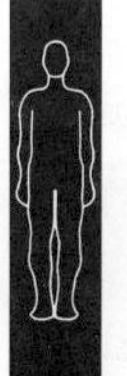

Primary structure: α1(III) chain

Ala	A	115	Cys	C	22	Asp	D	55	Glu	E	74
Phe	F	23	Gly	G	413	His	H	14	Ile	I	36
Lys	K	62	Leu	L	48	Met	M	17	Asn	N	41
Pro	P	281	Gln	Q	43	Arg	R	60	Ser	S	73
Thr	T	31	Val	V	36	Trp	W	7	Tyr	Y	15

Mol. wt (calc.) = 138 403 Residues = 1466

```
1     MMSFVQKGSW LLLALLHPTI ILAQQEAVEG GCSHLGQSYA DRDVWKPEPC
51    QICVCDSGSV LCDDIICDDQ ELDCPNPEIP FGECCAVCPQ PPTAPTRPPN
101   GQGPQGPKGD PGPPGIPGRN GDPGIPGQPG SPGSPGPPGI CESCPTGPQN
151   YSPQYDSYDV KSGVAVGGLA GYPGPAGPPG PPGPPGTSGH PGSPGSPGYQ
201   GPPGEPGQAG PSGPPGPPGA IGPSGPAGKD GESGRPGRPG ERGLPGPPGI
251   KGPAGIPGFP GMKGHRGFDG RNGEKGETGA PGLKGENGLP GENGAPGPMG
301   PRGAPGERGR PGLPGAAGAR GNDGARGSDG QPGPPGPPGT AGFPGSPGAK
351   GEVGPAGSPG SNGAPGQRGE PGPQGHAGAQ GPPGPPGING SPGGKGEMGP
401   AGIPGAPGLM GARGPPGPAG ANGAPGLRGG AGEPGKNGAK GEPGPRGERG
451   EAGIPGVPGA KGEDGKDGSP GEPGANGLPG AAGERGAPGF RGPAGPNGIP
501   GEKGPAGERG APGPAGPRGA AGEPGRDGVP GGPGMRGMPG SPGGPGSDGK
551   PGPPGSQGES GRPGPPGPSG PRGQPGVMGF PGPKGNDGAP GKNGERGGPG
601   GPGPQGPPGK NGETGPQGPP GPTGPGGDKG DTGPPGPQGL QGLPGTGGPP
651   GENGKPGEPG PKGDAGAPGA PGGKGDAGAP GERGPPGLAG APGLRGGAGP
701   PGPEGGKGAA GPPGPPGAAG TPGLQGMPGE RGGLGSPGPK GDKGEPGGPG
751   ADGVPGKDGP RGPTGPIGPP GPAGQPGDKG EGGAPGLPGI AGPRGSPGER
801   GETGPPGPAG FPGAPGQNGE PGGKGERGAP GEKGEGGPPG VAGPPGGSGP
851   AGPPGPQGVK GERGSPGGPG AAGFPGARGL PGPPGSNGNP GPPGPSGSPG
901   KDGPPGPAGN TGAPGSPGVS GPKGDAGQPG EKGSPGAQGP PGAPGPLGIA
951   GITGARGLAG PPGMPGPRGS PGPQGVKGES GKPGANGLSG ERGPPGPQGL
1001  PGLAGTAGEP GRDGNPGSDG LPGRDGSPGG KGDRGENGSP GAPGAPGHPG
1051  PPGPVGPAGK SGDRGESGPA GPAGAPGPAG SRGAPGPQGP RGDKGETGER
1101  GAAGIKGHRG FPGNPGAPGS PGPAGQQGAI GSPGPAGPRG PVGPSGPPGK
1151  DGTSGHPGPI GPPGPRGNRG ERGSEGSPGH PGQPGPPGPP GAPGPCCGGV
1201  GAAAIAGIGG EKAGGFAPYY GDEPMDFKIN TDEIMTSLKS VNGQIESLIS
1251  PDGSRKNPAR NCRDLKFCHP ELKSGEYWVD PNQGCKLDAI KVFCNMETGE
1301  TCISANPLNV PRKHWWTDSS AEKKHVWFGE SMDGGFQFSY GNPELPEDVL
1351  DVQLAFLRLL SSRASQNITY HCKNSIAYMD QASGNVKKAL KLMGSNEGEF
1401  KAEGNSKFTY TVLEDGCTKH TGEWSKTVFE YRTRKAVRLP IVDIAPYDIG
1451  GPDQEFGVDV GPVCFL
```

Structural and functional sites

Signal peptide: 1–23
N-Propeptide: 24–148
von Willebrand factor C repeat: 27–96
  COL2 domain: 103–141
N-Telopeptide: 149–167
Helical domain: 168–1196
C-Telopeptide: 1197–1205
C-Propeptide: 1206–1466
Interchain disulphide bond residues: 141, 144, 1196, 1197
Tyrosine sulphation sites: 151, 155, 158
Lysine/hydroxylysine cross-linking sites: 161, 263, 1106
Hydroxylysine glycosylation site: 263
Potential N-linked glycosylation site: 1367
N-Proteinase cleavage site: 148–149
C-Proteinase cleavage site: 1221–1222
Mammalian collagenase cleavage site: 951–952

Gene structure

The proα1(III) collagen chain is encoded by a single gene found on human chromosome 2 at locus q24.3–31. The gene probably contains 52 exons and, although the human gene has not been fully analysed, to date all the exons encoding the triple-helical domain are multiples of 9 bp corresponding to Gly-X-Y triplets (commonly 54 bp). The gene will exceed 38 kb [9–11].

References

1. Kadler, K.E. (1995) Extracellular matrix 1: fibril-forming collagens. In: Protein Profile, Vol. 2, ed. Shetcrline, P., Academic Press, London, pp. 491–619.
2. Kivirikko, K.I. (1993) Collagens and their abnormalities in a wide spectrum of diseases. Ann. Med. 25: 113–126.
3. Liu, X. et al. (1997) Type III collagen is crucial for collagen I fibrillogenesis and for normal cardiovascular development. Proc. Natl Acad. Sci. USA 94: 1852–1856.
4. Jukkola, A. et al. (1986) Incorporation of sulfate into type III procollagen by cultured human fibroblasts. Identification of tyrosine-O-sulfate. Eur. J. Biochem.154: 219–224.
5. Kuhn, K. (1987) The classical collagens: types I, II and III. In: Structure and Function of Collagen Types, ed. Mayne, R. and Burgeson, R.E., Academic Press, London, pp. 1–42.
6. Henkel, W. (1996) Cross-link analysis of the C-telopeptide domain from type III collagen. Biochem. J. 318: 497–503.
7. Byers, P.H. et al. (1974) Preparation of type III procollagen and collagen from rat skin. Biochemistry 13: 5243–5248.
8. Miller, E.J. and Rhodes, R.K. (1982) Preparation and characterization of the different types of collagen. Methods Enzymol. 82: 33–64.
9. Myers, J.C. and Dion, A.S. (1990) Types III and V procollagens: Homology in genetic organization and diversity in structure. In: Extracellular Matrix Genes, ed. Sandell, L.J. and Boyd, C.D., Academic Press, New York, pp. 57–78.
10. Sandell, L.J. and Boyd, C.D. (1990) Conserved and divergent sequence and functional elements within collagen genes. In: Extracellular Matrix Genes, ed. Sandell, L.J. and Boyd, C.D., Academic Press, New York, pp. 1–56.
11. Chu, M.L. and Prockop, D.J. (1993) Collagen: Gene structure. In: Connective Tissue and Its Heritable Disorders, ed. Steinmann, B. and Royce, P., Wiley-Liss, New York, pp. 149–165.

Collagen type IV

Type IV collagen is found exclusively in basement membranes where it provides the major structural support for this matrix. The molecule contains a long (approximately 350 nm) triple-helical domain containing approximately 20 short interruptions that are thought to introduce flexibility into the helix. Type IV collagen self-assembles into a meshwork by the anti-parallel interaction and extensive disulphide bonding of four molecules at their NH_2-termini to form the 7S domain, by the interaction of two molecules at their COOH-terminal non-collagenous (NC1) domains, and by lateral aggregations. The assembled type IV collagen meshwork provides a scaffold for the assembly of other basement membrane components through specific interactions with laminin, entactin/nidogen and heparan sulphate proteoglycan. In addition, this meshwork endows the basement membrane with a size-selective filtration property.

Molecular structure

Type IV collagen molecules are composed of three α chains selected from the translated products of six genetically distinct genes (the COL4A1–COL4A6 genes giving rise to the α1(IV)–α6(IV) chains, respectively). The most abundant form of type IV collagen has the composition $[\alpha 1(IV)]_2\alpha 2(IV)$. The composition of molecules containing the less abundant α3(IV)–α6(IV) chains is not certain, although the chains are expressed in a tissue-specific manner. Cysteine residues in the NC1 domains of all chains sequenced to date are conserved and participate in both intra-chain and inter-molecular disulphide bonding. The trimeric, disulphide-bonded CB3 cyanogen bromide fragment of type IV collagen has been shown to support cell adhesion via the integrins α1β1 and α2β1 [1–7].

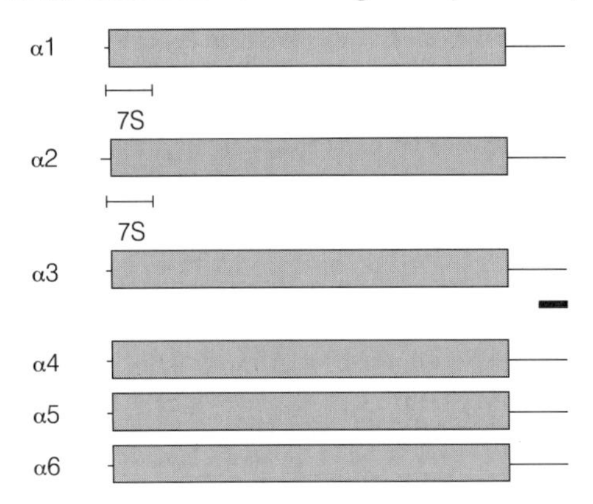

Isolation

Type IV collagen can be isolated from the Engelbreth–Holm–Swarm (EHS) tumour grown in lathyritic C57BL/6 mice [8].

Accession number

P02462

Primary structure: α1(IV) chain

Ala	A	59	Cys	C	20	Asp	D	58	Glu	E	69
Phe	F	46	Gly	G	478	His	H	16	Ile	I	58
Lys	K	94	Leu	L	92	Met	M	31	Asn	N	16
Pro	P	326	Gln	Q	73	Arg	R	45	Ser	S	70
Thr	T	43	Val	V	51	Trp	W	6	Tyr	Y	18

Mol. wt (calc.) = 160 423 Residues = 1669

```
1     MGPRLSVWLL LLPAALLLHE EHSRAAAKGG CAGSGCGKCD CHGVKGQKGE
51    RGLPGLQGVI GFPGMQGPEG PQGPPGQKGD TGEPGLPGTK GTRGPPGASG
101   YPGNPGLPGI PGQDGPPGPP GIPGCNGTKG ERGPLGPPGL PGFAGNPGPP
151   GLPGMKGDPG EILGHVPGML LKGERGFPGI PGTPGPPGLP GLQGPVGPPG
201   FTGPPGPPGP PGPPGEKGQM GLSFQGPKGD KGDQGVSGPP GVPGQAQVQE
251   KGDFATKGEK GQKGEPGFQG MPGVGEKGEP GKPGPRGKPG KDGDKGEKGS
301   PGFPGEPGYP GLIGRQGPAG EKGEAGPPGP PGIVIGTGPL GEKGERGYPG
351   TPGPRGEPGP KGFPGLPGQP GPPGLPVPGQ AGAPGFPGER GEKGDRGFPG
401   TSLPGPSGRD GLPGPPGSPG PPGQPGYTNG IVECQPGPPG DQGPPGIPGQ
451   PGFIGEIGEK GQKGESCLIC DIDGYRGPPG PQGPPGEIGF PGQPGAKGDR
501   GLPGRDGVAG VPGPQGTPGL IGQPGAKGEP GEFYFDLRLK GDKGDPGFPG
551   QPGMPGRAGS PGRDGHPGLP GPKGSPGSVG LKGERGPPGG VGFPGSRGDT
601   GPPGPPGYGP AGPIGDKGQA GFPGGPGSPG LPGPKGEPGK IVPLPGPPGA
651   EGLPGSPGFP GPQGDRGFPG TPGRPGLPGE KGAVGQPGIG FPGPPGPKGV
701   DGLPGDMGPP GTPGRPGFNG LPGNPGVQGQ KGEPGVGLPG LKGLPGLPGI
751   PGTPGEKGSI GVPGVPGEHG AIGPPGLQGI RGEPGPPGLP GSVGSPGVPG
801   IGPPGARGPP GGQGPPGLSG PPGIKGEKGF PGFPGLDMPG PKGDKGAQGL
851   PGITGQSGLP GLPGQQGAPG IPGFPGSKGE MGVMGTPGQP GSPGPVGAPG
901   LPGEKGDHGF PGSSGPRGDP GLKGDKGDVG LPGKPGSMDK VDMGSMKGQK
951   GDQGEKGQIG PIGEKGSRGD PGTPGVPGKD GQAGQPGQPG PKGDPGIKGT
1001  PGAPGLPGPP GKVGGMGLPG TPGEKGVPGI PGPQGSPGLP GDKGAKGEKG
1051  QAGPPGIGIP GLRGEKGDQG IAGFPGSPGE KGEKGSIGIP GMPGSPGLKG
1101  SPGSVGYPGS PGLPGEKGDK GLPGLDGIPG VKGEAGLPGT PGPTGPAGQK
1151  GEPGSDGIPG SAGEKGEPGL PGRGFPGFPG AKGDKGSKGE VGFPGLAGSP
1201  GIPGSKGEQG FMGPPGPQGQ PGLPGSPGHA TEGPKGDRGP QGQPGLPGLP
1251  GPMGPPGLPG IDGVKGDKGN PGWPGAPGVP GPKGDPGFQG MPGIGGSPGI
1301  TGSKGDMGPP GVPGFQGPKG LPGLQGIKGD QGDQGVPGAK GLPGPPGPPG
1351  PYDIIKGQPG LPGPEGPPGL KGLQGLPGPK GQQGVTGLVG IPGPPGIPGF
1401  DGAPGQKGEM GPAGPTGPRG FPGPPGPDGL PGSMGPPGTP SVDHGFLVTR
1451  HSQTIDDPQC PSGTKILYHG YSLLYVQGNE RAHGQDLGTA GSCLRKFSTM
1501  PFLFCNINNV CNFASRNDYS YWLSTPEPMP MSMAPITGEN IRPFISRCAV
1551  CEAPAMVMAV HSQTIQIPPC PSGWSSLWIG YSFVMHTSAG AEGSGQALAS
1601  PGSCLEEFRS APFIECHGRG TCNYYANAYS FWLATIERSE MFKKPTPSTL
1651  KAGELRTHVS RCQVCMRRT
```

Structural and functional sites

Signal peptide: 1–27
7S domain: 28–172
Triple-helical domain: 43–1440
COOH-terminal non-collagenous (NC1) domain: 1441–1669
Potential N-linked glycosylation site: 126
Proposed heparin/cell-binding sites: 531–543, 1263–1277

Accession number

P08572

Primary structure: α2(IV) chain

Ala	A	83	Cys	C	21	Asp	D	82	Glu	E	60
Phe	F	57	Gly	G	472	His	H	19	Ile	I	65
Lys	K	82	Leu	L	102	Met	M	24	Asn	N	18
Pro	P	286	Gln	Q	62	Arg	R	78	Ser	S	64
Thr	T	51	Val	V	49	Trp	W	9	Tyr	Y	28

Mol. wt (calc.) = 167 347 Residues = 1712

```
1     MGRDQRAVAG PALRRWLLLG TVTVGFLAQS VLAGVKKFDV PCGGRDCSGG
51    CQCYPEKGGR GQPGPVGPQG YNGPPGLQGF PGLQGRKGDK GERGAPGVTG
101   PKGDVGARGV SGFPGADGIP GHPGQGGPRG RPGYDGCNGT QGDSGPQGPP
151   GSEGFTGPPG PQGPKGQKGE PYALPKEERD RYRGEPGEPG LVGFQGPPGR
201   PGHVGQMGPV GAPGRPGPPG PPGPKGQQGN RGLGFYGVKG EKGDVGQPGP
251   NGIPSDTLHP IIAPTGVTFH PDQYKGEKGS EGEPGIRGIS LKGEEGIMGF
301   PGLRGYPGLS GEKGSPGQKG SRGLDGYQGP DGPRGPKGEA GDPGPPGLPA
351   YSPHPSLAKG ARGDPGFPGA QGEPGSQGEP GDPGLPGPPG LSIGDGDQRR
401   GLPGEMGPKG FIGDPGIPAL YGGPPGPDGK RGPPGPPGLP GPPGPDGFLF
451   GLKGAKGRAG FPGLPGSPGA RGPKGWKGDA GECRCTEGDE AIKGLPGLPG
501   PKGFAGINGE PGRKGDKGDP GQHGLPGFPG LKGVPGNIGA PGPKGAKGDS
551   RTITTKGERG QPGVPGVPGM KGDDGSPGRD GLDGFPGLPG PPGDGIKGPP
601   GDPGYPGIPG TKGTPGEMGP PGLGLPGLKG QRGFPGDAGL PGPPGFLGPP
651   GPAGTPGQID CDTDVKRAVG GDRQEAIQPG CIAGPKGLPG LPGPPGPTGA
701   KGLRGIPGFA GADGGPGPRG LPGDAGREGF PGPPGFIGPR GSKGAVGLPG
751   PDGSPGPIGL PGPDGPPGER GLPGEVLGAQ PGPRGDAGVP GQPGLKGLPG
801   DRGPPGFRGS QGMPGMPGLK GQPGLPGPSG QPGLYGPPGL HGFPGAPGQE
851   GPLGLPGIPG REGLPGDRGD PGDTGAPGPV GMKGLSGDRG DAGFTGEQGH
901   PGSPGFKGID GMPGTPGLKG DRGSPGMDGF QGMPGLKGRP GFPGSKGEAG
951   FFGIPGLKGL AGEPGFKGSR GDPGPPGPPP VILPGMKDIK GEKGDEGPMG
1001  LKGYLGAKGI QGMPGIPGLS GIPGLPGRPG HIKGVKGDIG VPGIPGLPGF
1051  PGVAGPPGIT GFPGFIGSRG DKGAPGRAGL YGEIGATGDF GDIGDTINLP
1101  GRPGLKGERG TTGIPGLKGF FGEKGTEGDI GFPGITGVTG VQGPPGLKGQ
1151  TGFPGLTGPP GSQGELGRIG LPGGKGDDGW PGAPGLPGFP GLRGIRGLHG
1201  LPGTKGFPGS PGSDIHGDPG FPGPPGERGD PGEANTLPGP VGVPGQKGDQ
1251  GAPGERGPPG SPGLQGFPGI TPPSNISGAP GDKGAPGIFG LKGYRGPPGP
1301  PGSAALPGSK GDTGNPGAPG TPGTKGWAGD SGPQGRPGVF GLPGEKGPRG
1351  EQGFMGNTGP TGAVGDRGPK GPKGDPGFPG APGTVGAPGI AGIPQKIAIQ
1401  PGTVGPQGRR GPPGAPGEIG PQGPPGEPGF RGAPGKAGPQ GRGGVSAVPG
1451  FRGDEGPIGH QGPIGQEGAP GRPGSPGLPG MPGRSVSIGY LLVKHSQTDQ
1501  EPMCPVGMNK LWSGYSLLYF EGQEKAHNQD LGLAGSCLAR FSTMPFLYCN
1551  PGDVCYYASR NDKSYWLSTT APLPMMPVAE DEIKPYISRC SVCEAPAIAI
1601  AVHSQDVSIP HCPAGWRSLW IGYSFLMHTA AGDEGGGQSL VSPGSCLEDF
1651  RATPFIECNG GRGTCHYYAN KYSFWLTTIP EQSFQGSPSA DTLKAGLIRT
1701  HISRCQVCMK NL
```

Structural and functional sites

Signal peptide: 1–25
7S domain: 26–183

Triple-helical domain: 58–1484
COOH-terminal non-collagenous (NC1) domain: 1485–1712
Potential N-linked glycosylation site: 138

Accession number

Q01955

Primary structure: α3(IV) chain

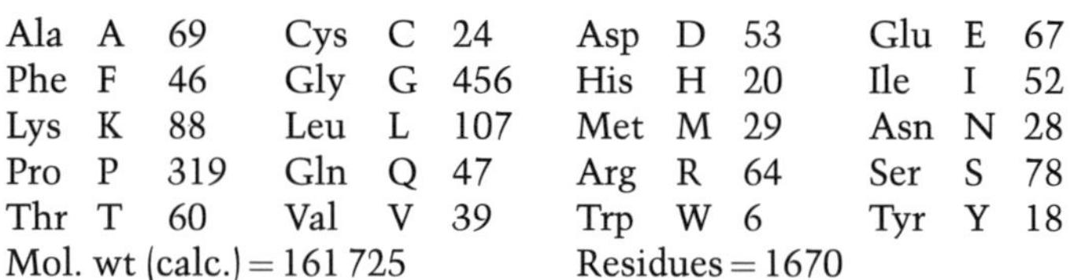

Ala	A	69	Cys	C	24	Asp	D	53	Glu	E	67
Phe	F	46	Gly	G	456	His	H	20	Ile	I	52
Lys	K	88	Leu	L	107	Met	M	29	Asn	N	28
Pro	P	319	Gln	Q	47	Arg	R	64	Ser	S	78
Thr	T	60	Val	V	39	Trp	W	6	Tyr	Y	18

Mol. wt (calc.) = 161 725 Residues = 1670

```
1     MSARTAPRPQ VLLLPLLLVL LAAAPAASKG CVCKDKGQCF CDGAKGEKGE
51    KGFPGPPGSP GQKGFTGPEG LTGPQGPKGF PGLPGLTGSK GVRGISGLPG
101   FSGSPGLPGT PGNTGPYGLV GVPGCSGSKG EQGFPGLPGT LGYPGIPGAA
151   GLKGQKGAPA KEEDIELDAK GDPGLPGAPG PQGLPGPPGF PGPVGPPGPP
201   GFFGFPGAMG PRGPKGHMGE RVIGHKGERG VKGLTGPPGP PGTVIVTLTG
251   PDNRTDLKGE KGDKGAMGEP GPPGPSGLPG ESYGSEKGAP GDPGLQGKPG
301   KDGVPGFPGS EGVKGNRGFP GLMGEDGIKG QKGDIGPPGF RGPTEYYDTY
351   QEKGDEGTPG PPGPRGARGP QGPSGPPGVP GSPGSSRPGL RGAPGWPGLI
401   GSKGERGRPG KDAMGTPGSP GCAGSPGLPG SPGPPGPPGD IGFRKGPPGD
451   HGLPGYLGSP GIPGVDGPKG EPGLLCTQCP YIPGPPGLPG LPGLHGVKGI
501   PGRQGAAGLK GSPGSPGNTG LPGFPGFPGA QGDPGLKGEK GETLQPEGQV
551   GVPGDPGLRG QPGRKGLDGI PGTPGVKGLP GPKGELALSG EKGDQGPPGD
601   PGSPGSPGPA GPAGPPGYGP QGEPGLQGTQ GVPGAPGPPG EAGPRGELSV
651   STPVPGPPGP PGPPGHPGPQ GPPGIPGSLG KCGDPGLPGP DGEPGIPGIG
701   FPGPPGPKGD QGFPGTKGSL GCPGKMGEPG LPGKPGLPGA KGEPAVAMPG
751   GPGTPGFPGE RGNSGEHGEI GLPGLPGLPG TPGNEGLDGP RGDPGQPGPP
801   GEQGPPGRCI EGPRGAQGLP GLNGLKGQQG RRGKTGPKGD PGIPGLDRSG
851   FPGETGSPGI PGHQGEMGPL GQRGYPGNPG ILGPPGEDGV IGMMGFPGAI
901   GPPGPPGNPG TPGHRGSPGI PGVKGQRGTP GAKGEQGDKG NPGPSEISHV
951   IGDKGEPGLK GFAGNPGEKG NRGVPGMPGL KGLKGLPGPA GPPGPRGDLG
1001  STGNPGEPGL LGIPGSMGNM GMPGSKGKRG TLGFPGRAGR PGLPGIHGLQ
1051  GDKGEPGYSE GTRPGPPGPT GDPGLPGDMG KKGEMGQPGP PGHLGPAGPE
1101  GAPGSPGSPG LPGKPGPHGD LGFKGIKGLL GPPGIRGPPG LPGFPGSPGP
1151  MGIRGDQGRD GIPGPAGEKG ETGLLRAPPG PRGNPGAQGA KGDRGAPGFP
1201  GLPGRKGAMG DAGPRGPTGI EGFPGPPGLP GAIIPGQTGN RGPPGSRGSP
1251  GAPGPPGPPG SHVIGIKGDK GSMGHPGPKG PPGTAGDMGP PGRLGAPGTP
1301  GLPGPRGDPG FQGFPGVKGE KGNPGFLGSI GPPGPIGPKG PPGVRGDPGT
1351  LKIISLPGSP GPPGTPGEPG MQGEPGPPGP PGNLGPCGPR GKPGKDGKPG
1401  TPGPAGEKGN KGSKGEPGPA GSDGLPGLKG KRGDSGSPAT WTTRGFVFTR
1451  HSQTTAIPSC PEGTVPLYSG FSFLFVQGNQ RAHGQDLGTL GSCLQRFTTM
1501  PFLFCNVNDV CNFASRNDYS YWLSTPALMP MNMAPITGRA LEPYISRCTV
1551  CEGPAIAIAV HSQTTDIPPC PHGWISLWKG FSFIMFTSAG SEGTGQALAS
1601  PGSCLEEFRA SPFLECHGRG TCNYYSNSYS FWLASLNPER MFRKPIPSTV
1651  KAGELEKIIS RCQVCMKKRH
```

Structural and functional sites

Signal peptide: 1–28
Amino-terminal non-collagenous domain: 29–42
Triple-helical domain: 43–1438
COOH-terminal non-collagenous (NC1) domain: 1439–1670
Potential N-linked glycosylation site: 253
Alternatively spliced domains: 1586–1670 (replaced by KAYSINCESWGIRKN NKSLSGVHEEKTLKLKKTAELVFFILKNKVMTEHAVI in form V), 1488–1670 (replaced by DALFVKVLRSP in form L5)

Accession number

P53420

Primary structure: α4(IV) chain

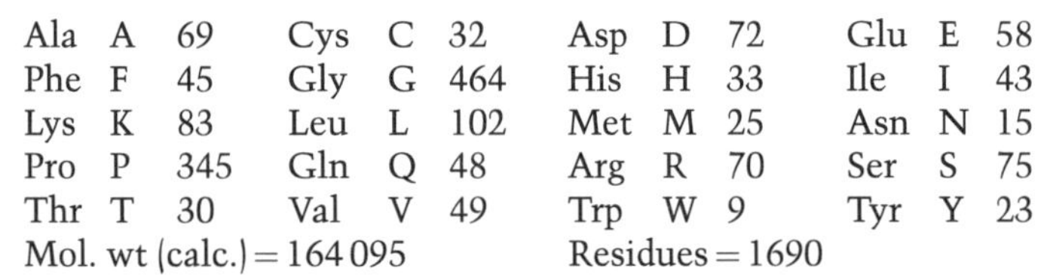

Ala	A	69	Cys	C	32	Asp	D	72	Glu	E	58
Phe	F	45	Gly	G	464	His	H	33	Ile	I	43
Lys	K	83	Leu	L	102	Met	M	25	Asn	N	15
Pro	P	345	Gln	Q	48	Arg	R	70	Ser	S	75
Thr	T	30	Val	V	49	Trp	W	9	Tyr	Y	23

Mol. wt (calc.) = 164 095   Residues = 1690

```
1     MWSLHIVLMR  CSFRLTKSLA  TGPWSLILIL  FSVQYVYGSG  KKYIGPCGGR
51    DCSVCHCVPE  KGSRGPPGPP  GPQGPIGPLG  APGPIGLSGE  KGMRGDRGPP
101   GAAGDKGDKG  PTGVPGFPGL  DGIPGHPGPP  GPRGKPGMSG  HNGSRGDPGF
151   PGGRGALGPG  GPLGHPGEKG  EKGNSVFILG  AVKGIQGDRG  DPGLPGLPGS
201   WGAGGPAGPT  GYPGEPGLVG  PPGQPGRPGL  KGNPGVGVKG  QMGDPGEVGQ
251   QGSPGPTLLV  EPPDFCLYKG  EKGIKGIPGM  VGLPGPPGRK  GESGIGAKGE
301   KGIPGFPGPR  GDPGSYGSPG  FPGLKGELGL  VGDPGLFGLI  GPKGDPGNRG
351   HPGPPGVLVT  PPLPLKGPPG  DPGFPGRYGE  TGDVGPPGPP  GLLGRPGEAC
401   AGMIGPPGPQ  GFPGLPGLPG  EAGIPGRPDS  APGKPGKPGS  PGLPGAPGLQ
451   GLPGSSVIYC  SVGNPGPQGI  KGKVGPPGGR  GPKGEKGNEG  LCACEPGPMG
501   PPGPPGLPGR  QGSKGDLGLP  GWLGTKGDPG  PPGAEGPPGL  PGKHGASGPP
551   GNKGAKGDMV  VSRVKGHKGE  RGPDGPPGFP  GQPGSHGRDG  HAGEKGDPGP
601   PGDHEDATPG  GKGFPGPLGP  PGKAGPVGPP  GLGFPGPPGE  RGHPGVPGHP
651   GVRGPDGLKG  QKGDTISCNV  TYPGRHGPPG  FDGPPGPKGF  PGPQGAPGLS
701   GSDGHKGRPG  TPGTAEIPGP  PGFRGDMGDP  GFGGEKGSSP  VGPPGPPGSP
751   GVNGQKGIPG  DPAFGHLGPP  GKRGLSGVPG  IKGPRGDPGC  PGAEGPAGIP
801   GFLGLKGPKG  REGHAGFPGV  PGPPGHSCER  GAPGIPGQPG  LPGYPGSPGA
851   PGGKGQPGDV  GPPGPAGMKG  LPGLPGRPGA  HGPPGLPGIP  GPFGDDGLPG
901   PPGPKGPRGL  PGFPGFPGER  GKPGAEGCPG  AKGEPGEKGM  SGLPGDRGLR
951   GAKGAIGPPG  DEGEMAIISQ  KGTPGEPGPP  GDDGFPGERG  DKGTPGMQGR
1001  RGELGRYGPP  GFHRGEPGEK  GQPGPPGPPG  PPGSTGLRGF  IGFPGLPGDQ
1051  GEPGSPGPPG  FSGIDGARGP  KGNKGDPASH  FGPPGPKGEP  GSPGCPGHFG
1101  ASGEQGLPGI  QGPRGSPGRP  GPPGSSGPPG  CPGDHGMPGL  RGQPGEMGDP
1151  GPRGLQGDPG  IPGPPGIKGP  SGSPGLNGLH  GLKGQKGTKG  ASGLHDVGPP
1201  GPVGIPGLKG  ERGDPGSPGI  SPPGPRGKKG  PPGPPGSSGP  PGPAGATGRA
1251  PKDIPDPGPP  GDQGPPGPDG  PRGAPGPPGL  PGSVDLLRGE  PGDCGLPGPP
1301  GPPGPPGPPG  YKGFPGCDGK  DGQKGPMGFP  GPQGPHGFPG  PPGEKGLPGP
```

```
1351  PGRKGPTGLP  GPRGEPGPPA  DVDDCPRIPG  LPGAPGMRGP  EGAMGLPGMR
1401  GPPGPGCKGE  PGLDGRRGVD  GVPGSPGPPG  RKGDTGEDGY  PGGPGPPGPI
1451  GDPGPKGFGP  GYLGGFLLVL  HSQTDQEPTC  PLGMPRLWTG  YSLLYLEGQE
1501  KAHNQDLGLA  GSCLPVFSTL  PFAYCNIHQV  CHYAQRNDRS  YWLASAAPLP
1551  MMPLSEEAIR  PYVSRCAVCE  APAQAVAVHS  QDQSIPPCPQ  TWRSLWIGYS
1601  FLMHTGAGDQ  GGGQALMSPG  SCLEDFRAAP  FLECQGRQGT  CHFFANKYSF
1651  WLTTVKADLQ  FSSAPAPDTL  KESQAQRQKI  SRCQVCVKYS
```

Structural and functional sites

Signal peptide: 1–38
Triple-helical domain: 42–1456
COOH-terminal non-collagenous (NC1) domain: 1457–1685
Potential N-linked glycosylation site: 125

Accession number

S19029

Primary structure: α5(IV) chain

Ala	A	46	Cys	C	20	Asp	D	54	Glu	E	61
Phe	F	41	Gly	G	477	His	H	13	Ile	I	69
Lys	K	80	Leu	L	113	Met	M	26	Asn	N	32
Pro	P	388	Gln	Q	74	Arg	R	39	Ser	S	65
Thr	T	38	Val	V	31	Trp	W	5	Tyr	Y	13

Mol. wt (calc.) = 160 998 Residues = 1685

```
1     MKLRGVSLAA  GLFLLALSLW  GQPAEAAACY  GCSPGSKCDC  SGIKGEKGER
51    GFPGLEGHPG  LPGFPGPEGP  PGPRGQKGDD  GIPGPPGPKG  IRGPPGLPGF
101   PGTPGLPGMP  GHDGAPGPQG  IPGCNGTKGE  RGFPGSPGFP  GLQGPPGPPG
151   IPGMKGEPGS  IIMSSLPGPK  GNPGYPGPPG  IQGLPGPTGI  PGPIGPPGPP
201   GLMGPPGPPG  LPGPKGNMGL  NFQGPKGEKG  EQGLQGPPGP  PGQISEQKRP
251   IDVEFQKGDQ  GLPGDRGPPG  PPGIRGPPGP  PGGEKGEKGE  QGEPGKRGKP
301   GKDGENGQPG  IPGLPGDPGY  PGEPGRDGEK  GQKGDTGPPG  PPGLVIPRPG
351   TGITIGEKGN  IGLPGLPGEK  GERGFPGIQG  PPGLPGPPGA  AVMGPPGPPG
401   FPGERGQKGD  EGPPGISIPG  PPGLDGQPGA  PGLPGPPGPG  SPHIPPSDEI
451   CEPGPPGPPG  SPGDKGLQGE  QGVKGDKGDT  CFNCIGTGIS  GPPGQPGLPG
501   LPGPPGSLGF  PGQKGEKGQA  GATGPKGLPG  IPGAPGAPGF  PGSKGEPGDI
551   LTFPGMKGDK  GELGSPGAPG  LPGLPGTPGQ  DGLPGLPGPK  GEPGGITFKG
601   ERGPPGNPGL  PGLPGNIGPM  GPPGLALQGP  VGEKGIQGVA  GNPGQPGIPG
651   PKGDPGQTIT  QPGKPGFRGN  PGRDGDVGLP  GDPGLPGQPG  LPGIPGSKGE
701   PGIPGIGLPG  PPGPKGFPGI  PGPPGAPGTP  GRIGLEGPPG  PPGFPGPKGE
751   PGFALPGPPG  PPGLPGFKGA  LGPKGDRGFP  GPPGPPGRTG  LDGLPGPKGD
801   VGPNGQPGPM  GPPGLPGIGV  QGPPGPPGIP  GPIGQPGLHG  IPGEKGDPGP
851   PGLDVPGPPG  ERGSPGIPGA  PGPIGPPGSP  GLPGKAGRSG  FPGTKGEMGM
901   MGPPGPPGPL  GIPGRSGVPG  LKGDDGLQGQ  PGLPGPTGEK  GSKGEPGLPG
951   PPGPMDPNLL  GSKGEKGEPG  LPGIPGVSGP  KGYQGLPGDP  GQPGLSGQPG
1001  LPGPPGPKGN  PGLPGQPGLI  GPPGLKGTIG  DMGFPGPQGV  EGPPGPSGVP
1051  GQPGSPGLPG  QKGDKGDPGI  SSIGLPGLPG  PKGEPGLPGY  PGNPGIKGSV
1101  GDPGLPGLPG  TPGAKGQPGL  PGFPGTPGPP  GPKGISGPPG  NPGLPGEPGP
```

1151 VGGGGHPGQP GPPGEKGKPG QDGIPGPAGQ KGEPGQPGFG NPGPPGLPGL
1201 SGQKGDGGLP GIPGNPGLPG PKGEPGFHGF PGVQGPPGPP GSPGPALEGP
1251 KGNPGPQGPP GRPGLPGPEG PPGLPGNGGI KGEKGNPGQP GLPGLPGLKG
1301 DQGPPGLQGN PGRPGLNGMK GDPGLPGVPG FPGMKGPSGV PGSAGPEGEP
1351 GLIGPPGPPG LPGPSGQSII IKGDAGPPGI PGQPGLKGLP GPQGPQGLPG
1401 PTGPPGDPGR NGLPGFDGAG GRKGDPGLPG QPGTRGLDGP PGPDGLQGPP
1451 GPPGTSSVAH GFLITRHSQT TDAPQCPQGT LQVYEGFSLL YVQGNKRAHG
1501 QDLGTAGSCL RRFSTMPFMF CNINNVCNFA SRNDYSYWLS TPEPMPMSMQ
1551 PLKGQSIQPF ISRCAVCEAP AVVIAVHSQT IQIPHCPQGW DSLWIGYSFM
1601 MHTSAGAEGS GQALASPGSC LEEFRSAPFI ECHGRGTCNY YANSYSFWLA
1651 TVDVSDMFSK PQSETLKAGD LRTRISRCQV CMKRT

Structural and functional sites

Signal peptide: 1–26
NH_2-terminal non-collagenous domain: 27–41
Triple-helical domain: 42–1456
COOH-terminal non-collagenous (NC1) domain: 1457–1685
Potential N-linked glycosylation site: 125

Accession number

A53404

Primary structure: α6(IV) chain

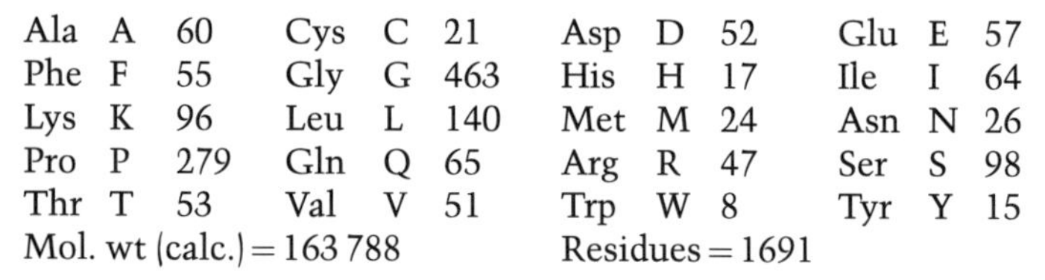

Ala	A	60	Cys	C	21	Asp	D	52	Glu	E	57
Phe	F	55	Gly	G	463	His	H	17	Ile	I	64
Lys	K	96	Leu	L	140	Met	M	24	Asn	N	26
Pro	P	279	Gln	Q	65	Arg	R	47	Ser	S	98
Thr	T	53	Val	V	51	Trp	W	8	Tyr	Y	15

Mol. wt (calc.) = 163 788 Residues = 1691

1 MLINKLWLLL VTLCLTEELA AAGEKSYGKP CGGQDCSGSC QCFPEKGARG
51 RPGPIGIQGP TGPQGFTGST GLSGLKGERG FPGLLGPYGP KGDKGPMGVP
101 GFLGINGIPG HPGQPGPRGP PGLDGCNGTQ GAVGFPGPDG YPGLLGPPGL
151 PGQKGSKGDP VLAPGSFKGI KGDPGLPGLD GITGPQGAPG FPGAVGPAGP
201 PGLQGPPGPP GPLGPDGNMG LGFQGEKGVK GDVGLPGPAG PPPSTGELEF
251 MGFPKGKKGS KGEPGPKGFP GISGPPGFPG LGTTGEKGEK GEKGIPGLPG
301 PRGPMGSEGV QGPPGQQGKK GTLGFPGLNG FQGIEGQKGD IGLPGPDVFI
351 DIDGAVISGN PGDPGVPGLP GLKGDEGIQG LRGPSGVPGL PALSGVPGAL
401 GPQGFPGLKG DQGNPGRTTI GAAGLPGRDG LPGPPGPPGP PSPEFETETL
451 HNKESGFPGL RGEQGPKGNL GLKGIKGDSG FCACDGGVPN TGPPGEPGPP
501 GPWGLIGLPG LKGARGDRGS GGAQGPAGAP GLVGPLGPSG PKGKKGEPIL
551 STIQGMPGDR GDSGSQGFRG VIGEPGKDGV PGLPGLPGLP GDGGQGFPGE
601 KGLPGLPGEK GHPGPPGLPG NGLPGLPGPR GLPGDKGKDG LPGQQGLPGS
651 KGITLPCIIP GSYGPSGFPG TPGFPGPKGS RGLPGTPGQP GSSGSKGEPG
701 SPGLVHLPEL PGFPGPRGEK GLPGFPGLPG KDGLPGMIGS PGLPGSKGAT
751 GDIFGAENGA PGEQGLQGLT GHKGFLGDSG LPGLKGVHGK PGLLGPKGER
801 GSPGTPGQVG QPGTPGSSGP YGIKGKSGLP GAPGFPGISG HPGKKGTRGK
851 KGPPGSIVKK GLPGLKGLPG NPGLVGLKGS PGSPGVAGLP ALSGPKGEKG

```
901   SVGFVGFPGI PGLPGIPGTR GLKGIPGSTG KMGPSGRAGT PGEKGDRGNP
951   GPVGIPSPRR PMSNLWLKGD KGSQGSAGSN GFPGPRGDKG EAGRPGPPGL
1001  PGAPGLPGII KGVSGKPGPP GFMGIRGLPG LKGSSGITGF PGMPGESGSQ
1051  GIRGSPGLPG ASGLPGLKGD NGQTVEISGS PGPKGQPGES GFKGTKGRDG
1101  LIGNIGFPGN KGEDGKVGVS GDVGLPGAPG FPGVAGMRGE PGLPGSSGHQ
1151  GAIGPLGSPG LIGPKGFPGF PGLHGLNGLP GTKGTHGTPG PSITGVPGPA
1201  GLPGPKGEKG YPGIGIGAPG KPGLRGQKGD RGFPGLQGPA GLPGAPGISL
1251  PSLIAGQPGD PGRPGLDGER GRPGPAGPPG PPGPSSNQGD TGDPGFPGIP
1301  GPKGPKGDQG IPGFSGLPGE LGLKGMRGEP GFMGTPGKVG PPGDPGFPGM
1351  KGKAGPRGSS GLQGDPGQTP TAEAVQVPPG PLGLPGIDGI PGLTGDPGAQ
1401  GPVGLQGSKG LPGIPGKDGP SGLPGPPGAL GDGLPGLQG  PPGFEGAPGQ
1451  QGPFGMPGMP GQSMRVGYTL VKHSQSEQVP PCPIGMSQLW VGYSLLFVEG
1501  QEKAHNQDLG FAGSCLPRFS TMPFIYCNIN EVCHYARRND KSYWLSTTAP
1551  IPMMPVSQTQ IPQYISRCSV CEAPSQAIAV HSQDITIPQC PLGWRSLWIG
1601  YSFLMHTAAG AEGGGQSLVS PGSCLEDFRA TPFIECSGAR GTCHYFANKY
1651  SFWLTTVEER QQFGELPVSE TLKAGQLHTR VSRCQVCMKS L
```

Structural and functional sites

Signal peptide: 1–21
NH_2-terminal non-collagenous domain: 22–46
Triple-helical domain: 47–1450
COOH-terminal non-collagenous (NC1) domain: 1451–1691
Potential N-linked glycosylation site: 127

Gene structure

The COL4A1 gene comprises 52 exons dispersed over more than 100 kb of genomic DNA. Exon sizes range from 27 to 192 bp. The first intron is at least 18 kb in length. The exon structure of the COL4A2 gene is not conserved with respect to the α1(IV) gene. The promoter(s) for the COL4A1 and COL4A2 genes are arranged head-to-head and the two gene products are transcribed from this common bidirectional promoter in opposite directions. The transcription start sites of the α1 and α2 chains are less than 150 bp apart. Both genes are localized to chromosome 13q34. The COL4A3 and COL4A4 genes have both been localized to human chromosome 2q35–q37 and are assumed to be organized in a similar head to head arrangement. Mutations in the COL4A3 and COL4A4 genes are associated with an autosomal recessive form of Alport syndrome and a mouse COL4A3 knockout model of the disease has been reported. A mutation in the COL4A4 gene has also been shown to cause benign familial haematuria. The COL4A5 and COL4A6 genes are localized to human chromosome Xq22 and are arranged in a head to head manner. The COL4A5 gene spans >140 kb and contains 51 exons varying in size from 27 to 1254 bp. Mutations in the COL4A5 gene cause Alport syndrome. The COL4A6 spans 425 kb and contains 46 exons varying in size from 11 to 287 bp. Intron 2 is estimated to be approximately 340 kb. Deletions involving both the COL4A6 and COL4A5 genes have been found in individuals with Alport syndrome associated with a diffuse leiomyomatosis [6,9–14].

References

1. Soininen, R. et al. (1987) Complete primary structure of the α1-chain of human basement membrane (type IV) collagen. FEBS Lett. 225: 188–194.
2. Hostikka, S.L. and Tryggvason, K. (1988) The complete structure of the α2 chain of human type IV collagen and comparison with the α1(IV) chain. J. Biol. Chem. 263: 19488–19493.
3. Mariyama, M. et al. (1994) Complete primary structure of the human α3(IV) chain. Coexpression of the α3(IV) and α4(IV) collagen chains in human tissue. J. Biol. Chem. 269: 23013–23017.
4. Leinonen, A. et al. (1994) Complete primary structure of the human type IV collagen α4(IV) chain. Comparison with structure and expression of the other a(IV) chains' tissue. J. Biol. Chem. 269: 26172–26177.
5. Zhou, J. et al. (1992) Complete amino acid sequence of the human α5(IV) collagen chain and identification of a single-base mutation in exon 23 converting glycine 521 in the collagenous domain to cysteine in an Alport Syndrome patient. J. Biol. Chem. 267: 12475–12481.
6. Oohashi, T. et al. (1994) Identification of a new collagen IV chain, α6(IV), by cDNA isolation and assignment of the gene to chromosome Xq22, which is the same locus for COL4A5. J. Biol. Chem. 269: 7520–7526.
7. Vandenberg, P. et al. (1991) Characterization of a type IV collagen major cell binding site with affinity to the $\alpha1\beta1$ and $\alpha2\beta1$ integrins. J. Cell Biol. 113: 1475–1483.
8. Furuto, D.K. and Miller, E.J. (1987) Isolation and characterization of collagens and procollagens. Methods Enzymol. 144: 41–61.
9. Soininen, R. et al. (1988) The structural genes for the α1 and α2 chains of type IV collagen are divergently encoded on opposite DNA strands and have an overlapping promoter region. J. Biol. Chem. 263: 17217–17220.
10. Morrison, K.E. et al. (1991) Sequence and localization of a partial cDNA encoding the human α3 chain of type IV collagen. Am. J. Hum. Genet. 49: 545–554.
11. Mariyama, M. et al. (1992) Colocalisation of the genes for the α3(IV) and the α4(IV) chains of type IV collagen to chromosome 2 bands q35–q37. Genomics 13: 809–813.
12. Zhou. J. et al. (1994) Structure of the human type IV collagen COL4A5 gene. J. Biol. Chem. 269: 6608–6614.
13. Barker, D.F. et al. (1990) Identification of mutations in the COL4A5 collagen gene in Alport syndrome. Science 248: 1224–1227.
14. Zhang, X. et al. (1996) Structure of the human type IV collagen COL4A6 gene, which is mutated in Alport syndrome-associated leiomyomatosis. Genomics 33: 473–479.

Collagen type V

Type V collagen is a quantitatively minor fibrillar collagen and forms heterotypic fibrils with type I or types I and III collagens in bone, tendon, cornea, skin, blood vessels and the more compliant tissues, liver, lung and placenta. It also forms cross-type heterotrimers with type XI collagen chains, and collagens V and XI are now classed as one type V/XI family[1]. Mutations in the COL5A1 gene have been linked to the more common Ehlers–Danlos syndrome types I and II, characterized by joint hypermobility, fragile skin, atrophic scars and in which there is a general abnormality in the architecture of the major collagen fibrils[2,3]. Mutations in the COL5A2 gene have been generated that lead to homozygous mice producing normal amounts of structurally abnormal type V collagen. These mice have a high rate of perinatal mortality due to spinal deformities and both skin and ocular (notably corneal) abnormalities. There is a general disorganization of type I collagen-containing fibrils[4].

Molecular structure

Type V collagen is synthesized in both homotrimeric and heterotrimeric procollagen forms. The major form comprises two proα1(V) chains and one proα2(V) chain but in some tissues (notably uterus and placenta) a second form, comprising proα1(V), proα2(V) and proα3(V) chains, is present. A homotrimer of proα1(V) chains has been observed in cell culture and the fully processed homotrimer has been identified in tissues. In addition, the proα1(XI) and proα2(XI) chains can substitute for the proα1(V) and proα2(V) chains, respectively, forming cross-type heterotrimers. The procollagen forms are processed extracellularly at the COOH-terminus by C-proteinase. However, both partial and complete processing occurs at the NH_2-terminal end of the proα1(V) chain depending on whether it forms heterotrimers or homotrimers, respectively. The NH_2-terminal domain of the proα2(V) chain is not processed. Type V collagen forms thin fibrils, and controls the diameter of type I collagen fibrils as a copolymerized constituent. The type V collagen

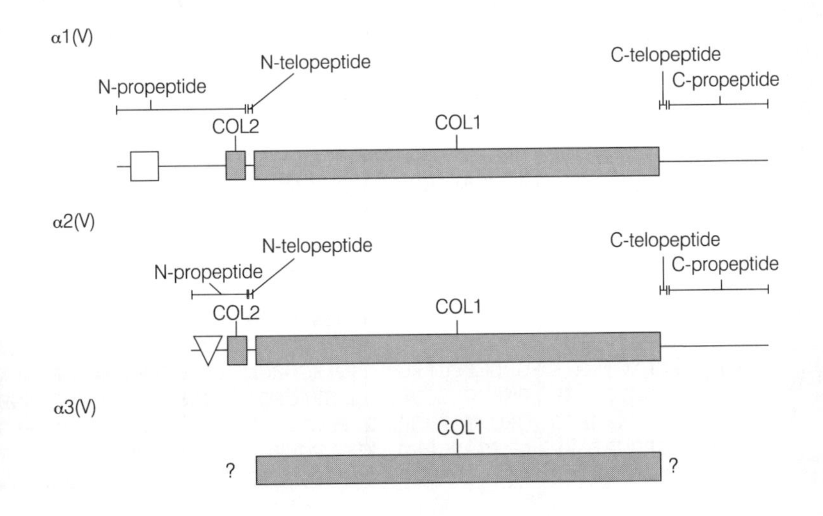

molecules are staggered by 4D (D = 67 nm) and are primarily cross-linked to each other by head-to-tail bonds involving the N-telopeptides and the COOH-terminus of the helix. The α1(I) C-telopeptide is also cross-linked laterally to the amino terminus of the α1(V) helix. Native type V collagen resists mammalian collagenase but is cleaved by the 92 kDa gelatinase (MMP-9) at two sites, one near the amino terminus, the other in the triple helix. Type V collagen has also been shown to interact *in vitro* with a number of extracellular macromolecules including thrombospondin, heparin, heparan sulphate, decorin and biglycan [1,5–12].

Isolation

Type V collagen can be prepared as the partially processed form by NaCl extraction of fetal skin [13] or as the shortened triple-helical form by pepsin digestion of placental tissues [14].

Accession number

P20908

Primary structure: α1(V) chain

Ala	A	92	Cys	C	12	Asp	D	105	Glu	E	118
Phe	F	39	Gly	G	428	His	H	17	Ile	I	52
Lys	K	98	Leu	L	97	Met	M	23	Asn	N	34
Pro	P	334	Gln	Q	74	Arg	R	72	Ser	S	72
Thr	T	70	Val	V	54	Trp	W	7	Tyr	Y	40

Mol. wt (calc.) = 183 415 Residues = 1838

```
1     MDVHTRWKAR  SALRPGAPLL  PPLLLLLLWA  PPPSRAAQPA  DLLKVLDFHN
51    LPDGITKTTG  FCATRRSSKG  PDVAYRVTKD  AQLSAPTKQL  YPASAFPEDF
101   SILTTVKAKK  GSQAFLVSIY  NEQGIQQIGL  ELGRSPVFLY  EDHTGKPGPE
151   DYPLFRGINL  SDGKWHRIAL  SVHKKNVTLI  LDCKKKTTKF  LDRSDHPMID
201   INGIIVFGTR  ILDEEVFEGD  IQQLLFVSDH  RAAYDYCEHY  SPDCDTAVPD
251   TPQSQDPNPD  EYYTEGDGEG  ETYYYEYPYY  EDPEDLGKEP  TPSKKPVEAA
301   KETTEVPEEL  TPTPTEAAPM  PETSEGAGKE  EDVGIGDYDY  VPSEDYYTPS
351   PYDDLTYGEG  EENPDQPTDP  GAGAEIPTST  ADTSNSSNPA  PPPGEGADDL
401   EGEFTEETIR  NLDENYYDPY  YDPTSSPSEI  GPGMPANQDT  IYEGIGGPRG
451   EKGQKGEPAI  IEPGMLIEGP  PGPEGPAGLP  GPPGTMGPTG  QVGDPGERGP
501   PGRPGLPGAD  GLPGPPGTML  MLPFRFGGGG  DAGSKGPMVS  AQESQAQAIL
551   QQARLALRGP  AGPMGLTGRP  GPVGPPGSGG  LKGEPGDVGP  QGPRGVQGPP
601   GPAGKPGRRG  RAGSDGARGM  PGQTGPKGDR  GFDGLAGLPG  EKGHRGDPGP
651   SGPPGPPGDD  GERGDDGEVG  PRGLPGKPGP  RGLLGPKGPP  GPPGPPGVTG
701   MDGQPGPKGN  VGPQGEPGPP  GQQGNPGAQG  LPGPQGAIGP  PGEKGPLGKP
751   GLPGMPGADG  PPGHPGKEGP  PGEKGGQGPP  GPQGPIGYPG  PRGVKGADGI
801   RGLKGTKGEK  GEDGFPGFKG  DMGIKGDRGE  IGPPGPRGED  GPEGPKGRGG
851   PNGDPGPLGP  PGEKGKLGVP  GLPGYPGRQG  PKGSIGFPGF  PGANGEKGGR
901   GTPGKPGPRG  QRGPTGPRGE  RGPRGITGKP  GPKGNSGGDG  PAGPPGERGP
951   NGPQGPTGFP  GPKGPPGPPG  KDGLPGHPGQ  RGETGFQGKT  GPPGPPGVVG
1001  PQGPTGETGP  MGERGHPGPP  GPPGEQGLPG  LAGKEGTKGD  PGPAGLPGKD
```

```
1051  GPPGLRGFPG  DRGLPGPVGA  LGLKGNEGPP  GPPGPAGSPG  ERGPAGAAGP
1101  IGIPGRPGPQ  GPPGPAGEKG  APGEKGPQGP  AGRDGLQGPV  GLPGPAGPVG
1151  PPGEDGDKGE  IGEPGQKGSK  GDKGEQGPPG  PTGPQGPIGQ  PGPSGADGEP
1201  GPRGQQGLFG  QKGDEGPRGF  PGPPGPVGLQ  GLPGPPGEKG  ETGDVGQMGP
1251  PGPPGPRGPS  GAPGADGPQG  PPGGIGNPGA  VGEKGEPGEA  GEPGPSGRSG
1301  PPGPKGERGE  KGESGPSGAA  GPPGPKGPPG  DDGPKGSPGP  VGFPGDPGPP
1351  GEPGPAGQDG  PPGDKGDDGE  PGQTGSPGPT  GEPGPSGPPG  KRGPPGPAGP
1401  EGRQGEKGAK  GEAGLEGPPG  KTGPIGPQGA  PGKPGPDGLR  GIPGPVGEQG
1451  LPGSPGPDGP  PGPMGPPGLP  GLKGDSGPKG  EKGHPGLIGL  IGPPGEQGEK
1501  GDRGLPGPQG  SSGPKGEQGI  TGPSGPIGPP  GPPGLPGPPG  PKGAKGSSGP
1551  TGPRGEAGHP  GPPGPPGPPG  EVIQPLPIQA  SRTRRNIDAS  QLLDDGNGEN
1601  YVDYADGMEE  IFGSLNSLKL  EIEQMKRPLG  TQQNPARTCK  DLQLCHPDFP
1651  DGEYWVDPNQ  GCSRDSFKVY  CNFTAGGSTC  VFPDKKSEGA  RITSWPKENP
1701  GSWFSEFKRG  KLLSYVDAEG  NPVGVVQMTF  LRLLSASAHQ  NVTYHCYQSV
1751  AWQDAATGSY  DKALRFLGSN  DEEMSYDNNP  YIRALVDGCA  TKKGYQKTVL
1801  EIDTPKVEQV  PIVDIMFNDF  GEASQKFGFE  VGPACFMG
```

Structural and functional sites

Signal peptide: 1–37
N-Propeptide: 38–558
PARP repeat: 38–244
COL2 domain: 469–519 [17 triplets similar to COL2 domains of proα1(XI) and proα2(XI)]
Helical domain: 559–1572
C-Telopeptide: 1573–1605
C-Propeptide: 1606–1838
Lysine/hydroxylysine cross-linking sites: 535, 642, 1482
Potential N-linked glycosylation sites: 156, 1259, 1397
C-Proteinase cleavage site: 1605–1606
MMP-9 cleavage site (helix): 997–998

Accession number

P05997

Primary structure: α2(V) chain

Ala	A	80	Cys	C	17	Asp	D	67	Glu	E	75
Phe	F	20	Gly	G	403	His	H	19	Ile	I	37
Lys	K	63	Leu	L	66	Met	M	24	Asn	N	38
Pro	P	269	Gln	Q	59	Arg	R	72	Ser	S	62
Thr	T	51	Val	V	56	Trp	W	6	Tyr	Y	12

Mol. wt (calc.) = 144 560　　Residues = 1496

```
1     MMANWAEARP  LLILIVLLGQ  FVSIKAQEED  EDEGYGEEIA  CTQNGQMYLN
51    RDIWKPAPCQ  ICVCDNGAIL  CDKIECQDVL  DCADPVTPPG  ECCPVCSQTP
101   GGGNTNFGRG  RKGQKGEPGL  VPVVTGIRGR  PGPAGPPGSQ  GPRGERGPKG
151   RPGPRGPQGI  DGEPGVPGQP  GAPGPPGHPS  HPGPDGLSRP  FSAQMAGLDE
201   KSGLGSQVGL  MPGSVGPVGP  RGPQGLQGQQ  GGAGPTGPPG  EPGDPGPMGP
```

```
251   IGSRGPEGPP GKPGEDGEPG RNGNPGEVGF AGSPGARGFP GAPGLPGLKG
301   HRGHKGLEGP KGEVGAPGSK GEAGPTGPMG AMGPLGPRGM PGERGRLGPQ
351   GAPGQRGAHG MPGKPGPMGP LGIPGSSGFP GNPGMKGEAG PTGARGPEGP
401   QGQRGETGPP GPVGSPGLPG AIGTDGTPGP KGPTGSPGTS GPPGSAGPPG
451   SPGPQGSTGP QGNSGLPGDP GFKGEAGPKG EPGPHGIQGP IGPPGEEGKR
501   GPRGDPGTLG PPGPVGERGA PGNRGFPGSD GLPGPKGAQG ERGPVGSSGP
551   KGSQGDPGRP GEPGLPGARG LTGNPGVQGP EGKLGPLGAP GEDGRPGPPG
601   SIGIKGQPGT MGLPGPKGSN GDPGKPGEAG NPGVPGQRGA PGKDGKVGPY
651   GPPGPPGLRG ERGEQGPPGP TGFQGHPGPP GPPGEGGKPG DQGVPGGPGA
701   VGPLGPRGER GNPGERGEPG ITGLPGEKGM AGGHGPDGPK GSPGPSGTPG
751   DTGPPGLQGM PGERGIAGTP GPKGDRGGIG EKGAEGTAGN DGAGGLPGPL
801   GPPGPAGLLG EKGEPGPRGL VGPPGSRGNP GSRGENGPTG AVGFAGPQGS
851   DGQPGVKGEP GEPGQKGDAG SPGPQGLAGS PGPHGPNGVP GLKGGRGTQG
901   PPGATGFPGS AGRVGPPGPA GAPGPAGPLG EPGKEGPPGP RGDPGSHGRV
951   GVRGPAGPPG GPGDKGDPGE DGQPGPDGPP GPAGTTGQRG IVGMPGQRGE
1001  RGMPGLPGPA GTPGKVGPTG ATGDKGPPGP VGPPGSNGPV GEPGPEGPAG
1051  NDGTPGRDGA VGERGDRGDP GPAGLPGSQG APGTPGPVGA PGDAGQRGDP
1101  GSRGPIGHLG RAGKRGLPGP QGPRGDKGDH GDRGDRGQKG HRGFTGLQGL
1151  PGPPGPNGEQ GSAGIPGPFG PRGPPGPVGP SGKEGNPGPL GPLGPPGVRG
1201  SVGEAGPEGP PGEPGPPGPP GPPGHLTAAL GDIMGHYDES MPDPLPEFTE
1251  DQAAPDDKNK TDPGVHATLK SLSSQIETMR SPDGSKKHPA RTCDDLKLCH
1301  SAKQSGEYWI DPNQGSVEDA IKVYCNMETG ETCISANPSS VPRKTWWASK
1351  SPDNKPVWYG LDMNRGSQFA YGDHQSPNTA ITQMTFLRLL SKEASQNITY
1401  ICKNSVGYMD DQAKNLKKAV VLKGANDLDI KAEGNIRFRY IVLQDTCSKR
1451  NGNVGKTVFE YRTQNVARLP IIDLAPVDVG GTDQEFGVEI GPVCFV
```

Structural and functional sites

Signal peptide: 1–26
N-Propeptide: 27–212
von Willebrand factor C repeat: 36–104
COL2 domain: 126–179 [18 triplets in contrast to the COL2 domain in proα1(V), proα1(XI) and proα2(XI)]
Helical domain: 213–1223
C-Telopeptide: 1224–1226
C-Propeptide: 1227–1496
Lysine/hydroxylysine cross-linking sites: 201, 1127
Potential N-linked glycosylation sites: 1259, 1397
C-Proteinase cleavage site: 1226–1227
MMP-9 cleavage site (helix): 657–658

Primary structure: α3(V) chain

Only the triple-helical domain has been sequenced (Accession: P25940 [9]).

Gene structure

The proα1(V) and proα2(V) collagen chains are encoded by single genes found on human chromosomes 9 (locus q34.2–34.3) and 2 (locus q24.3–31), respectively. The COL5A1 gene consists of 66 exons and spans approximately 150 kb (excluding the first intron which itself is greater than 600 kb). Fourteen exons encode the signal peptide, the N-propeptide and the beginning of the triple-helical domain, as in the case of the COL11A2 gene. The remainder of the triple helix is encoded by exons 15 to 62 which comprise mostly 54 or 45 bp. The C-propeptide is encoded by exons 63 to 66. The final exon is identical in size (144 bp) to that of the genes encoding the major fibrillar collagen types I to III. The proα2(V) gene is evolutionary closely related to, and syntenic with, the proα1(III) gene [15–17].

References

[1] Fichard, A. et al. (1995) Another look at collagen V and XI molecules. Matrix Biol. 14: 515–531.
[2] Burrows, N.P. et al. (1996) The gene encoding collagen alpha-1(V) (COL5A1) is linked to mixed Ehlers–Danlos syndrome type I/II. J. Invest. Dermatol. 106: 1273–1276.
[3] DePaepe, A. et al. (1997) Mutations in the COL5A1 gene are causal in the Ehlers–Danlos syndromes I and II. Amer. J. Hum. Genet. 60: 547–554.
[4] Andrikopoulos, K. et al. (1995) Targeted mutation in the col5a2 gene reveals a regulatory role for type V collagen during matrix assembly. Nature Genetics 9: 31–36.
[5] Weil, D. et al. (1987) The pro-α2(V) collagen gene is evolutionarily related to the major fibril-forming collagens. Nucleic Acids Res. 15: 181–198.
[6] Woodbury, D. et al. (1989) Amino-terminal propeptide of human pro-α2(V) collagen conforms to the structural criteria of a fibrillar procollagen molecule. J. Biol. Chem. 264: 2735–2738.
[7] Greenspan, D.S. et al. (1991) The pro-α1(V) collagen chain. Complete primary structure, distribution of expression, and comparison with the pro-α1(XI) collagen chain. J. Biol. Chem. 266: 24727–24733.
[8] Takahara, K. et al. (1991) Complete primary structure of human collagen α1(V) chain. J. Biol. Chem. 266: 13124–13129.
[9] Mann, K. (1992) Isolation of the α3-chain of human type V collagen and characterisation by partial sequencing. Hoppe-Seyler Z. Biol. Chem. 373: 69–75.
[10] Moradi-Ameli, M. et al. (1994) Diversity in the processing events at the N-terminus of type-V collagen. Eur. J. Biochem. 221: 987–995.
[11] Niyibizi, C and Eyre, D.R. (1994) Structural characteristics of the cross-linking sites in type V collagen of bone. Chain specificities and heterotypic links to type I collagen. Eur. J. Biochem. 224: 943–950.
[12] Niyibizi, C. et al. (1994) A 92 kDa gelatinase (MMP-9) cleavage site in native type V collagen. Biochem. Biophys. Res. Commun. 202: 328–333.
[13] Elstow, S.F. and Weiss, J.B. (1983) Extraction, isolation and characterization of neutral salt soluble type V collagen from fetal calf skin. Collagen Related Res. 3: 181–193.
[14] Abedin, M.Z. et al. (1982) Isolation and native characterization of cysteine-rich collagens from bovine placental tissues and uterus and their relationship to types IV and V collagens. Biosci. Rep. 2: 493–502.
[15] Greenspan, D.S. et al. (1992) Human collagen gene COL5A1 maps to the q34.2–q34.3 region of chromosome-9, near the locus for nail-patella syndrome. Genomics 12: 836–837.

[16] Myers, J.C. and Dion, A.S. (1990) Types III and V procollagens: homology in genetic organization and diversity in structure. In: Extracellular Matrix Genes, ed. Sandell, L.J. and Boyd, C.D., Academic Press, New York, pp. 57–78.
[17] Takahara, K. et al. (1995) Complete structural organization of the human alpha-1(V) collagen gene (COL5A1) – divergence from the conserved organization of other characterized fibrillar collagen genes. Genomics 29: 588–597.

Collagen type VI

Type VI collagen is essentially a glycoprotein with a short collagenous central domain and is present in most, possibly all, connective tissues. It assembles into a unique supramolecular structure of 5 nm-diameter microfibrils with a periodicity of approximately 100 nm [1]. Mutations in the genes COL6A1, COL6A2 and COL6A3 encoding the three distinct α chains, have been linked to the autosomal dominant disorder of Bethlem myopathy, characterized by contractures of multiple joints and generalized muscular weakness and wasting [2].

Molecular structure

Type VI collagen is synthesized largely as a heterotrimer comprising three genetically distinct α1(VI), α2(VI) and α3(VI) chains, although there is evidence for the existence of alternative but less stable assemblies comprised of either α3(VI) or α1(VI)/α2(VI) chains. Each chain consists of a short (105 nm) triple helix at the ends of which are large globular carboxy-terminal (NC1) and amino-terminal (NC2) domains. The α1(VI) and α2(VI) chains are similar in size, but the α3(VI) chain is much larger owing to its prominent NC2 domain. The NC1 and NC2 domains of the α1(VI) and α2(VI) chains are comprised of two and one repeats, respectively, that are homologous to the A-type repeats of von Willebrand factor. The NC1 domain of the α3(VI) chain also contains two A-type repeats as well as three other repeats: a lysine/proline-rich repeat homologous to several salivary proteins and containing threonine repeats which may be O-glycosylated, a fibronectin type III repeat and a repeat that is highly homologous to Kunitz-type serine proteinase inhibitors. The NC2 domain of the α3(VI) chain is comprised of ten von Willebrand factor A repeats and is subject to multiple alternative splicing, resulting in at least four mutually exclusive isoforms. Three additional variants in the NC1 domain of the α2(VI) chain also arise by alternative splicing and involve the mutually exclusive use of the two most COOH-terminal exons, together with the selective use of an internal acceptor splice site in the penultimate exon. Monomers are assembled intracellularly into anti-parallel dimers and then into tetramers prior to secretion from the cell. The α3(VI) chain is partially processed at the NC1 domain with loss of the

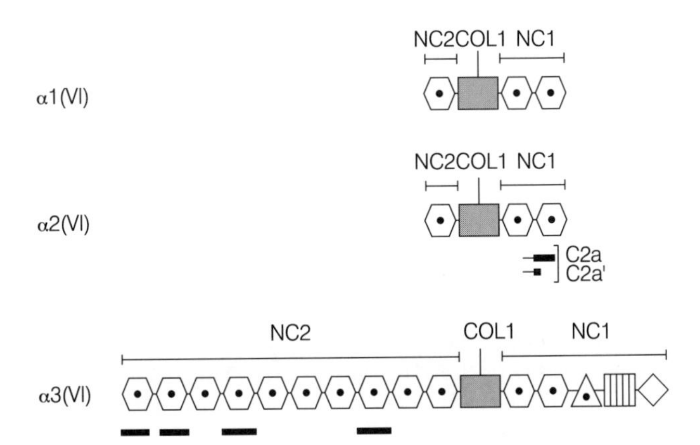

Kunitz module and possibly other adjacent modules. The tetramers are then assembled into microfibrils by end to end aggregation. Type VI collagen tetramers are stabilized by intra- and inter-molecular disulphide bonds, but are not cross-linked by lysine/hydroxylysine-derived bonds. The end to end association of tetramers within the microfibrils occurs by strong non-covalent interactions. Colocalization and ligand binding studies *in vitro* indicate an interaction between type VI collagen and other macromolecules including hyaluronan, type II and XIV collagens, decorin, biglycan and the cell membrane-bound chondroitin sulphate proteoglycan NG2 receptor [1,3–9].

Isolation

Type VI collagen can be extracted completely and in an intact form from most tissues by selective glycosidase and/or guanidine–HCl treatment or bacterial collagenase digestion [10,11].

Accession number

S05377; A31952

Primary structure: α1(VI) chain

Ala	A	78	Cys	C	20	Asp	D	71	Glu	E	69
Phe	F	33	Gly	G	155	His	H	14	Ile	I	42
Lys	K	54	Leu	L	71	Met	M	11	Asn	N	28
Pro	P	92	Gln	Q	41	Arg	R	61	Ser	S	50
Thr	T	41	Val	V	65	Trp	W	4	Tyr	Y	28

Mol. wt (calc.) = 108 522 Residues = 1028

```
1     MRAARALLPL  LLQACWTAAQ  DEPETPRAVA  FQDCPVDLFF  VLDTSESVAL
51    RLKPYGALVD  KVKSFTKRFI  DNLRDRYYRC  DRNLVWNAGA  LHYSDEVEII
101   QGLTRMPGGR  DALKSSVDAV  KYFGKGTYTD  CAIKKGLEQL  LVGGSHLKEN
151   KYLIVVTDGH  PLEGYKEPCG  GLEDAVNEAK  HLGVKVFSVA  ITPDHLEPRL
201   SIIATDHTYR  RNFTAADWGQ  SRDAEEAISQ  TIDTIVDMIK  NNVEQVCCSF
251   ECQPARGPPG  LRGDPGFEGE  RGKPGLPGEK  GEAGDPGRPG  DLGPVGYQGM
301   KGEKGSRGEK  GSRGPKGYKG  EKGKRGIDGV  DGVKGEMGYP  GLPGCKGSPG
351   FDGIQGPPGP  KGDPGAFGLK  GEKGEPGADG  EAGRPGARGP  SGDEGPAGEP
401   GPPGEKGEAG  DEGNPGPDGA  PGERGGPGER  GPRGTPGTRG  PRGDPGEAGP
451   QGDQGREGPV  GVPGDPGEAG  PIGPKGYRGD  EGPPGSEGAR  GAPGPAGPPG
501   DPGLMGERGE  DGPAGNGTEG  FPGFPGYPGN  RGAPGINGTK  GYPGLKGDEG
551   EAGDPGDDNN  DIAPRGVKGA  KGYRGPEGPQ  GPPGHQGPPG  PDECEILDII
601   MKMCSCCECK  CGPIDLLFVL  DSSESIGLQN  FEIAKDFVVK  VIDRLSRDEL
651   VKFEPGQSYA  GVVQYSHSQM  QEHVSLRSPS  IRNVQELKEA  IKSLQWMAGG
701   TFTGEALQYT  RDQLLPPSPN  NRIALVITDG  RSDTQRDTTP  LNVLCSPGIQ
751   VVSVGIKDVF  DFIPGSDQLN  VISCQGLAPS  QGRPGLSLVK  ENYAELLEDA
801   FLKNVTAQIC  IDKKCPDYTC  PITFSSPADI  TILLEPPPDV  GSHNFDTTKR
851   FAKRLAERFL  TAGRTDPAHD  VRVAVVQYSG  TGQQRPERAS  LQFLQNYTAL
901   ASAVDAMDFI  NDATDVNDAL  GYVTRFYREA  SSGAAKKRLL  LFSDGNSQGA
951   TPAAIEKAVQ  EAQRAGIEIF  VVVVGRQVNE  PHIRVLVTGK  TAEYDVAYGE
1001  SHLFRVPSYQ  ALLRGVFHQT  VSRKVALG
```

Structural and functional sites

Signal peptide: 1–19 (probable)
NC2 domain: 20–256
Helical domain: 257–592
NC1 domain: 593–1028
von Willebrand factor A repeats: 30–216, 609–783, 801–1003
Hydroxylysine glycosylation sites: All residues in the Y position of Gly-X-Y triplets
Potential N-linked glycosylation sites: 212, 516 (determined), 537, 804, 896
Imperfections in Gly-X-Y triplets: 515–516, 559–565 (required for supercoiling of dimers)
Cysteine involved in dimer formation: 345

Accession number

B31952; S05378; S09646

Primary structure: α2(VI) chain

Major variant:

Ala	A	60	Cys	C	21	Asp	D	70	Glu	E	66
Phe	F	39	Gly	G	154	His	H	21	Ile	I	42
Lys	K	51	Leu	L	66	Met	M	13	Asn	N	33
Pro	P	89	Gln	Q	44	Arg	R	67	Ser	S	54
Thr	T	45	Val	V	60	Trp	W	5	Tyr	Y	18

Mol. wt (calc.) = 108 354 Residues = 1018

C2a variant:

Ala	A	52	Cys	C	20	Asp	D	63	Glu	E	63
Phe	F	32	Gly	G	145	His	H	13	Ile	I	40
Lys	K	51	Leu	L	55	Met	M	12	Asn	N	26
Pro	P	91	Gln	Q	40	Arg	R	55	Ser	S	42
Thr	T	45	Val	V	48	Trp	W	6	Tyr	Y	18

Mol. wt (calc.) = 97 219 Residues = 917

C2a′ variant:

Ala	A	42	Cys	C	20	Asp	D	57	Glu	E	55
Phe	F	27	Gly	G	142	His	H	12	Ile	I	36
Lys	K	49	Leu	L	46	Met	M	11	Asn	N	25
Pro	P	82	Gln	Q	35	Arg	R	50	Ser	S	41
Thr	T	35	Val	V	42	Trp	W	4	Tyr	Y	16

Mol. wt (calc.) = 87 092 Residues = 827

```
1     MLQGTCSVLL  LWGILGAIQA  QQQEVISPDT  TERNNNCPEK  TDCPIHVYFV
51    LDTSESVTMQ  SPTDILLFHM  KQFVPQFISQ  LQNEFYLDQV  ALSWRYGGLH
101   FSDQVEVFSP  PGSDRASFIK  NLQGISSFRR  GTFTDCALAN  MTEQIRQDRS
151   KGTVHFAVVI  TDGHVTGSPC  GIKLQAERAR  EEGIRLFAVA  PNQNLKEQGL
201   RDIASTPHEL  YRNDYATMLP  DSTEINQDTI  NRIIKVMKHE  AYGECYKVSC
```

```
251   LEIPGPSGPK  GYRGQKGAKG  NMGEPGEPGQ  KGRQGDPGIE  GPIGFPGPKG
301   VPGFKGEKGE  FGADGRKGAP  GLAGKNGTDG  QKGKLGRIGP  PGCKGDPGNR
351   GPDGYPGEAG  SPGERGDQGG  KGDPGRPGRR  GPPGEIGAKG  SKGYQGNNGA
401   PGSPGVKGAK  GGPGPRGPKG  EPGRRGDPGT  KGSPGSDGPK  GEKGDPGPEG
451   PRGLAGEVGN  KGAKGDRGLP  GPRGPQGALG  EPGKQGSRGD  PGDAGPRGDS
501   GQPGPSGDPG  RPGFSYPGPR  GAPGEKGEPG  PRGPEGGRGD  FGLKGEPGRK
551   GEKGEPADPG  PPGEPGPRGP  RGVPGPEGEP  GPPGDPGLTE  CDVMTYVRET
601   CGCCDCEKRC  GALDVVFVID  SSESIGYTNF  TLEKNFVINV  VNRLGAIAKD
651   PKSETGTRVG  VVQYSHEGTF  EAIQLDDEHI  DSLSSFKEAV  KNLEWIAGGT
701   WTPSALKFAY  DRLIKESRRQ  KTRVFAVVIT  DGRHDPRDDD  LNLRALCDRD
751   VTVTAIGIGD  MFHEKHESEN  LYSIACDKPQ  QVRNMTLFSD  LVAEKFIDDM
801   EDVLCPDPQI  VCPDLPCQT
```

Major variant continues:

```
820                        E  LSVAQCTQRP  VDIVFLLDGS  ERLGEQNFHK
850   ARRFVEQVAR  RLTLARRDDD  PLNARVALLQ  FGGPGEQQVA  FPLSHNLTAI
901   HEALETTQYL  NSFSHVGAGV  VHAINAIVRS  PRGGARRHAE  LSFVFLTDGV
951   TGNDSLHESA  HSMRNENVVP  TVLALGSDVD  MDVLTTLSLG  DRAAVFHEKD
1001  YDSLAQPGFF  DRFIRWIC
```

C2a variant continues:

```
820                        D  APWPGGEPPV  TFLRTEEGPD  ATFPRTIPLI
850   QQLLNATELT  QDPAAYSQLV  AVLVYTAERA  KFATGVERQD  WMELFIDTFK
901   LVHRDIVGDP  ETALALC
```

C2a′ variant continues:

```
820                        G  LDGAVLC
```

Structural and functional sites

Signal peptide: 1–20
NC2 domain: 21–254
Helical domain: 255–589
NC1 domain: 590–1018 (main variant); 590–917 (C2a variant); 590–827 (C2a′ variant)
von Willebrand factor A repeats: 36–218, 606–783, 818–995
Hydroxylysine glycosylation sites: All residues in the Y position of Gly-X-Y triplets
Potential N-linked glycosylation sites: 140 (determined), 326, 629 (determined), 784, 855 (C2a variant only), 896, 953
Imperfections in Gly-X-Y triplets: 516–517, 557–559 (required for supercoiling of dimers)
Cysteine involved in dimer formation: 343

Accession number

S13679

Primary structure: $\alpha3$(VI) chain

Ala	A	234	Cys	C	30	Asp	D	180	Glu	E	183
Phe	F	150	Gly	G	303	His	H	43	Ile	I	154
Lys	K	158	Leu	L	283	Met	M	34	Asn	N	130
Pro	P	211	Gln	Q	146	Arg	R	189	Ser	S	215
Thr	T	169	Val	V	295	Trp	W	6	Tyr	Y	62

Mol. wt (calc.) = 342 960 Residues = 3175

```
1     MRKHRHLPLV AVFCLFLSGF PTTHAVKNGA AADIIFLVDS SWTIGEEHFQ
51    LVREFLYDVV KSLAVGENDF HFALVQFNGN PHTEFLLNTY RTKQEVLSHI
101   SNMSYIGGTN QTGKGLEYIM AKHLTKAAGS LAGDGVPQVI VVLTDGHSKD
151   GLALPSAELK SADVNVFAIG VEDADEGALK EIASEPLNMH MFNLENFTSL
201   HDIVGNLVSC VHSSVSPERA GDTETLKDIT QQQQAAQDSA DIIFLIDGSN
251   NTGSVNFAVI LDFLVNLLEK LPIGTQQIRV GVVQFSDEPR TMFSLDTYST
301   KAQVLGAVKA LGFAGGELAN IGLALDFVVE NHFTRAGGSR VEEGVPQVLV
351   LISAGPSSDE IRYGVVALKQ ASVFSFGLGA QAASRAELQH IATDDNLVFT
401   VPEFRSFGDL QEKLLPYIVG VAQRHIVLKP PTIVTQVIEV NKRDIVFLVD
451   GSSALGLANF NAIRDFIAKV IQRLEIGQDL IQVAVAQYAD TVRPEFYFNT
501   HPTKREVITA VRKMKPLDGS ALYTGSALDF VRNNLFTSSA GYRAAEGIPK
551   LLVLITGGKS LDEISQPAQE LKRSSIMAFA IGNKGADQAE LEEIAFDSSL
601   VFIPAEFRAA PLQGMLPGLL APLRTLSGTP EVHSNKRDII FLLDGSANVG
651   KTNFPYVRDF VMNLVNSLDI GNDNIRVGLV QFSDTPVTEF SLNTYQTKSD
701   ILGHLRQLQL QGGSGLNTGS ALSYVYANHF TEAGGSRIRE HVPQLLLLLT
751   AGQSEDSYLQ AANALTRAGI LTFCVGASQA NKAELEQIAF NPSLVYLMDD
801   FSSLPALPQQ LIQPLTTYVS GGVEEVPLAQ PESKRDILFL FDGSANLVGQ
851   FPVVRDFLYK IIDELNVKPE GTRIAVAQYS DDVKVESRFD EHQSKPEILN
901   LVKRMKIKTG KALNLGYALD YAQRYIFVKS AGSRIEDGVL QFLVLLVAGR
951   SSDRVDGPAS NLKQSGVVPF IFQAKNADPA ELEQIVLSPA FILAAESLPK
1001  IGDLHPQIVN LLKSVHNGAP APVSGEKDVV FLLDGSEGVR SGFPLLKEFV
1051  QRVVESLDVG QDRVRVAVVQ YSDRTRPEFY LNSYMNKQDV VNAVRQLTLL
1101  GGPTPNTGAA LEFVLRNILV SSAGSRITEG VPQLLIVLTA DRSGDDVRNP
1151  SVVVKRGGAV PIGIGIGNAD ITEMQTISFI PDFAVAIPTF RQLGTVQQVI
1201  SERVTQLTRE ELSRLQPVLQ PLPSPGVGGK RDVVFLIDGS QSAGPEFQYV
1251  RTLIERLVDY LDVGFDTTRV AVIQFSDDPK AEFLLNAHSS KDEVQNAVQR
1301  LRPKGGRQIN VGNALEYVSR NIFKRPLGSR IEEGVPQFLV LLISSGKDDE
1351  VVVPAVELKQ FGVAPFTIAR NADQEELVKI SLSPEYVFSV STFRELPSLE
1401  QKLLTPITTL TSEQIQKLLA STRYPPPAVE SDAADIVFLI DSSEGVRPDG
1451  FAHIRDFVSR IVRRLNIGPS KVRVGVVQFS NDVFPEFYLK TYRSQAPVLD
1501  AIRRLRLRGG SPLNTGKALE FVARNLFVKS AGSRIEDGVP QHLVLVLGGK
1551  SQDDVSRFAQ VIRSSGIVSL GVGDRNIDRT ELQTITNDPR LVFTVREFRE
1601  LPNIEERIMN SFGPSAATPA PPGVDTPPPS RPEKKKADIV FLLDGSINFR
1651  RDSFQEVLRF VSEIVDTVYE DGDSIQVGLV QYNSDPTDEF FLKDFSTKRQ
1701  IIDAINKVVY KGGRHANTKV GLEHLRVNHF VPEAGSRLDQ RVPQIAFVIT
1751  GGKSVEDAQD VSLALTQRGV KVFAVGVRNI DSEEVGKIAS NSATAFRVGN
1801  VQELSELSEQ VLETFDDAID ETLCPGVTDA AKACNLDVIL GFDGSRDQNV
1851  FVAQKGFESK VDAILNRISQ MHRVSCSGGR SPTVRVSVVA NTPSGPVEAF
1901  DFDEYQPEML EKFRNMRSQH PYVLTEDTLK VYLNKFRQSS PDSVKVVIHF
1951  TDGADGDLAD LHRASENLRQ EGVRALILVG LERVVNLERL MHLEFGRGFM
2001  YDRPLRLNLL DLDYELAEQL DNIAEKACCG VPCKCSGQRG DRGPIGSIGP
2051  KGIPGEDGYR GYPGDEGGPG ERGPPGVNGT QGFQGCPGQR GVKGSRGFPG
2101  EKGEVGEIGL DGLDGEDGDK GLPGSSGEKG NPGRRGDKGP RGEKGERGDV
2151  GIRGDPGNPG QDSQERGPKG ETGDLGPMGV PGRDGVPGGP GETGKNGGFG
2201  RRGPPGAKGN KGGPGQPGFE GEQGTRGAQG PAGPAGPPGL IGEQGISGPR
```

```
2251 GSGGARGAPG ERGRTGPLGR KGEPGEPGPK GGIGNPGPRG ETGDDGRDGV
2301 GSEGRRGKKG ERGFPGYPGP KGNPGEPGLN GTTGPKGIRG RRGNSGPPGI
2351 VGQKGRPGYP GPAGPRGNRG DSIDQCALIQ SIKDKCPCCY GPLECPVFPT
2401 ELAFALDTSE GVNQDTFGRM RDVVLSIVNV LTIAESNCPT GARVAVVTYN
2451 NEVTTEIRFA DSKRKSVLLD KIKNLQVALT SKQQSLETAM SFVARNTFKR
2501 VRNGFLMRKV AVFFSNTPTR ASPQLREAVL KLSDAGITPL FLTRQEDRQL
2551 INALQINNTA VGHALVLPAG RDLTDFLENV LTCHVCLDIC NIDPSCGFGS
2601 WRPSFRDRRA AGSDVDIDMA FILDSAETTT LFQFNEMKKY IAYLVRQLDM
2651 SPDPKASQHF ARVAVVQHAP SESVDNASMP PVKVEFSLTD YGSKEKLVDF
2701 LSRGMTQLQG TRALGSAIEY TIENVFESAP NPRDLKIVVL MLTGEVPEQQ
2751 LEEAQRVILQ AKCKGYFFVV LGIGRKVNIK EVYTFASEPN DVFFKLVDKS
2801 TELNEEPLMR FGRLLPSFVS SENAFYLSPD IRKQCDWFQG DQPTKNLVKF
2851 GHKQVNVPNN VTSSPTSNPV TTTKPVTTTK PVTTTTKPVT TTTKPVTIIN
2901 QPSVKPAAAK PAPAKPVAAK PVATKTATVR PPVAVKPATA AKPVAAKPAA
2951 VRPPAAAAKP VATKPEVPRP QAAKPAATKP ATTKPVVKML REVQVFEITE
3001 NSAKLHWERP EPPGPYFYDL TVTSAHDQSL VLKQNLTVTD RVIGGLLAGQ
3051 TYHVAVVCYL RSQVRATYHG SFSTKKSQPP PPQPARSASS STINLMVSTE
3101 PLALTETDIC KLPKDEGTCR DFILKWYYDP NTKSCARFWY GGCGGNENKF
3151 GSQKECEKVC APVLAKPGVI SVMGT
```

Structural and functional sites

Signal peptide: 1–25
NC2 domain: 26–2036
Helical domain: 2037–2372
NC1 domain: 2373–3175
von Willebrand factor A repeats: 26–230, 239–425, 432–632, 636–827, 834–1020, 1026–1212, 1230–1419, 1433–1625, 1636–1823, 1835–2027, 2399–2582, 2616–2863
Lysine/proline-rich repeat: 2864–2985
Fibronectin type III repeat: 2986–3074
Kunitz-type serine proteinase inhibitor repeat: 3110–3160
Alternatively spliced repeats: 1433–1625 (N3), 636–827 (N7), 239–425 (N9), 26–230 (N10)
Hydroxylysine glycosylation sites: All residues in the Y position of Gly-X-Y triplets
Potential N-linked glycosylation sites: 102, 110, 196, 250, 791, 1149, 2078, 2330, 2557, 2676, 2860, 3035
Imperfections in Gly-X-Y triplets: 2163–2166, 2299–2300 (required for super coiling of dimers)
Cysteine involved in tetramer formation: 2086

Gene structure

The COL6A1 and COL6A2 genes span 29 kb and 36 kb, respectively, and are separated by 150 kb on chromosome 21 (locus q22.3) in a head-to-tail organization (5′ COL6A1 3′–5′ COL6A2 3′). COL6A2 contains 30 exons, two of which (1/1A and 28A/28) are alternatively used. Transcription of exons 1 and 1A are controlled by two promoters P1 and P2 located upstream of exon 1 and 1A, respectively. However, both exons code only for the 5′-untranslated region. The three NC1 variants of the α2(VI) chain arise by alternative splicing of the penultimate exon 28A and the final exon 28. The exon structure encoding the triple-helical domain of both the α1(VI) and α2(VI) chains is similar and based on multiples of 9 bp (27, 36, 45, 54, 63, and 90 bp) except for interrupted regions. The COL6A3 gene is located on human chromosome 2 at locus q37. Separate exons encode each of the repetitive nine most terminal α3 NC2 subdomains and three of these are spliced in or out in a mutually exclusive manner. The exons encoding the NC2 domain alone span 26 kb [3,12–15].

References

[1] Timpl, R. and Chu, M.-L. (1994) Microfibrillar collagen type VI. In: Extracellular Matrix Assembly and Structure, ed. Yurchenco, P.D. et al., Academic Press, Orlando, Fl., pp. 207–242.

[2] Jobsis, G.J. et al. (1996) Type VI collagen mutations in Bethlem myopathy, an autosomal dominant myopathy with contractures. Nature Genetics 14: 113–115.

[3] Weil, D. et al. (1988) Cloning and chromosomal localization of human genes encoding the three chains of type VI collagen. Am. J. Hum. Genet. 42: 435–445.

[4] Chu, M.-L. et al. (1989) Sequence analysis of α1(VI) and α2(VI) chains of human type VI collagen reveals internal triplication of globular domains similar to the A domains of von Willebrand factor and two α2(VI) chain variants that differ in the carboxy terminus. EMBO J. 8: 1939–1946.

[5] Chu, M.-L. et al. (1990) Mosaic structure of globular domains in the human type VI collagen α3 chain: Similarity to von Willebrand factor, salivary proteins and aprotinin type protease inhibitors. EMBO J. 9: 385–393.

[6] Saitta, B. et al. (1990) Alternative splicing of the human α2(VI) collagen gene generates multiple mRNA transcripts which predict three protein variants with distinct carboxy termini. J. Biol. Chem. 265: 6473–6480.

[7] Stokes, D.G. et al. (1991) Human α3(VI) collagen gene. Characterization of exons coding for the amino-terminal globular domain and alternative splicing in normal and tumor cells. J. Biol. Chem. 266: 8626–8633.

[8] Mayer, U. et al. (1994) Recombinant expression and properties of the Kunitz-type protease-inhibitor module from human type VI collagen α3(VI) chain. Eur. J. Biochem. 225: 573–580.

[9] Zanussi, S. et al. (1992) The human type VI collagen gene. mRNA and protein variants of the α3 chain generated by alternative splicing of an additional 5′-end exon. J. Biol. Chem. 267: 24082–24089.

[10] Wu, J.-J. et al. (1987) Type VI collagen of the intervertebral disc. Biochemical and electron-microscopic characterization of the native protein. Biochem. J. 248: 373–381.

[11] Kielty, C.M. et al. (1991) Isolation and ultrastructural analysis of microfibrillar structures from foetal bovine elastic tissues. Relative abundance and supramolecular architecture of type VI collagen assemblies and fibrillin. J. Cell Sci. 99: 797–807.

[12] Saitta, B. et al. (1992) Human α2(VI) collagen gene. Heterogeneity at the 5′-untranslated region generated by an alternative exon. J. Biol. Chem. 267: 6188–6196.

[13] Saitta, B. et al. (1991) The exon organization of the triple-helical coding regions of the human $\alpha1$(VI) and $\alpha2$(VI) collagen genes is highly similar. Genomics 11: 145–153.
[14] Heiskanen, M. et al. (1995) Head to tail organization of the human COL6A1 and COL6A2 genes by Fiber-FISH. Genomics 29: 801–803.
[15] Saitta, B. and Chu, M.-L. (1994) Two promoters control the transcription of the human $\alpha2$(VI) collagen gene. Eur. J. Biochem. 223: 675–682.

Collagen type VII

Type VII collagen has a highly specific tissue distribution. It is the major component of anchoring fibrils that are elaborated by specialized epithelia (e.g. epidermis and intestinal sub-mucosa) and that anchor the basement membrane to the underlying stromal tissue. Mutations in the COL7A1 gene result in the heritable diseases known as the dystrophic forms of epidermolysis bullosa in which the anchoring fibrils are abnormal in morphology, reduced in number or completely absent, leading to skin fragility and severe blistering[1,2].

Molecular structure

Type VII collagen is a homotrimer of three $\alpha 1$(VII) chains. Each chain comprises a triple helix, approximately 424 nm in length, a large NH_2-terminal globular domain (NC2) and a small COOH-terminal domain (NC1). The NC2 domain comprises a partial von Willebrand factor A repeat, followed by nine consecutive fibronectin type III repeats, a von Willebrand factor A repeat and a cysteine/proline-rich domain. The triple helix contains 19 interruptions, one major interruption in the centre of the helix being 39 amino acids in length. The small NC1 domain is cysteine-rich and contains a module homologous to the Kunitz proteinase inhibitor-type repeat found in the NC1 domain of the $\alpha 3$(VI) chain. Limited processing occurs at the end of the NC1 domain. The processed monomers assemble into anti-parallel dimers in the tissue with an NC1 to NC1 domain overlap of 60 nm, resulting in an overall triple helix of length 785 nm. The dimers then associate laterally to produce 'segment long spacing (SLS)-like' structures that insert into the basement membrane via the large NC2 domains[1,3–5]. (N.B. While the COL and NC domain designations follow the usual convention and are numbered from C to N, it should be noted that in references 1, 2, 7 and 8, the domains are numbered from N to C.)

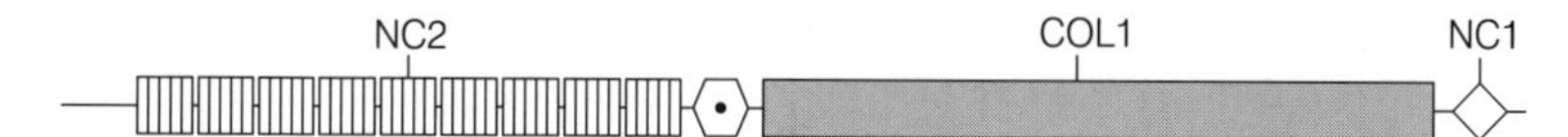

Isolation

Type VII collagen can be isolated in an intact form either from the culture medium of specialized epithelial cells or from guanidine–HCl extracts of amniotic membranes[4]. The triple helical domain can be isolated in larger quantities by pepsin digestion[3].

Accession number

L02870

Primary structure: α1(VII) chain

Ala	A	193	Cys	C	17	Asp	D	145	Glu	E	187
Phe	F	39	Gly	G	627	His	H	28	Ile	I	56
Lys	K	93	Leu	L	212	Met	M	18	Asn	N	27
Pro	P	426	Gln	Q	105	Arg	R	217	Ser	S	166
Thr	T	132	Val	V	201	Trp	W	19	Tyr	Y	36

Mol. wt (calc.) = 295 219 Residues = 2944

```
1     MTLRLLVAAL  CAGILAEAPR  VRAQHRERVT  CTRLYAADIV  FLLDGSSSIG
51    RSNFREVRSF  LEGLVLPFSG  AASAQGVRFA  TVQYSDDPRT  EFGLDALGSG
101   GDVIRAIREL  SYKGGNTRTG  AAILHVADHV  FLPQLARPGV  PKVCILITDG
151   KSQDLVDTAA  QRLKGQGVKL  FAVGIKNADP  EELKRVASQP  TSDFFFFVND
201   FSILRTLLPL  VSRRVCTTAG  GVPVTRPPDD  STSAPRDLVL  SEPSSQSLRV
251   QWTAASGPVT  GYKVQYTPLT  GLGQPLPSER  QEVNVPAGET  SVRLRGLRPL
301   TEYQVTVIAL  YANSIGEAVS  GTARTTALEG  PELTIQNTTA  HSLLVAWRSV
351   PGATGYRVTW  RVLSGGPTQQ  QELGPGQGSV  LLRDLEPGTD  YEVTVSTLFG
401   RSVGPATSLM  ARTDASVEQT  LRPVILGPTS  ILLSWNLVPE  ARGYRLEWRR
451   ETGLEPPQKV  VLPSDVTRYQ  LDGLQPGTEY  RLTLYTLLEG  HEVATPATVV
501   PTGPELPVSP  VTDLQATELP  GQRVRVSWSP  VPGATQYRII  VRSTQGVERT
551   LVLPGSQTAF  DLDDVQAGLS  YTVRVSARVG  PREGSASVLT  VRREPETPLA
601   VPGLRVVVSD  ATRVRVAWGP  VPGASGFRIS  WSTGSGPESS  QTLPPDSTAT
651   DITGLQPGTT  YQVAVSVLRG  REEGPAAVIV  ARTDPLGPVR  TVHVTQASSS
701   SVTITWTRVP  GATGYRVSWH  SAHGPEKSQL  VSGEATVAEL  DGLEPDTEYT
751   VHVRAHVAGV  DGPPASVVVR  TAPEPVGRVS  RLQILNASSD  VLRITWVGVT
801   GATAYRLAWG  RSEGGPMRHQ  ILPGNTDSAE  IRGLEGGVSY  SVRVTALVGD
851   REGTPVSIVV  TTPPEAPPAL  GTLHVVQRGE  HSLRLRWEPV  PRAQGFLLHW
901   QPEGGQEQSR  VLGPELSSYH  LDGLEPATQY  RVRLSVLGPA  GEGPSAEVTA
951   RTESPRVPSI  ELRVVDTSID  SVTLAWTPVS  RASSYILSWR  PLRGPGQEVP
1001  GSPQTLPGIS  SSQRVTGLEP  GVSYIFSLTP  VLDGVRGPEA  SVTQTPVCPR
1051  GLADVVFLPH  ATQDNAHRAE  ATRRVLERLV  LALGPLGPQA  VQVGLLSYSH
1101  RPSPLFPLNG  SHDLGIILQR  IRDMPYMDPS  GNNLGTAVVT  AHRYMLAPDA
1151  PGRRQHVPGV  MVLLVDEPLR  GDIFSPIREA  QASGLNVVML  GMAGADPEQL
1201  RRLAPGMDSV  QTFFAVDDGP  SLDQAVSGLA  TALCQASFTT  QPRPEPCPVY
1251  CPKGQKGEPG  EMGLRGQVGP  PGDPGLPGRT  GAPGPQGPPG  SATAKGERGF
1301  PGADGRPGSP  GRAGNPGTPG  APGLKGSPGL  PGPRGDPGER  GPRGPKGEPG
1351  APGQVIGGEG  PGLPGRKGDP  GPSGPPGPRG  PLGDPGPRGP  PGLPGTAMKG
1401  DKGDRGERGP  PGPGEGGIAP  GEPGLPGLPG  SPGPQGPVGP  PGKKGEKGDS
1451  EDGAPGLPGQ  PGSPGEQGPR  GPPGAIGPKG  DRGFPGPLGE  AGEKGERGPP
1501  GPAGSRGLPG  VAGRPGAKGP  EGPPGPTGRQ  GEKGEPGRPG  DPAVVGPAVA
1551  GPKGEKGDVG  PAGPRGATGV  QGERGPPGLV  LPGDPGPKGD  PGDRGPIGLT
1601  GRAGPPGDSG  PPGEKGDPGR  PGPPGPVGPR  GRDGEVGEKG  DEGPPGDPGL
1651  PGKAGERGLR  GAPGVRGPVG  EKGDQGDPGE  DGRNGSPGSS  GPKGDRGEPG
1701  PPGPPGRLVD  TGPGAREKGE  PGDRGQEGPR  GPKGDPGLPG  APGERGIEGF
1751  RGPPGPQGDP  GVRGPAGEKG  DRGPPGLDGR  SGLDGKPGAA  GPSGPNGAAG
1801  KAGDPGRDGL  PGLRGEQGLP  GPSGPPGLPG  KPGEDGKPGL  NGKNGEPGDP
1851  GEDGRKGEKG  DSGASGREGR  DGPKGERGAP  GILGPQGPPG  LPGPVGPPGQ
1901  GFPGVPGGTG  PKGDRGETGS  KGEQGLPGER  GLRGEPGSVP  NVDRLLETAG
1951  IKASALREIV  ETWDESSGSF  LPVPERRRGP  KGDSGEQGPP  GKEGPIGFPG
2001  ERGLKGDRGD  PGPQGPPGLA  LGERGPPGPS  GLAGEPGKPG  IPGLPGRAGG
2051  VGEAGRPGER  GERGEKGERG  EQGRDGPPGL  PGTPGPPGPP  GPKVSVDEPG
```

```
2101 PGLSGEQGPP GLKGAKGEPG SNGDQGPKGD RGVPGIKGDR GEPGPRGQDG
2151 NPGLPGERGM AGPEGKPGLQ GPRGPPGPVG GHGDPGPPGA PGLAGPAGPQ
2201 GPSGLKGEPG ETGPPGRGLT GPTGAVGLPG PPGPSGLVGP QGSPGLPGQV
2251 GETGKPGAPG RDGASGKDGD RGSPGVPGSP GLPGPVGPKG EPGPTGAPGQ
2301 AVVGLPGAKG EKGAPGGLAG DLVGEPGAKG DRGLPGPRGE KGEAGRAGEP
2351 GDPGEDGQKG APGPKGFKGD PGVGVPGSPG PPGPPGVKGD LGLPGLPGAP
2401 GVVGFPGQTG PRGEMGQPGP SGERGLAGPP GREGIPGPLG PPGPPGSVGP
2451 PGASGLKGDK GDPGVGLPGP RGERGEPGIR GEDGRPGQEG PRGLTGPPGS
2501 RGERGEKGDV GSAGLKGDKG DSAVILGPPG PRGAKGDMGE RGPRGLDGDK
2551 GPRGDNGDPG DKGSKGEPGD KGSAGLPGLR GLLGPQGQPG AAGIPGDPGS
2601 PGKDGVPGIR GEKGDVGFMG PRGLKGERGV KGACGLDGEK GDKGEAGPPG
2651 RPGLAGHKGE MGEPGVPGQS GAPGKEGLIG PKGDRGFDGQ PGPKGDQGEK
2701 GERGTPGIGG FPGPSGNDGS AGPPGPPGSV GPRGPEGLQG QKGERGPPGE
2751 RVVGAPGVPG APGERGEQGR PGPAGPRGEK GEAALTEDDI RGFVRQEMSQ
2801 HCACQGQFIA SGSRPLPSYA ADTAGSQLHA VPVLRVSHAE EEERVPPEDD
2851 EYSEYSEYSV EEYQDPEAPW DSDDPCSLPL DEGSCTAYTL RWYHRAVTGS
2901 TEACHPFVYG GCGGNANRFG TREACERRCP PRVVQSQGTG TAQD
```

Structural and functional sites

Signal peptide: 1–16 (probable)
NC2 domain: 17–1253
Helical domain: 1254–2784
NC1 domain: 2785–2944
Fibronectin type III repeats: 231–325, 327–413, 419–502, 506–593, 598–683, 684–771, 776–862, 864–952, 954–1045
von Willebrand factor A repeat: 1052–1188 (partial)
Cysteine/proline-rich repeat: 1189–1253
Kunitz-type serine proteinase inhibitor repeat: 2876–2929
Potential *N*-glycosylation sites: 337, 786, 1109
Imperfections/insertions in Gly-X-Y triplets: 1293–1295, 1356–1361, 1398–1399, 1415–1420, 1451–1452, 1549–1550, 1581–1582, 1709–1718, 1902–1903, 1940–1978 (major interruption in centre of triple helix), 2021, 2095–2101, 2216–2217, 2302–2303, 2319–2323, 2371–2373, 2464–2465, 2523–2526, 2752–2753
Cysteines involved in anti-parallel dimer formation: 2634 with 2802 or 2804

Gene structure

The type VII collagen gene COL7A1 is approximately 31 kb and is located on the short arm of chromosome 3 at locus 3p21.3. It comprises 118 exons, which is more than for any other characterized gene. Exon 1 encodes the 5′-untranslated region and the signal peptide and exons 2–28 encode the NC2 domain: exons 2–5 encode the cartilage matrix protein-like region; exons 6–23 encode the nine fibronectin type III repeats; exons 24–26 encode the von Willebrand factor A repeat and exons 27 and 28 encode the cysteine/proline-rich domain. Exons 29–112 encode the triple helical region and most are multiples of 9 bp. Exon 112 is a junctional exon that also encodes the start of the NC1 domain. Exons 113–118 encode the major part of the NC1 domain and the 3′-untranslated region, the Kunitz-like motif being encoded by exon 117 [6–8].

References
[1] Christiano, A.M. et al. (1994) Cloning of human type VII collagen. Complete primary sequence of the $\alpha 1$(VII) chain and identification of intragenic polymorphisms. J. Biol. Chem. 269: 20256–20262.
[2] Christiano, A.M. et al. (1996) Glycine substitution in the triple-helical region of type VII collagen result in a spectrum of dystrophic epidermolysis bullosa phenotypes and patterns of inheritance. Amer. J. Hum. Genet. 58: 671–681.
[3] Bentz, H. et al. (1983) Isolation and partial characterization of a new human collagen with extended triple-helical structural domain. Proc. Natl Acad. Sci. USA 80: 3168–3172.
[4] Lunstrum, G.P. et al. (1986) Large complex globular domains of type VII procollagen contribute to the structure of anchoring fibrils. J. Biol. Chem. 261: 9042–9048.
[5] Burgeson, R.E. (1987) Type VII collagen. In: Structure and Function of Collagen Types, ed. Mayne, R. and Burgeson, R.E., Academic Press, London, pp. 145–172.
[6] Parente, M.G. et al. (1991) Human type VII collagen: cDNA cloning and chromosomal mapping of the gene. Proc. Natl Acad. Sci. USA 88: 6931–6935.
[7] Greenspan, D.S. (1993) The carboxy-terminal half of type VII collagen, including the non-collagenous NC-2 domain and intron/exon organization of the corresponding region of the COL7A1 gene. Hum. Molec. Genet. 2: 273–278.
[8] Christiano, A.M. et al. (1994) Structural organization of the human type VII collagen gene (COL7A1), composed of more exons than any previously characterized gene. Genomics 21: 169–179.

Collagen type VIII

Type VIII collagen is a major component of Descemet's membrane, the specialized basement membrane elaborated by corneal endothelial cells, but it is also synthesized by vascular endothelial cells and epithelial and mesenchymal cells of other tissues. It assembles into the hexagonal lattice that is a characteristic feature of Descemet's membrane.

Molecular structure

Two α chains, α1(VIII) and α2(VIII), have been characterized and there is evidence to suggest that in Descemet's membrane the molecule is a heterotrimer comprising two α1(VIII) chains and one α2(VIII) chain. The COOH-terminal three-quarters of each NC1 domain is homologous to the corresponding region of the α1(X) chain. The lengths of the triple-helical and NC1 domains are similar to those of type X collagen, but the NC2 domain is 2–3 times larger[1–6].

α1(VIII)

NC2 COL1 NC1

α2(VIII)

NC2 COL1 NC1

?

Isolation

Type VIII collagen can be isolated in an intact form from the medium of cultured endothelial cells[7] or as its triple-helical domain by pepsin digestion[8].

Accession number

S15435

Primary structure: α1(VIII) chain

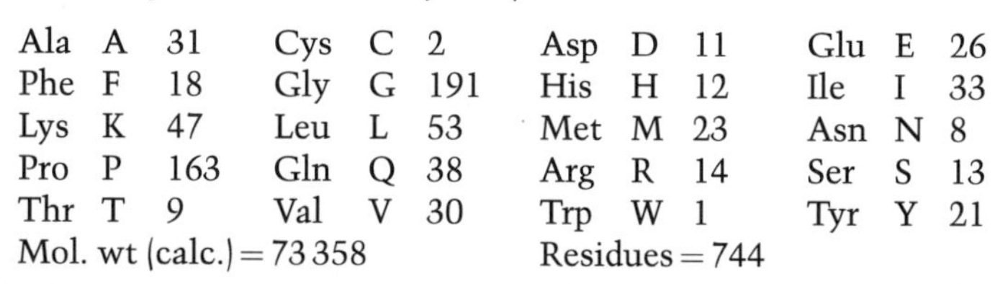

Ala	A	31	Cys	C	2	Asp	D	11	Glu	E	26
Phe	F	18	Gly	G	191	His	H	12	Ile	I	33
Lys	K	47	Leu	L	53	Met	M	23	Asn	N	8
Pro	P	163	Gln	Q	38	Arg	R	14	Ser	S	13
Thr	T	9	Val	V	30	Trp	W	1	Tyr	Y	21

Mol. wt (calc.) = 73 358 Residues = 744

```
1     MAVLPGPLQL  LGVLLTISLS  SIRLIQAGAY  YGIKPLPPQI  PPQMPPQIPQ
51    YQPLGQQVPH  MPLAKDGLAM  GKEMPHLQYG  KEYPHLPQYM  KEIQPAPRMG
101   KEAVPKKGKE  IPLASLRGEQ  GPRGEPGPRG  PPGPPGLPGH  GIPGIKGKPG
151   PQGYPGVGKP  GMPGMPGKPG  AMGMPGAKGE  IGQKGEIGPM  GIPGPQGPPG
```

201 PHGLPGIGKP GGPGLPGQPG PKGDRGPKGL PGPQGLRGPK GDKGFGMPGA

251 PGVKGPPGMH GLPGPVGLPG VGKPGVTGFP GPQGPLGKPG APGEPGRQGP

301 IGVPGVQGPP GIPGIGKPGQ DGIPGQPGFP GGKGEQGLPG LPGAPGLPGI

351 GKPGFPGPKG DRGMGGVPGA LGPRGEKGPI GSPGIGGSPG EPGLPGIPGP

401 MGPPGAIGFP GPKGEGGIVG PQGPPGPKGE PGLQGFPGKP GFLGEVGPPG

451 MRGFPGPIGP KGEHGQKGVP GLPGVPGLLG PKGEPGIPGD QGLQGPPGIP

501 GIGGPSGPIG PPGIPGPKGE PGLPGPPGFP GIGKPGVAGL HGPPGKPGAL

551 GPQGQPGLPG PPGPPGPPGP PAVMPPTPPP QGEYLPDMGL GIDGVKPPHA

601 TGAKKGKNGG PAYEMPAFTA ELTAPFPPVG GPVKFNKLLY NGRQNYNPQT

651 GIFTCEVPGV YYFAYHVHCK GGNVWVALFK NNEPVMYTYD EYKKGFLDQA

701 SGSAVLLLRP GDRVFLQMPS EQAAGLYAGQ YVHSSFSGYL LYPM

Structural and functional sites

Signal peptide: 1–28

NC2 domain: 29–117

Helical domain: 118–571

NC1 domain: 572–744

Imperfections in Gly-X-Y triplets: 139–140, 156–157, 206–207, 244–245, 270–271, 314–315, 349–350, 531–532 [same relative locations as imperfections in α2(VIII) and α1(X)]

Mammalian collagenase cleavage sites: 206–207, 531–532

Accession number

P25067

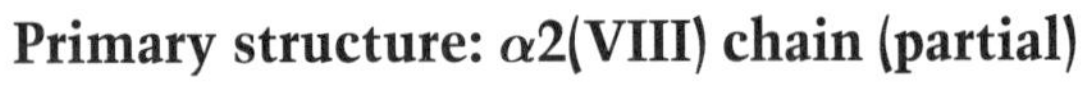

Primary structure: α2(VIII) chain (partial)

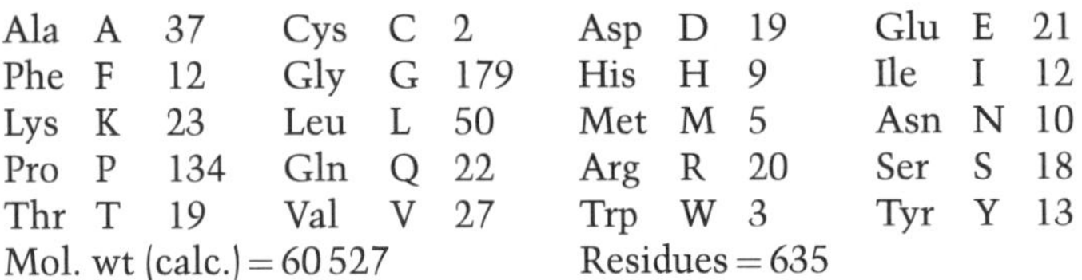

Ala	A	37	Cys	C	2	Asp	D	19	Glu	E	21
Phe	F	12	Gly	G	179	His	H	9	Ile	I	12
Lys	K	23	Leu	L	50	Met	M	5	Asn	N	10
Pro	P	134	Gln	Q	22	Arg	R	20	Ser	S	18
Thr	T	19	Val	V	27	Trp	W	3	Tyr	Y	13

Mol. wt (calc.) = 60 527 Residues = 635

1 MPLPLLPMDL KGEPGPPGKP GPWGPPGPPG FPGKPGHGKP GLHGQPGPAG

51 PPGFSRMGKA GPPGLPGNVG PPGQPGLRGE PGIRGDQGLR GPPGPPGLPG

101 PSGITIPGKP GAQGVPGPPG FQGEPGPQGE PGPPGDRGLK GDNGVGQPGL

151 PGAPGQGGAP GPPGLPGPAG LGKPGLDGLP GAPGDKGESG PPGVPGPRGE

201 PGAVGPKGPP GVDGVGVPGA AGLPGPQGPS GAKGEPGTRG PPGLIGPTGY

251 GMPGLPGPKG DRGPAGVPGL LGDRGEPGED GDPGEQGPQG LGGPPGLPGS

301 AGLPGRRGPP GLRGEAGPGG PPGVPGIRGD QGPSGLAGKP GVPGERGLPG

351 AHGPPGPTGP KGEPGFTGRP GGPGVAGALG QKGDLGLPGQ PGLRGPSGIP

401 GLQGPAGPIG PQGLPGLKGE PGLPGPPGEG RAGEPGTAGP RGPPGVPGSP

451 GITGPPGLPG PPGAPGAFDE TGIAGLHLPN GGVEGAVLGK GGKPQFGLGE

501 LSAHATPAFT AVLTSPLPAS GMPVKFDRTL YNGHSGYNPA TGIFTCPVGG

551 VYYFAYHVHV KGTNVWVALY KNNVPATYTY DEYKKGYLDQ ASGGAVLQLR

601 PNDQVWVQIP SDQANGLYST EYIHSSFSGF LLCPT

Structural and functional sites

Signal peptide: Not determined
NC2 domain: <1–11
Helical domain: 12–468
NC1 domain: 469–635
Imperfections in Gly-X-Y triplets: 36–37, 56–57, 106–107, 144–145, 170–171, 214–215, 249–250, 428–429 [same relative locations as in α1(VIII) and α1(X)]

Gene structure

The α1(VIII) collagen gene has been localized to the long arm of human chromosome 3 at locus q12–13.1. The α2(VIII) collagen gene is located on the short arm of chromosome 1 at locus p32.3–34.3. The genes encoding both α chains have a condensed structure similar to that of type X collagen. The gene for the rabbit α1(VIII) chain comprises four exons, exons 1 and 2 coding for the 5′-untranslated region, exon 3 encoding most of the NC2 domain and exon 4 encoding the remainder of the NC2 domain, the entire triple-helical domain, the NC1 domain and the 3′-untranslated region [4,9].

References

1. Shuttleworth, C.A. (1997) Type VIII collagen. Int. J. Biochem. Cell Biol. 28: 1–4.
2. Mann, K. et al. (1990) The primary structure of a triple-helical domain of collagen type VIII from bovine Descemet's membrane. FEBS Lett. 273: 168–172.
3. Sawada, H. et al. (1990) Characterization of the collagen in the hexagonal lattice of Descemet's membrane: Its relation to type VIII collagen. J. Cell Biol. 110: 219–227.
4. Muragaki, Y. et al. (1991) The complete primary structure of the human α1(VIII) chain and assignment of its gene (COL8A1) to chromosome 3. Eur. J. Biochem. 197: 615–622.
5. Yamaguchi, N. et al. (1991) The α1(VIII) collagen gene is homologous to the α1(X) collagen gene and contains a large exon encoding the entire triple helical and carboxyl-terminal non-triple helical domains of the α1(VIII) polypeptide. J. Biol. Chem. 266: 4508–4513.
6. Muragaki, Y. et al. (1992) α1(VIII) collagen gene transcripts encode a short chain collagen polypeptide and are expressed by various epithelial, endothelial and mesenchymal cells in newborn mouse tissues. Eur. J. Biochem. 207: 895–902.
7. Benya, P.D. and Padilla, S.R. (1986) Isolation and characterization of type VIII collagen synthesised by cultured rabbit corneal endothelial cells. A conventional structure replaces the interrupted-helix model. J. Biol. Chem. 261: 4160–4169.
8. Kapoor, R. et al. (1986) Type VIII collagen from bovine Descemet's membrane: Structural characterization of a triple-helical domain. Biochemistry, 25: 3930–3937.
9. Muragaki, Y. et al. (1991) The α2(VIII) collagen gene. A novel member of the short chain collagen family located on the human chromosome 1. J. Biol. Chem. 266: 7721–7727.

Collagen type IX

Type IX collagen is the prototype of a sub-family of collagens called FACIT collagens that include types XII, XIV, XVI and XIX. Type IX collagen associates specifically with the surface of, and participates in the formation of, type II collagen fibrils. It is found in cartilage, intervertebral disc and vitreous humour. As well as linking type II collagen molecules, it may also serve to bridge collagen fibrils with other matrix macromolecules [1]. Mutations in the genes for the α1(IX) and α2(IX) chains have been linked to certain chondrodysplasias and early onset of osteoarthritis. In particular, mutations in the COL9A2 gene are associated with multiple epiphyseal dysplasia (EDM2) [2–4]. The homozygous knockout mice for COL9A1 develop normally but show signs of osteoarthritis by 4 months and subsequently have flattened articulating surfaces and epiphyses [2].

Molecular structure

Type IX collagen is synthesized as a disulphide-bonded heterotrimer comprising three distinct α1(IX), α2(IX) and α3(IX) chains. The molecule is not processed prior to its deposition in the extracellular matrix. The molecule comprises three collagenous domains (COL1–3) and four non-collagenous domains (NC1–4). The α2(IX) chain can have a chondroitin/dermatan sulphate glycosaminoglycan covalently attached at the NC3 domain; there are therefore both proteoglycan and non-proteoglycan forms of type IX collagen. The length of the glycosaminoglycan is both species- and tissue-dependent. Another form of type IX collagen lacks the large NC4 domain at the NH_2-terminus of the α1(IX) chain owing to the use of an alternative promoter in the α1(IX) gene. The expression of this variant is tissue-specific and developmentally regulated. The COL1 domain of type IX collagen is homologous to the COL1 domains of type XII, XIV and XVI collagens. The NC4 domain contains a PARP repeat and shows homology to the NC3 domain of types XII and XIV collagen. The NH_2-terminal region of the COL2 domain of all three chains is cross-linked to the N-telopeptides of type II collagen; the COL2 domain of α3(IX) is cross-linked to the C-telopeptide of type II collagen. The two cross-link sites on the α3(IX) chain span precisely the gap zone of type II collagen fibrils and, as a consequence, the type IX collagen molecules are anti-parallel to the type II collagen molecules in the

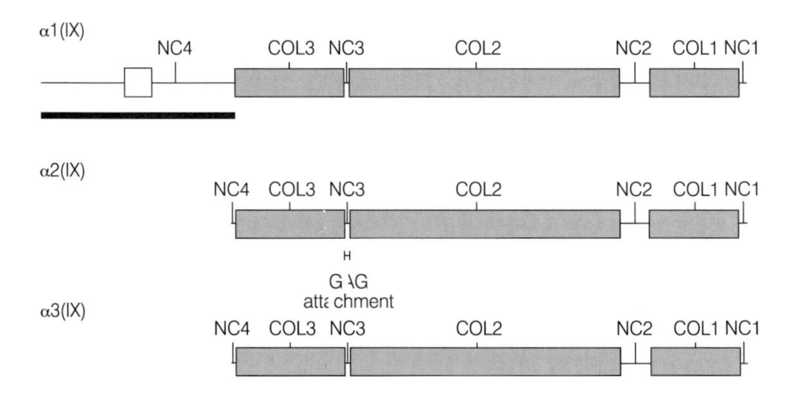

fibrils. The α3(IX) NC1 domain also cross-links with the COL2 domain of α1(IX) and α3(IX) chains. The NC4 domain of the long form of the α1(IX) chain may interact with the glycosaminoglycan chains of proteoglycans[5–13].

Isolation

The intact type IX molecule has been isolated from fetal cartilage by 1 M NaCl or 4 M guanidine–HCl extraction[13,14]. Large quantities of the cleaved COL1–3 domains are prepared by pepsin digestion of the guanidine–HCl-insoluble cartilage residue[15].

Accession number

P20849

Primary structure: α1(IX) chain

Ala	A	49	Cys	C	11	Asp	D	38	Glu	E	49
Phe	F	21	Gly	G	216	His	H	9	Ile	I	35
Lys	K	43	Leu	L	65	Met	M	12	Asn	N	19
Pro	P	149	Gln	Q	38	Arg	R	53	Ser	S	45
Thr	T	29	Val	V	37	Trp	W	7	Tyr	Y	6

Mol. wt (calc.) = 92 867 Residues = 931

```
1    MKTCWKIPVF FFVCSFLEPW ASAAVKRRPR FPVNSNSNGG NELCPKIRIG
51   QDDLPGFDLI SQFQVDKAAS RRAIQRVVGS ATLQVAYKLG NNVDFRIPTR
101  NLYPSGLPEE YSFLTTFRMT GSTLKKNWNI WQIQDSSGKE QVGIKINGQT
151  QSVVFSYKGL DGSLQTAAFS NLSSLFDSQW HKIMIGVERS SATLFVDCNR
201  IESLPIKPRG PIDIDGFAVL GKLADNPQVS VPFELQWMLI HCDPLRPRRE
251  TCHELPARIT PSQTTDERGP PGEQGPPGAS GPPGVPGIDG IDGDRGPKGP
301  PGPPGPAGEP GKPGAPGKPG TPGADGLTGP DGSPGSIGSK GQKGEPGVPG
351  SRGFPGRGIP GPPGPPGTAG LPGELGRVGP VGDPGRRGPP GPPGPPGPRG
401  TIGFHDGDPL CPNACPPGRS GYPGLPGMRG HKGAKGEIGE PGRQGHKGEE
451  GDQGELGEVG AQGPPGAQGL RGITGLVGDK GEKGARGLDG EPGPQGLPGA
501  PGDQGQRGPP GEAGPKGDRG AEGARGIPGL PGPKGDTGLP GVDGRDGIPG
551  MPGTKGEPGK PGPPGDAGLQ GLPGVPGIPG AKGVAGEKGS TGAPGKPGQM
601  GNSGKPGQQG PPGEVGPRGP QGLPGSRGEL GPVGSPGLPG KLGSLGSPGL
651  PGLPGPPGLP GMKGDRGVVG EPGPKGEQGA SGEEGEAGER GELGDIGLPG
701  PKGSAGNPGE PGLRGPEGSR GLPGVEGPRG PPGPRGVQGE QGATGLPGVQ
751  GPPGRAPTDQ HIKQVCMRVI QEHFAEMAAS LKRPDSGATG LPGRPGPPGP
801  PGPPGENGFP GQMGIRGLPG IKGPPGALGL RGPKGDLGEK GERGPPGRGP
851  NGLPGAIGLP GDPGPASYGK NGRDGERGPP GLAGIPGVPG PPGPPGLPGF
901  CEPASCTMQL VSEHLTKGLT LERLTAAWLS A
```

Structural and functional sites

Signal peptide: 1–23
NC4 domain: 24–268
PARP repeat 28–267
COL3 domain: 269–405

NC3 domain: 406–417
COL2 domain: 418–756
Lysine/hydroxylysine cross-linking site: 432
NC2 domain: 757–786
COL1 domain: 787–901
NC1 domain: 902–931
Potential N-linked glycosylation site: 171
Inter-chain disulphide bond residues: 411, 415, 901, 906
Imperfections in Gly-X-Y triplets: 356–360, 847–851, 864–868
Stromelysin cleavage site: 780–781

Primary structure: α1(IX) chain (alternative short form)

```
1     MAWTARDRGA  LGLLLLGLCL  CAAQR
```

Structural and functional sites

Signal peptide: 1–23
NC4 domain: 24–25
COL3–NC1: 26–688 (identical in sequence to the long form)

Accession number

M95610; Z22923

Primary structure: α2(IX) chain

Ala	A	40	Cys	C	4	Asp	D	24	Glu	E	31
Phe	F	5	Gly	G	203	His	H	8	Ile	I	23
Lys	K	38	Leu	L	27	Met	M	11	Asn	N	6
Pro	P	139	Gln	Q	34	Arg	R	27	Ser	S	17
Thr	T	15	Val	V	30	Trp	W	0	Tyr	Y	6

Mol. wt (calc.) = 65 130 Residues = 688

```
1     MTAVPAPRSL  FVLLQVVVLA  LAQIRGPPGE  RGPPGPPGPP  GVPGSDGIDG
51    DKGPPGKAGP  PGPKGEPGKA  GPDGPDGKPG  IDGLTGAKGE  PGPMGIPGVK
101   GQPGLPGPPG  LPGPGFAGPP  GPPGPVGLPG  EIGIRGPKGD  PGPDGPSGPP
151   GPPGKPGRPG  TIQGLEGSAD  FLCPTNCPPG  MKGPPGLQGV  KGHAGKRGIL
201   GDPGHQGKPG  PKGDVGASGE  QGIPGPPGPQ  GIRGYPGMAG  PKGETGPHGY
251   KGMVGAIGAT  GPPGEEGPRG  PPGRAGEKGD  EGSPGIRGPQ  GITGPKGATG
301   PPGINGKDGT  PGTPGMKGSA  GQAGQPGSPG  HQGLAGVPGQ  PGTKGGPGDQ
351   GEPGPQGLPG  FSGPPGKEGE  PGPRGEIGPQ  GIMGQKGDQG  ERGPVGQPGP
401   QGRQGPKGEQ  GPPGIPGPQG  LPGVKGDKGS  PGKTGPRGKV  GDPGVAGLPG
451   EKGEKGESGE  PGPKGQQGVR  GEPGYPGPSG  DAGAPGVQGY  PGPPGPRGLA
501   GNRGVPGQPG  RQGVEGRDAT  DQHIVDVALK  MLQEQLAEVA  VSAKREALGA
551   VGMMGPPGPP  GPPGYPGKQG  PHGHPGPRGV  PGIVGAVGQI  GNTGPKGKRG
601   EKGDPGEVGR  GHPGMPGPPG  IPGLPGRPGQ  AINGKDGDRG  SPGAPGEAGR
651   PGLPGPVGLP  GFCEPAACLG  ASAYASARLT  EPGSIKGP
```

Structural and functional sites

Signal peptide: 1–22
NC4 domain: 23–25
COL3 domain: 26–162
NC3 domain: 163–179
COL2 domain: 180–518
NC2 domain: 519–548
COL1 domain: 549–663
NC1 domain: 664–678
GAG attachment site: 168
Lysine/hydroxylysine cross-linking site: 182
Inter-chain disulphide bond residues: 173, 177, 663, 668
Imperfections in Gly-X-Y triplets: 113–114, 612–613, 632–633
Stromelysin cleavage site: 542–543

Accession number

L41162

Primary structure: α3(IX) chain

Ala	A	44	Cys	C	6	Asp	D	28	Glu	E	32
Phe	F	4	Gly	G	210	His	H	4	Ile	I	14
Lys	K	29	Leu	L	49	Met	M	6	Asn	N	5
Pro	P	137	Gln	Q	26	Arg	R	33	Ser	S	24
Thr	T	11	Val	V	20	Trp	W	0	Tyr	Y	2

Mol. wt (calc.) = 63 742 Residues = 684

```
1    MAGPRACAPL LLLLLLGQLL AAAGAQRVGL PGPPGPPGRP GKPGQDGIDG
51   EAGPPGLPGP PGPKGAPGKP GKPGEAGLPG LPGVDGLTGR DGPPGPKGAP
101  GERGSLGPPG PPGLGGKGLP GPPGEAGVSG PPGGIGLRGP PGPPGLPGLP
151  GPPGPPGPPG HPGVLPEGAT DLQCPSICPP GPPGPPGMPG FKGPTGYKGE
201  QGEVGKDGEK GDPGPPGPAG LPGSVGLQGP RGLRGLPGPL GPPGDRGPIG
251  FRGPPGIPGA PGKAGDRGER GPEGFRGPKG DLGRPGPKGT PGVAGPSGEP
301  GMPGKDGQNG VPGLDGQKGE AGRNGAPGEK GPNGLPGLPG RAGSKGEKGE
351  RGRAGELGEA GPSGEPGVPG DAGMPGERGE AGHRGSAGAL GPQGPPGAPG
401  VRGFQGQKGS MGDPGLPGPQ GLRGDVGDRG PGGAEGPKGD QGIAGSDGLP
451  GDKGELGPSG LVGPKGESGS RGELGPKGTQ GPNGTSGVQG VPGPPGPLGL
501  QGVPGVPGIT GKPGVPGKEA SEQRIRELCG GMISEQIAQL AAHLRKPLAP
551  GSIGRPGPAG PPGPPGPPGS IGHPGARGPP GYRGPTGELG DPGPRGNQGD
601  RGDKGAAGAG LDGPEGDQGP QGPQGVPGTS KDGQDGAPGE PGPPGDPGLP
651  GAIGAQGTPG ICDTSACQGA VLGGVGEKSG SRSS
```

Structural and functional sites

Signal peptide: 1–25
NC4 domain: 26–28
COL3 domain: 29–165
NC3 domain: 166–180

COL2 domain: 181–519
NC2 domain: 520–550
COL1 domain: 551–662
NC1 domain: 663–684
Cross-linking site to COL2 domains of α1(IX) and α3(IX) chains: 678
Lysine/hydroxylysine cross-linking sites: 192, 330
Potential N-linked glycosylation site: 483
Inter-chain disulphide bond residues: 174, 178, 529, 662, 667
Imperfections in Gly-X-Y triplets: 116–117, 608–609, 631–632
Stromelysin cleavage site: 543–544

Gene structure

The use of the alternative promoter in the intron between exons 6 and 7 of the α1(IX) gene causes the loss of polypeptide encoded by exons 1–6. The new exon 1* encodes a different amino acid sequence in the shortened NC4 domain of the α1(IX) chain. The α1(IX) chain is encoded by a single gene found on human chromosome 6 at locus q12–13. The gene contains 19 exons and spans 100 kb [5–7]. The α2(IX) gene has 32 exons spanning approximately 16 kb and maps to human chromosome 1 at locus p32.3–p33 [8–10]. The gene for the α3(IX) chain is located on human chromosome 20q13.3 [12].

References

1 Reichenberger, E. and Olsen, B.R. (1996) Collagens as organizers of the extracellular matrix during morphogenesis. Semin. Cell Develop. Biol. 7: 631–638.

2 Faessler, R. et al. (1994) Mice lacking α1(IX) collagen develop non-inflammatory degenerative joint disease. Proc. Natl Acad. Sci. USA 91: 5070–5074.

3 Muragaki, Y. et al. (1996) A mutation in the gene encoding the α2(IX) chain of the fibril-associated collagen IX, COL9A2, causes multiple epiphyseal dysplasia (EDM2). Nature Genet. 12: 103–105.

4 Briggs, M. et al. (1994) Genetic mapping of a locus for multiple epiphyseal dysplasia (EDM2) to a region of chromosome 1 containing a type IX collagen gene. Am. J. Hum. Genet. 55: 678–684.

5 Kimura, T. et al. (1989) The complete primary structure of two distinct forms of human α1(IX) collagen chains. Eur. J. Biochem. 179: 71–78.

6 Nishimura, I. et al. (1989) Tissue-specific forms of type IX collagen–proteoglycan arise from the use of two widely separated promoters. J. Biol. 264: 20033–20041.

7 Muragaki, Y. et al. (1990) Molecular cloning of rat and human type IX collagen cDNA and localization of the α1(IX) gene on the human chromosome 6. Eur. J. Biochem. 192: 703–708.

8 Perala, M. et al. (1994) The exon structure of the mouse α2(IX) collagen gene shows unexpected divergence from the chick gene. J. Biol. Chem. 269: 5064–5071.

9 Perala, M. et al. (1993) Molecular cloning of human α2(IX) collagen cDNA and assignment of the human COL9A2 gene to chromosome 1. FEBS Lett. 319: 177–180.

[10] Warman, M.L et al. (1994) The genes encoding alpha 2(IX) collagen (COL9A2) map to human chromosome 1p32.3–p33 and mouse chromosome 4. Genomics 23: 158–162.
[11] Ayad, S. et al. (1991) Mammalian cartilage synthesizes both proteoglycan and non-proteoglycan forms of type IX collagen. Biochem. J. 278: 441–445.
[12] Brewton, R.G. et al. (1995) Molecular cloning of the α3 chain of human type IX collagen: Linkage of the gene COL9A3 to chromosome 20q13.3. Genomics 30: 329–336.
[13] Diab, M. et al. (1996) Collagen type IX from human cartilage: a structural profile of intermolecular cross-linking sites. Biochem. J. 314: 327–332.
[14] Ayad, S. et al. (1989) Bovine cartilage types VI and IX collagens. Characterization of their forms *in vivo*. Biochem. J. 262: 753–761.
[15] Grant, M.E. et al. (1988) The structure and synthesis of cartilage collagens. In: The Control of Tissue Damage, ed. Glauert, A.M., Elsevier, Amsterdam, pp. 3–28.

Collagen type X

Type X collagen is a short-chain collagen which is both temporally and spatially regulated during fetal development. It is synthesized predominantly by hypertrophic chondrocytes during endochondral bone formation and therefore has a very restricted tissue distribution in the calcifying cartilage that is eventually replaced by bone. Mutations within the COL10A1 gene encoding the NC1 domain of the α1(X) chain have been linked to metaphyseal chondrodysplasia type Schmid (SMCD)[1,2]. Mouse COL10A1 gene null mutants have a phenotype that partly resembles SMCD and, in particular, develop *coxa vara*. Surprisingly, the depletion of type X collagen protein does not affect the hypertrophic zone of the growth plate but rather other zones, where there is an altered distribution of matrix components. The thickness of both the growth plate resting zone and articular cartilage is reduced and the trabecular structure is abnormal with retention of a cartilaginous matrix[3].

Molecular structure

Type X collagen is a homotrimer comprising three identical α1(X) chains. It has a short triple helix approximately 132 nm in length, a small NH_2-terminal domain and a large COOH-terminal globular domain. It is deposited in the cartilage matrix without apparent processing and, although its macromolecular organization has not been determined *in vivo*, it may form a hexagonal type lattice as in the case of type VIII collagen. The COOH-terminal three-quarters of the NC1 domain is homologous to that of the two type VIII collagen chains. The lengths of the triple-helical and NC1 domains are similar to those of type VIII collagen, but the NC2 domain is shorter[4–6].

Isolation

Type X collagen is isolated in its intact form from the medium of cultured chondrocytes[7] or as its triple-helical domain by pepsinization[8].

Accession number

S18249

Primary structure: α1(X) chain

Ala	A	36	Cys	C	1	Asp	D	12	Glu	E	22
Phe	F	15	Gly	G	175	His	H	10	Ile	I	27
Lys	K	35	Leu	L	35	Met	M	10	Asn	N	14
Pro	P	145	Gln	Q	23	Arg	R	19	Ser	S	27
Thr	T	25	Val	V	26	Trp	W	2	Tyr	Y	21

Mol. wt (calc.) = 66 053 Residues = 680

```
1     MLPQIPFLLL VSLNLVHGVF YAERYQTPTG IKGPLPNTKT QFFIPYTIKS
51    KGIAVRGEQG TPGPPGPAGP RGHPGPSGPP GKPGYGSPGL QGEPGLPGPP
101   GPSAVGKPGV PGLPGKPGER GPYGPKGDVG PAGLPGPRGP PGPPGIPGPA
151   GISVPGKPGQ QGPTGAPGPR GFPGEKGAPG VPGMNGQKGE MGYGAPGRPG
201   ERGLPGPQGP TGPSGPPGVG KRGENGVPGQ PGIKGDRGFP GEMGPIGPPG
251   PQGPPGERGP EGIGKPGAAG APGQPGIPGT KGLPGAPGIA GPPGPPGFGK
301   PGLPGLKGER GPAGLPGGPG AKGEQGPAGL PGKPGLTGPP GNMGPQGPKG
351   IPGSHGLPGP KGETGPAGPA GYPGAKGERG SPGSDGKPGY PGKPGLDGPK
401   GNPGLPGPKG DPGVGGPPGL PGPVGPAGAK GMPGHNGEAG PRGAPGIPGT
451   RGPIGPPGIP GFPGSKGDPG SPGPPGPAGI ATKGLNGPTG PPGPPGPRGH
501   SGEPGLPGPP GPPGPPGQAV MPEGFIKAGQ RPSLSGTPLV SANQGVTGMP
551   VSAFTVILSK AYPAIGTPIP FDKILYNRQQ HYDPRTGIFT CQIPGIYYFS
601   YHVHVKGTHV WVGLYKNGTP VMYTYDEYTK GYLDQASGSA IIDLTENDQV
651   WLQLPNAESN GLYSSEYVHS SFSGFLVAPM
```

Structural and functional sites

Signal peptide: 1–18
NC2 domain: 19–56
Helical domain: 57–519
NC1 domain: 520–680
Lysine/hydroxylysine cross-linking sites: Cross-links present but sites unknown
Potential N-linked glycosylation site: 617
Imperfections in Gly-X-Y triplets: 84–85, 101–105, 151–155, 192–193, 218–219, 222–223, 297–298, 479–480 [same relative locations as imperfections in α1(VIII) and α2(VIII)]
Mammalian collagenase cleavage sites: 151–152, 479–480

Gene structure

The human type X collagen gene spans approximately 7 kb and is located on the long arm of chromosome 6 at locus q21–22.3. Except for the type VIII collagen gene, its structure is unique in comparison with other known vertebrate collagen genes in that it is condensed, comprising only three exons. Exon 1 encodes a large portion of the 5′-untranslated region, exon 2 encodes the remaining 5′-untranslated region and most of the NC2 domain and exon 3 encodes the remainder of the NC2 domain, the entire triple helix, the NC1 domain and part of the 3′-untranslated region [6,9–11].

References

[1] McIntosh, I. et al. (1995) Concentration of mutations causing Schmid metaphyseal chondrodysplasia in the C-terminal noncollagenous domain of type X collagen. Hum. Mutat. 5: 121–125.

[2] Wallis, G.A. et al. (1996) Mutations within the gene encoding the alpha-1(X) chain of type X collagen (COL10A1) cause metaphyseal chondrodysplasia type Schmid but not several other forms of metaphyseal chondrodysplasia. J. Med. Genet. 33: 450–457.

[3] Kwan, K.M. et al. (1997) Abnormal compartmentalization of cartilage matrix components in mice lacking collagen X: implications for function. J. Cell Biol. 136: 459–471.

[4] Schmid, T.M. and Linsenmayer, T.F. (1987) Type X collagen. In: Structure and Function of Collagen Types, ed. Mayne, R. and Burgeson, R.E., Academic Press, London, pp. 223–259.
[5] Kwan, A.P.L. et al. (1991) Macromolecular organization of chicken type X collagen in vitro. J. Cell Biol. 114: 597–604.
[6] Thomas, J.T. et al. (1991) The human collagen X gene. Complete primary translated sequence and chromosomal localization. Biochem. J. 280: 617–623.
[7] Marriott, A. et al. (1992) The synthesis of type X collagen by bovine and human growth-plate chondrocytes. J. Cell Sci. 99: 641–649.
[8] Kielty, C.M. et al. (1984) Embryonic chick cartilage collagens. Differences in the low-Mr species present in sternal cartilage and tibiotarsal articular cartilage. FEBS Lett. 169: 179–184.
[9] Reichenberger, E. et al. (1992) Genomic organization and full-length cDNA sequence of human collagen X. FEBS Lett. 282: 393–396.
[10] Apte, S. et al. (1991) Cloning of human $\alpha 1$(X) collagen DNA and localization of the COL10A1 gene to the q21–q22 region of human chromosome 6. FEBS Lett. 282: 393–396.
[11] Thomas, J.T. et al. (1995) Sequence comparison of three mammalian type-X collagen promoters and preliminary functional analysis of the human promoter. Gene 160: 291–296.

Collagen type XI

Type XI collagen is a quantitatively minor fibrillar collagen and forms heterotypic fibrils with types II and IX collagens mainly in cartilaginous tissues. However, it is also expressed by non-chondrogenic tissues, particularly during embryonic development, and in these tissues cross-type heterotrimers of type V and XI collagens frequently exist. Type V and XI collagens should therefore be classed as one type V/XI collagen family [1]. Mutations in the COL11A1 and COL11A2 genes have been linked to Stickler syndrome characterized by numerous skeletal deformities. However, only mutations in the COL11A1 gene cause additional ocular abnormalities. Mice that are homozygous for the mutation in the COL11A1 gene that causes autosomal recessive chondrodysplasia (*cho*) are essentially functionally equivalent to homozygous knockouts for COL11A1. *Cho/cho* mice die at birth with numerous skeletal defects: shortened snout, cleft palate, protruding tongue, short limbs with widened metaphyses and a shortened spine. The chondrocytes within the growth plates are also highly disorganized [1,2].

Molecular structure

Type XI collagen is synthesized as a heterotrimeric procollagen comprising three distinct proα1(XI), proα2(XI) and proα3(XI) chains, although alternative assemblies may exist. The proα1(V) and proα2(V) chains can substitute for the proα1(XI) and proα2(XI) chains, respectively, forming cross-type heterotrimers. The proα3(XI) chain appears identical to the alternatively spliced form (lacking exon 2) of the proα1(II) chain but the triple-helical domain has more hydroxylysine glycosides than the α1(II) chain. The proα3(XI) chain is also not cleaved by *N*-proteinase or mammalian collagenase, suggesting differences between the two chains either in amino acid sequence or enzyme accessibility. The procollagen forms are processed extracellularly at the COOH-terminus, but only partial processing occurs at the NH_2-terminus for the proα1(XI) and proα2(XI) chains and the proα3(XI) is not processed further. Complex alternative splicing occurs in the primary mRNAs encoding the (variable) region between the PARP and COL2 domains of both the proα1(XI) and proα2(XI) chains. Three exons are involved for both chains and the resulting variants are differentially expressed according to tissue source. In the case of the proα1(XI) chain, two adjacent exons encoding an acidic 39 amino acid and basic 51 amino acid region are mutually exclusive. A third exon encoding an acidic 85 amino acid region can also be present or absent. In the case of the proα2(XI) chain exons 6, 7 and 8 can be alternatively spliced. Two further potential splice variants within the intron between exons 6 and 7 are also possible, but the sequence is not known. The bulk of the type XI collagen triple helix is copolymerized in the interior of the type II collagen containing fibrils. However, the portion of aminopropeptide that remains after processing projects from the fibril surface and is available for interaction with other extracellular matrix components, the interactions varying according to which alternatively spliced forms are present. In cartilage, the type XI collagen molecules are staggered by 4D (D = 67 nm) and are primarily cross-linked to each other by head-to-tail bonds involving the N-telopeptides and COOH-terminus of the helix. The α1(II) C-telopeptide is also linked to the amino-terminus of the α1(XI) helix [1–10].

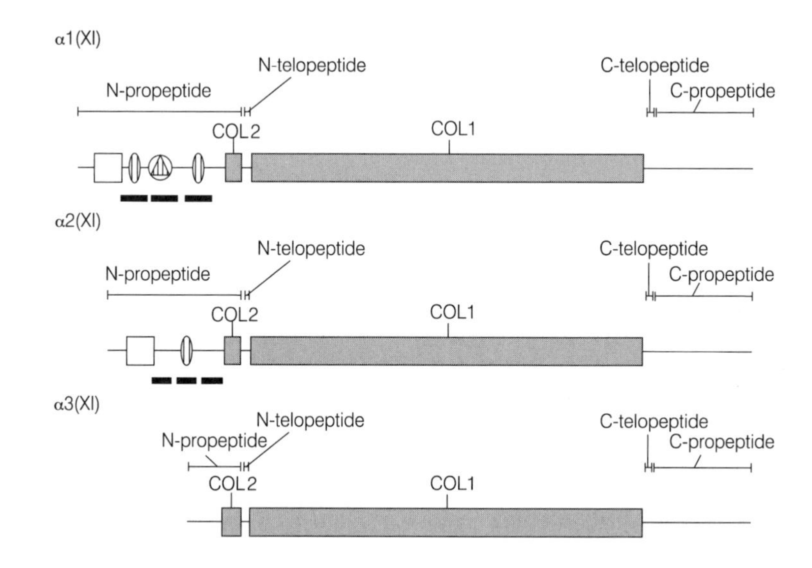

Isolation

Small amounts of the intact form of type XI collagen can be obtained from chondrocyte culture medium or from 1 M NaCl or 4 M-guanidine–HCl extracts of cartilage. Larger quantities of its shorter triple-helical form are prepared from cartilage by pepsin digestion [5,6].

Accession number

P12107; A56371

Primary structure: α1(XI) chain

Ala	A	101	Cys	C	12	Asp	D	97	Glu	E	124
Phe	F	50	Gly	G	424	His	H	16	Ile	I	56
Lys	K	125	Leu	L	81	Met	M	25	Asn	N	37
Pro	P	288	Gln	Q	82	Arg	R	71	Ser	S	81
Thr	T	73	Val	V	65	Trp	W	10	Tyr	Y	37

Mol. wt (calc.) = 186 539 Residues =1857

```
1    MEPWSSRWKT KRWLWDFTVT TLALTFLFQA REVRGAAPVD VLKALDFHNS
51   PEGISKTTGF CTNRKNSKGS DTAYRVSKQA QLSAPTKQLF PGGTFPEDFS
101  ILFTVKPKKG IQSFLLSIYN EHGIQQIGVE VGRSPVFLFE DHTGKPAPED
151  YPLFRTVNIA DGKWHRVAIS VEKKTVTMIV DCKKKTTKPL DRSERAIVDT
201  NGITVFGTRI LDEEVFEGDI QQFLITGDPK AAYDYCEHYS PDCDSSAPKA
251  AQAQEPQIDE YAPEDIIEYD YEYGEAEYKE AESVTEGPTV TEETIAQTEK
301  KKSNFKKKMR TVATKSKEKS KKFTPPKSEK FSSKKKKSYQ ASAKAKLGVK
351  ANIVDDFQEY NYGTMESYQT EAPRHVSGTN EPNPVEEIFT EEYLTGEDYD
401  SQRKNSEDTL YENKEIDGRD SDLLVDGDLG EYDFYEYKEY EDKPTSPPNE
```

```
451   EFGPGVPAET DITETSINGH GAYGEKGQKG EPAVVEPGML VEGPPGPAGP
501   AGIMGPPGLQ GPTGPPGDPG DRGPPGRPGL PGADGLPGPP GTMLMLPFRY
551   GGDGSKGPTI SAQEAQAQAI LQQARIALRG PPGPMGLTGR PGPVGGPGSS
601   GAKGESGDPG PQGPRGVQGP PGPTGKPGKR GRPGADGGRG MPGEPGAKGD
651   RGFDGLPGLP GDKGHRGERG PQGPPGPPGD DGMRGEDGEI GPRGLPGEAG
701   PRGLLGPRGT PGAPGQPGMA GVDGPPGPKG NMGPQGEPGP PGQQGNPGPQ
751   GLPGPQGPIG PPGEKGPQGK PGLAGLPGAD GPPGHPGKEG QSGEKGALGP
801   PGPQGPIGXP GPRGVKGADG VRGLKGSKGE KGEDGFPGFK GDMGLKGDRG
851   EVGQIGPRGX DGPEGPKGRA GPTGDPGPSG QAGEKGKLGV PGLPGYPGRQ
901   GPKGSTGFPG FPGANGEKGA RGVAGKPGPR GQRGPTGPRG SRGARGPTGK
951   PGPKGTSGGD GPPGPPGERG PQGPQGPVGF PGPKGPPGPP GRMGCPGHPG
1001  QRGETGFQGK TGPPGPGGVV GPQGPTGETG PIGERGYPGP PGPPGEQGLP
1051  GAAGKEGAKG DPGPQGISGK DGPAGLRGFP GERGLPGAQG APGLKGGEGP
1101  QGPPGPVGSP GERGSAGTAG PIGLRGRPGP QGPPGPAGEK GAPGEKGPQG
1151  PAGRDGVQGP VGLPGPAGPA GSPGEDGDKG EIGEPGQKGS KGGKGENGPP
1201  GPPGLQGPVG APGIAGGDGE PGPRGQQGMF GQKGDEGARG FPGPPGPIGL
1251  QGLPGPPGEK GENGDVGPWG PPGPPGPRGP QGPNGADGPQ GPPGSVGSVG
1301  GVGEKGEPGE AGNPGPPGEA GVGGPKGERG EKGEAGPPGA AGPPGAKGPP
1351  GDDGPKGNPG PVGFPGDPGP PGELGPAGQD GVGGDKGEDG DPGQPGPPGP
1401  SGEAGPPGPP GKRGPPGAAG AEGRQGEKGA KGEAGAEGPP GKTGPVGPQG
1451  PAGKPGPEGL RGIPGPVGEQ GLPGAAGQDG PPGPMGPPGL PGLKGDPGSK
1501  GEKGHPGLIG LIGPPGEQGE KGDRGLPGTQ GSPGAKGDGG IPGPAGPLGP
1551  PGPPGLPGPQ GPKGNKGSTG PAGQKGDSGL PGPPGPPGPP GEVIQPLPIL
1601  SSKKTRRHTE GMQADADDNI LDYSDGMEEI FGSLNSLKQD IEHMKFPMGT
1651  QTNPARTCKD LQLSHPDFPD GEYWIDPNQG CSGDSFKVYC NFTSGGETCI
1701  YPDKKSEGVR ISSWPKEKPG SWFSEFKRGK LLSYLDVEGN SINMVQMTFL
1751  KLLTASARQN FTYHCHQSAA WYDVSSGSYD KALRFLGSND EEMSYDNNPF
1801  IKTLYDGCTS RKGYEKTVIE INTPKIDQVP IVDVMISDFG DQNQKFGFEV
1851  GPVCFLG
```

Structural and functional sites

Signal peptide: 1–36 (probable)
N-Propeptide: 37–579
PARP repeat 37–259
Alternatively spliced domains: 261–299 (acidic) and 300–350 (basic) are mutually exclusive; 382–466 (acidic)
Constant region: 467–492
COL2 domain: 493–543 [17 triplets similar to the COL2 domain in proα2(XI), proα1(V), and proα2(V)]
Helical domain: 580–1593
C-Telopeptide: 1594–1614
C-Propeptide: 1615–1857
Lysine/hydroxylysine cross-linking sites: 556, 663, 1503
Potential N-linked glycosylation sites: 1691, 1760
C-Proteinase cleavage site: 1614–1615

Accession number

U32169; U41065–41069

Primary structure: α2(XI) chain

Ala	A	99	Cys	C	14	Asp	D	79	Glu	E	110
Phe	F	31	Gly	G	428	His	H	20	Ile	I	30
Lys	K	72	Leu	L	113	Met	M	16	Asn	N	15
Pro	P	293	Gln	Q	89	Arg	R	102	Ser	S	66
Thr	T	59	Val	V	74	Trp	W	3	Tyr	Y	23

Mol. wt (calc.) = 172 031   Residues = 1736

```
1     MERCSRCHRL LLLLPLVLGL SAAPGWAGAP PVDVLRALRF PSLPDGVRRA
51    KGICPADVAY RVARPAQLSA PTRQLFPGGF PKDFSLLTVV RTRPGLQAPL
101   LTLYSAQGVR QLGLELGRPV RFLYEDQTGR PQPPSQPVFR GLSLADGKWH
151   RVAVAVKGQS VTLIVDCKKR VTRPLPRSAR PVLDTHGVII FGARILDEEV
201   FEGDVQELAI VPGVQAAYES CEQKELECEG GQRERPQNQQ PHRAQRSPQQ
251   QPSRLHRPQN QEPQSQPTES LYYDYEPPYY DVMTTGTTPD YQDPTPGEEE
301   EILESSLLPP LEEEQTDLQV PPTADRFQAE EYGEGGTDPP EGPYDYTYGY
351   GDDYREETEL GPALSAETAH SGAAAHGPRG LKGEKGEPAV LEPGMLVEGP
401   PGPEGPAGLI GPPGIQGNPG PVGDPGERGP PGRAGLPGSD GAPGPPGTSL
451   MLPFRFGSGG GDKGPVVAAQ EAQAQAILQQ ARLALRGPPG PMGYTGRPGP
501   LGQPGSPGLK GESGDLGPQG PRGPQGLTGS LGKAGRRGRA GADGARGMPG
551   DPGVKGDRGF DGLPGLPGEK GHRGDTGAQG LPGPPGEDGE RGDDGEIGPR
601   GLPGESGPRG LLGPKGPPGI PGPPGVRGMD GPQGPKGSLG PQGEPGPPGQ
651   QGTPGTQGLP GPQGAIGPHG EKGPQGKPGL PGMPGSDGPP GHPGKEGPPG
701   TKGNQGPSGP QGPLGYPGPR GVKGVDGIRG LKGHKGEKGE DGFPGFKGDI
751   GVKGDRGEVG VPGSRGEDGP EGPKGRTGPT GDPGPPGLMG EKGKLGVPGL
801   PGYPGRQGPK GSLGFPGFPG ASGEKGARGL SGKSGPRGER GPTGPRGQRG
851   PRGATGKSGA KGTSGGDGPH GPPGERGLPG PQGPNGFPGP KGPLGPPGKD
901   GLPGHPGQRG EVGFQGKTGP PGPPGVVGPQ GAAGETGPMG ERGHPGPPGP
951   PGEQGLPGTA GKEGTKGDPG PPGAPGKDGP AGLRGFPGER GLPGTAGGPG
1001  LKGNEGPSGP PGPAGSPGER GAAGSGGPIG RQGRPGPQGP PGAAGEKGVP
1051  GEKGPIGPTG RDGVQGPVGL PGPAGPPGVA GEDGDKGEVG DPGQKGTKGN
1101  KGEHGPPGPP GPIGPVGQPG AAGADGEPGA RGPQGHFGAK GDEGTRGFNG
1151  PPGPIGLQGL PGPSGEKGET GDVGPMGPPG PPGPRGPAGP NGADGPQGPP
1201  GGVGNLGPPG EKGEPGESGS PGIQGEPGVK GPRGERGEKG ESGQPGEPGP
1251  PGAKGPQGDD GPKGNPGPVG FPGDPGPPGE GGPRGQDGAK GDRGEDGEPG
1301  QPGSPGPTGE NGPPGPLGKR GPAGSPGSEG RQGGKGAKGD PGAIGAPGKT
1351  GPVGPAGPAG KPGPDGLRGL PGSVGQQGRP GATGQAGPPG PVGPPGLPGL
1401  RGDAGAKGEK GHPGLIGLIG PPGEQGEKGD RGLPGPQGSP GQKGEMGIPG
1451  ASGPIGPGGP PGLPGPAGPK GAKGATGPGG PKGEKGVQGP PGHPGPPGEV
1501  IQPLPIQMPK KTRRSVDGSR LMQEDEAIPT GGAPGSPGGL EEIFGSLDSL
1551  REEIEQMRRP TGTQDSPART CQDLKLCHPE LPDGEYWVDP NQGCARDAFR
1601  VFCNFTAGGE TCVTPRDDVT QFSYVDSEGS PVGVVQLTFL RLLSVSAHQD
1651  VSYPCSGAAR DGPLRLRGAN EDELSPETSP YVKEFRDGCQ TQQGRTVLEV
1701  RTPVLEQLPV LDASFSDLGA PPRRGGVLLG PVCFMG
```

Structural and functional sites

Signal peptide: 1–22 (probable)
N-Propeptide: 23–486
PARP repeat: 23–245
Alternatively spliced domains: 267–292 (exon 6); 293–313 (acidic amino acids) (exon 7); 314–373 (exon 8)

Constant region (type V/XI family): 374–398
COL2 domain: 399–449 [17 triplets similar to the COL2 domain in proα1(XI), proα1(V), and proα2(V)]
Helical domain: 487–1500
C-Telopeptide: 1501–1524
C-Propeptide: 1525–1736
Lysine/hydroxylysine cross-linking sites: N- and helical-sites: 463, 1410
Potential N-linked glycosylation sites: 1605
C-Proteinase cleavage site: 1524–1525

Primary structure: α3(XI) chain

The proα3(XI) chain is the alternatively spliced form of the proα1(II) chain but the main triple-helical domain is more glycosylated on hydroxylysine residues.

Structural and functional sites

Lysine/hydroxylysine cross-linking sites: 190 and 1130 [compare with α1(II) chain].

Gene structure

The proα1(XI) and proα2(XI) collagen chains are encoded by single genes found on human chromosomes 1 (locus p21) and 6 (locus p21.3), respectively. The COL11A2 gene is 30.5 kb with a minimum of 62 exons and is located within the major histocompatibility complex (MHC) locus, approximately 45 kb centromeric to the human leukocyte antigen DPB2 class II gene. Two non-class II genes, the retinoid X receptor β (RXRβ) gene and KE5, are located 1.1 kb upstream of, and within, the COL11A2 gene (between exons 6 and 7), respectively. The amino propeptide is encoded by 14 exons. The proα3(XI) chain is probably encoded by the same gene as the proα1(II) chain (chromosome 12, locus q13.11–12) but is modified post-translationally[3,4,9].

References

1 Fichard, A. et al. (1995) Another look at collagen V and XI molecules. Matrix Biol. 14: 515–531.
2 Li, Y. et al. (1995) A fibrillar collagen gene, Col11a1, is essential for skeletal morphogenesis. Cell 80: 425–430.
3 Bernard, M. et al. (1988) Cloning and sequencing of pro-α1(XI) collagen cDNA demonstrates that type XI belongs to the fibrillar class of collagens and reveals that the expression of the gene is not restricted to cartilaginous tissue. J. Biol. Chem. 263: 17159–17166.
4 Kimura, T. et al. (1989) The human α2(XI) collagen (COL11A2) chain. Molecular cloning of cDNA and genomic DNA reveals characteristics of a fibrillar collagen with differences in genomic organisation. J. Biol. Chem. 264: 13910–13916.
5 Rousseau, J.-C. et al. (1996) Processing of type XI collagen. Determination of the matrix forms of the α1(XI) chain. J. Biol. Chem. 271: 23743–23748.

[6] Grant, M.E. et al. (1988) The structure and synthesis of cartilage collagens. In: The Control of Tissue Damage. ed. Glauert, A.M., Elsevier, Amsterdam, pp. 3–28.
[7] Thom, J. et al. (1995) Alternative exon splicing within the amino-terminal non-triple helical domain of the rat pro-α1(XI) collagen chain generates multiple forms of the mRNA transcript which exhibit tissue-dependent variation. J. Biol. Chem. 270: 9478–9485.
[8] Zhidkova, N.I. et al. (1995) Alternative mRNA processing occurs in the variable region of the pro-α1(XI) and pro-α2(XI) collagen chains. J. Biol. Chem. 270: 9486–9493.
[9] Lui, V.C.H. et al. (1996) The human α2(XI) collagen gene (COL11A2): Completion of coding information, identification of the promoter sequence, and precise localization within the major histocompatibility complex reveal overlap with the KE5 gene. Genomics 32: 401–412.
[10] Wu, J.-J. and Eyre, D.R. (1995) Structural analysis of cross-linking domains in cartilage type XI collagen. Insights on polymeric assembly. J. Biol. Chem. 270: 18865–18870.

Collagen type XII

Type XII collagen is a member of the FACIT collagens that include collagen types IX, XIV, XVI and XIX. Type XII collagen is found mainly in tissues rich in type I collagen but is also present in cartilaginous tissues that contain type II collagen. It is thought to modulate the biomechanical properties of tissues.

Molecular structure

Type XII collagen is synthesized as a disulphide-bonded homotrimeric molecule comprising three α1(XII) chains. The molecule is not processed prior to deposition in the extracellular matrix. Each chain consists of two collagenous triple-helical domains (COL1–2) and three non-collagenous domains (NC1–3). The COL1 domain is homologous to the COL1 domains of type IX, XIV and XVI collagens. Chain selection, registration and assembly of FACIT collagens are directed by the COL1 domain and the adjacent five residues of the NC1 domain. Type XII collagen does not possess a domain homologous to COL2(IX) and therefore is not cross-linked either to itself or to other collagens by lysine/hydroxylysine-derived bonds. Two type XII collagen variants that share a common signal peptide but differ in the size of their NC3 domains arise by alternative splicing. The NC3 domain of the long form is composed of 18 fibronectin type III repeats, four von Willebrand factor A repeats and a PARP repeat. The short form lacks the first two von Willebrand factor A and the first eight fibronectin type III repeats. The amino acid sequence of the human α1(XII) chain shows 92% and 78% identity to the mouse and chicken chains, respectively. Three potential sites for glycanation are present in the fibronectin 4, 5 and 6 repeats, and are also conserved in both mouse and chicken. In some tissues, e.g. fetal bovine cartilage, these sites are utilized and the larger type XII collagen variant is a proteoglycan. Heterotrimers comprising both long and short form chains are also possible which, together with the variable glycantion of the long form, allows a large number of possible isoforms. Type XII collagen interacts *in vitro* with the glycosaminoglycan chain of decorin and the protein core of fibromodulin[1–9].

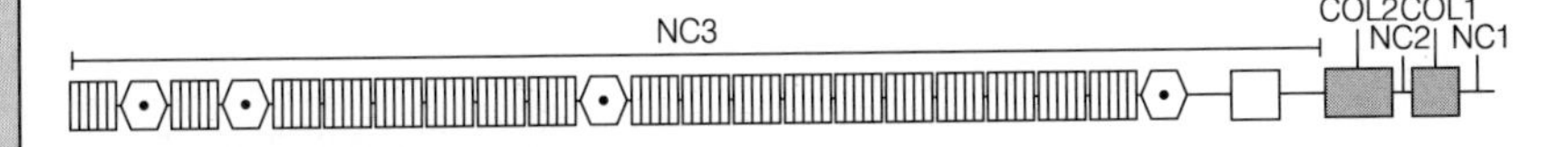

Isolation

Type XII collagen variants are readily extracted from tissues by low-molarity NaCl solutions since they are not covalently cross-linked within the extracellular matrix[3,5].

Accession number

U73778–73779

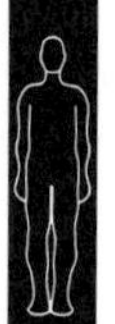

Primary structure: α1(XII) chain

Ala	A	158	Cys	C	22	Asp	D	172	Glu	E	195
Phe	F	91	Gly	G	281	His	H	31	Ile	I	148
Lys	K	155	Leu	L	203	Met	M	45	Asn	N	118
Pro	P	261	Gln	Q	105	Arg	R	158	Ser	S	242
Thr	T	266	Val	V	270	Trp	W	30	Tyr	Y	112

Mol. wt (calc.) = 333 193   Residues = 3063

```
1     MRSRLPPALA ALGAALLLSS IEAEVDPPSD LNFKIIDENT VHMSWAEPVD
51    PIVGYRITVD PTTDGPTKEF TLSASTTETL LSELVPETEY VVTITSYDEV
101   EESVPVIGQL TIQTGSSTKP VEKKPGKTEI QKCSVSAWTD LVFLVDGSWS
151   VGRNNFKYIL DFIAALVSAF DIGEEKTRVG VVQYSSDTRT EFNLNQYYQR
201   DELLAAIKKI PYKGGNTMTG DAIDYLVKNT FTESAGARVG FPKVAIIITD
251   GKSQDEVEIP ARELRNVGVE VFSLGIKAAD AKELKQIAST PSLNHVFNVA
301   NFDAIVDIQN EIISQVCSGV DEQLGELVSG EEVVEPPSNL IAMEVSSKYV
351   KLNWNPSPSP VTGYKVILTP MTAGSRQHAL SVGPQTTTLS VRDLSADTEY
401   QISVSAMKGM TSSEPISIME KTQPMKVQVE CSRGVDIKAD IVFLVDGSYS
451   IGIANFVKVR AFLEVLVKSF EISPNRVQIS LVQYSRDPHT EFTLKKFTKV
501   EDIIEAINTF PYRGGSTNTG KAMTYVREKI FVPSKGSRSN VPKVMILITD
551   GKSSDAFRDP AIKLRNSDVE IFAVGVKDAV RSELEAIASP PAETHVFTVE
601   DFDAFQRISF ELTQSICLRI EQELAAIKKK AYVPPKDLSF SEVTSYGFKT
651   NWSPAGENVF SYHITYKEAA GDDEVTVVEP ASSTSVVLSS LKPETLYLVN
701   VTAEYEDGFS IPLAGEETTE EVKGAPRNLK VTDETTDSFK ITWTQAPGRV
751   LRCRIIYRPV AGGESREVTT PPNQRRRTLE NLIPDTKYEV SVIPEYFSGP
801   GTPLTGNAAT EEVRGNPRDL RVSDPTTSTM KLSWSGAPGK VKQYLVTYTP
851   VAGGETQEVT VRGDTTNTVL QGLKEGTQYA LSVTALYASG AGDALFGEGT
901   TLEERGSPQD LVTKDITDTS IGAYWTSAPG MVRGYRVSWK SLYDDVDTGE
951   KNLPEDAIHT MIENLQPETK YRISVFATYS SGEGEPLTGD ATTELSQDSK
1001  TLKVDEETEN TMRVTWKPAP GKVVNYRVVY RPHGRGKQMV AKVPPTVTST
1051  VLKRLQPQTT YDITVLPIYK MGEGKLRQGS GTTASRFKSP RNLKTSDPTM
1101  SSFRVTWEPA PGEVKGYKVT FHPTGDDRRL GELVVGPYDN TVVLEELRAG
1151  TTYKVNVFGM FDGGESSPLV GQEMTTLSDT TVMPILSSGM ECLTRAEADI
1201  VLLVDGSWSI GRANFRTVRS FISRIVEVFD IGPKRVQIAL AQYSGDPRTE
1251  WQLNAHRDKK SLLQAVANLP YKGGNTLTGM ALNFIRQQNF RTQAGMRPRA
1301  RKIGVLITDG KSQDDVEAPS KKLKDEGVEL FAIGIKNADE VELKMIATDP
1351  DTHDYNVAD  FESLSRIVDD LTINLCNSVK GPGDLEAPSN LVISERTHRS
1401  FRVSWTPPSD SVDRYKVEYY PVSGGKRQEF YVSRMETSTV LKDLKPETEY
1451  VVNVYSVVED EYSEPLKGTE KTLPVPVVSL NIYDVGPTTM HVQWQPVGGA
1501  TGYILSYKPV KDTEPTRPKE VRLGPTVNDM QLTDLVPNTE YAVTVQAVLH
1551  DLTSEPVTVR EVTLPLPRPQ DLKLRDVTHS TMNVFWEPVP GKVRKYIVRY
1601  KTPEEDVKEV EVDRSETSTS LKDLFSQTLY TVSVSAVHDE GESPPVTAQE
1651  TTRPVPAPTN LKITEVTSEG FRGTWDHGAS DVSLYRITWG PFGSSDKMET
1701  ILNGDENTLV FENLNPNTIY EVSITAIYAD ESESDDLIGS ERTLPILTTQ
1751  APKSGPRNLQ VYNATSNSLT VKWDPASGRV QKYRITYQPS TGEGNEQTTT
1801  IGGRQNSVVL QKLKPDTPYT ITVSSLYPDG EGGRMTGRGK TKPLNTVRNL
1851  RVYDPSTSTL NVRWDHAEGN PRQYKLFYAP AAGGPEELVP IPGNTNYAIL
1901  RNLQPDTSYT VTVVPVYTEG DGGRTSDTGR TLMRGLARNV QVYNPTPNRL
1951  GVRWDPAPGP VLQYRVVYSP VDGTRPSESI VVPGNTRMVH LERLIPDTLY
2001  SVNLVALYSD GEGNPSPAQG RTLPRSGPRN LRVFGETTNS LSVAWDHADG
2051  PVQQYRIIYS PTVGDPIDEY TTVPGRRNNV ILQPLQPDTP YKITVIAVYE
```

```
2101 DGDGGHLTGN GRTVGLLPPQ NIHISDEWYT RFRVSWDPSP SPVLGYKIVY
2151 KPVGSNEPME AFVGEMTSYT LHNLNPSTTY DVNVYAQYDS GLSVPLTDQG
2201 TTLYLNVTDL KTYQIGWDTF CVKWSPHRAA TSYRLKLSPA DGTRGQEITV
2251 RGSETSHCFT GLSPDTDYGV TVFVQTPNLE GPGVSVKEHT TVKPTEAPTE
2301 PPTPPPPPTI PPARDVCKGA KADIVFLTDA SWSIGDDNFN KVVKFIFNTV
2351 GGFDEISPAG IQVSFVQYSD EVKSEFKLNT YNDKALALGA LQNIRYRGGN
2401 TRTGKALTFI KEKVLTWESG MRKNVPKVLV VVTDGRSQDE VKKAALVIQQ
2451 SGFSVFVVGV ADVDYNELAN IASKPSERHV FIVDDFESFE KIEDNLITFV
2501 CETATSSCPL IYLDGYTSPG FKMLEAYNLT EKNFASVQGV SLESGSFPSY
2551 SAYRIQKNAF VNQPTADLHP NGLPPSYTII LLFRLLPETP SDPFAIWQIT
2601 DRDYKPQVGV IADPSSKTLS FFNKDTRGEV QTVTFDTEEV KTLFYGSFHK
2651 VHIVVTSKSV KIYIDCYEII EKDIKEAGNI TTDGYEILGK LLKGERKSAA
2701 FQIQSFDIVC SPVWTSRDRC CDIPSRRDEG KCPAFPNSCT CTQDSVGPPG
2751 PPGPAGGPGA KGPRGERGIS GAIGPPGPRG DIGPPGPQGP PGPQGPNGLS
2801 IPGEQGRQGM KGDAGEPGLP GRTGTPGLPG PPGPMGPPGD RGFTGKDSAM
2851 GPRGPPGRPG SPGSPGVTGP SGKPGKPGDH GRPGPSGLKG EKGDRGDIAS
2901 QNMMRAVARQ VCEQLISGQM NRFNQMLNQI PNDYQSSRNQ PGPPGPPGPP
2951 GSAGARGEPG PGGRPGFPGT PGMQGPPGER GLPGEKGERG TGSSGPRGLP
3001 GPPGPQGESR TGPPGSTGSR GPPGPPGRPG NSGIQGPPGP PGYCDSSQCA
3051 SIPYNGQSYP GSG
```

Structural and functional sites

Signal peptide: 1–24
NC3 domain: 25–2746
Alternatively spliced domain: 25–1188
COL2 domain: 2747–2898
NC2 domain: 2899–2941
COL1 domain: 2942–3044
NC1 domain: 3045–3063
Fibronectin type III repeats: 25–115, 332–420, 630–721, 722–812, 813–903, 904–996, 999–1084, 1087–1178, 1384–1470, 1476–1564, 1565–1648, 1655–1751, 1752–1842, 1843–1932, 1933–2023, 2024–2124, 2115–2202, 2203–2290
von Willebrand factor A repeats: 130–320, 427–620, 1189–1410, 2293–2504
PARP repeat: 2505–2720
Potential N-linked glycosylation sites: 651, 700, 1763, 2206, 2528, 2679
Potential glycanation sites: 798–802, 889–892, 981–984
Imperfections in Gly-X-Y triplets: 2990–2991, 3010–3011 (conserved in the COL1 domains of type IX, XIV and XVI collagens)
Cysteines 3044, 3049 conserved in type IX, XIV, XVI and XIX collagens

Gene structure

The gene for the human α1(XII) chain has been assigned to chromosome 6 at locus q12–q13, close to the locus for the genes encoding the α1(IX) and α1(XIX) chains[6,10].

References

[1] Yamagata, M. et al. (1991) The complete primary structure of type XII collagen shows a chimeric molecule with reiterated fibronectin type III motifs, von Willebrand factor A motifs, a domain homologous to a noncollagenous region of type IX collagen, and short collagenous domains with an arg-gly-asp site. J. Cell Biol. 115: 209–221.

[2] Koch, M. et al. (1992) A major oligomeric fibroblast proteoglycan identified as a novel large form of type XII collagen. Eur. J. Biochem. 207: 847–856.

[3] Lunstrum, G.P. et al. (1992) Identification and partial characterization of a large variant form of type XII collagen. J. Biol. Chem. 267: 20087–20092.

[4] Trueb, J. and Trueb, B. (1992) The splice variants of collagen XII share a common 5′ end. Biochim. Biophys. Acta. 1171: 97–98.

[5] Watt, S.L. et al. (1992) Characterization of collagen types XII and XIV from fetal bovine cartilage. J. Biol. Chem. 267: 20093–20099.

[6] Gerecke, D.R. et al. (1997) Complete primary structure of two splice variants of collagen XII, and assignment of α1(XII) collagen (COL12A1), α1(IX) collagen (COL9A1), and α1(XIX) collagen (COL19A1) to human chromosome 6q12–q13. Genomics 41: 236–242.

[7] Bohme, K. et al. (1995) Primary structure of the long and short splice variants of mouse collagen XII and their tissue-specific expression during embryonic development. Dev. Dynamics 204: 432–445.

[8] Koch, M. et al. (1995) Large and small splice variants of collagen XII: differential expression and ligand binding. J. Cell Biol. 130: 1005–1014.

[9] van der Rest, M. and Dublet, B. (1996) Type XII and type XIV collagens: interfibrillar constituents of dense connective tissues. Semin. Cell Develop. Biol. 7: 639–648.

[10] Oh, S.P. et al. (1992) The mouse alpha 1(XII) and human alpha 1(XII)-like collagen genes are localized on mouse chromosome 9 and human chromosome 6. Genomics 14: 225–231.

Collagen type XIII

Type XIII is a short-chain, non-fibrillar collagen that is expressed in skin, intestine, placenta, bone, cartilage and striated muscle. It has, however, only been characterized at the genomic and cDNA levels and the protein has yet to be isolated. Recent studies in the mouse indicate that it is a transmembrane component of focal adhesion sites.

Molecular structure

Type XIII is probably synthesized as a homotrimer of three α1(XIII) chains, but whether the three chains exhibit identical splicing is not known. Each chain consists of three collagenous domains (COL1–3) and four non-collagenous domains (NC1–4). Complex alternative splicing of the primary transcript occurs in both collagenous and non-collagenous domains. Alternatively spliced forms may vary between 516 and 623 amino acids and are generated by exon skipping (exons 3B, 4A, 4B and 5 in COL3; exons 12 and 13 in NC3; exons 29 and 33 in COL1, and exon 37 at the COL1/NC1 junction site). At least 12 mRNA species exist through the alternations of exons 3B-5, 12 and 13, and distinct differences in the proportions of the variant mRNAs were observed in four cultured cell lines and seven tissues. Furthermore, four combinations of alternatively spliced exons 29–37 have been found among cDNA clones isolated from human endothelial cells and HT1080 cells, but additional combinations are predicted to occur[1–4]. Mouse cDNA clones indicate a potential transmembrane region of 23 amino acids in the amino-terminal NC4 domain, which is predicted to be in a type II orientation similar to that in type XVII collagen[5]. (N.B. While these COL and NC domain designations follow the usual convention and are numbered from C to N, it should be noted that in references 1 to 5, the domains are numbered from N to C.)

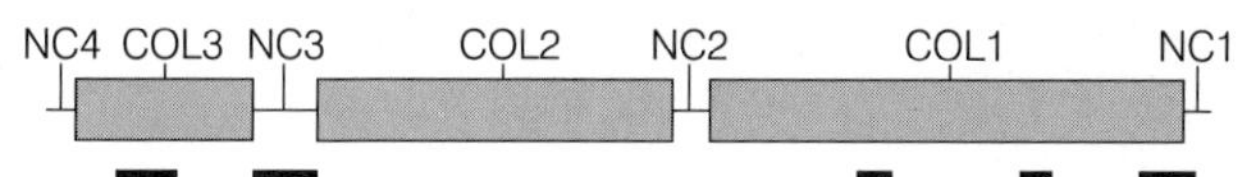

Isolation

Type XIII collagen protein has not been isolated but peptide antibodies have identified polypeptides ranging from 50 to 180 kDa on Western blots of human skin fibroblasts.

Accession number

A38298; B38298; C38298

Primary structure: α1(XIII) chain

Ala	A	29	Cys	C	5	Asp	D	23	Glu	E	36
Phe	F	5	Gly	G	178	His	H	9	Ile	I	17
Lys	K	45	Leu	L	38	Met	M	11	Asn	N	11
Pro	P	119	Gln	Q	25	Arg	R	24	Ser	S	16
Thr	T	13	Val	V	14	Trp	W	2	Tyr	Y	3

Mol. wt (calc.) = 60 104 Residues = 623

```
1     METAILGRVN  QLLDEKWKLH  SRRRREAPKT  SPGCNCPPGP  PGPTGRPGLP
51    GDKGAIGMPG  RVGVKGQPGE  KGSPGDAGLS  IIGPRGPPGQ  PGTRGFPGFP
101   GPIGLDGFPG  HPGPKGDMGL  TGPPGQPGPQ  GQKGEKGQCG  EYPHRLLPLL
151   NSVRLAPPPV  IKRRTFQGEQ  SQASIQGPPG  PPGPPGPSGP  LGHPGLPGPM
201   GPPGLPGPPG  PKGDPGIQGY  HGRKGERGMP  GMPGKHGAKG  APGIAVAGMK
251   GEPGIPGTKG  EKGAEGSPGL  PGLLGQKGEK  GDAGNSIGGG  RGEPGPPGLP
301   GPPGPKGEAG  VDGQVGPPGQ  PGDKGERGAA  GEQGPDGPKG  SKGEPGKGEM
351   VDYNGNINEA  LQEIRTLALM  GPPGLPGQIG  PPGAPGIPGQ  KGEIGLLGPL
401   GHDGKGPRGK  PGDMGPPGPQ  GPPGKDGPPG  VKGENGHPGS  PGEKGEKGET
451   GQAGSPGEKG  EAGEKGNPGA  EVPGLPGPEG  PPGPPGLQGV  PGPKGEAGLD
501   GAKGEKGFQG  EKGDRGPLGL  PGASGLDGRP  GPPGTPGPIG  VPGPAGPKGE
551   RGSKGDPGMT  GPTGAAGLPG  LHGPPGDKGN  RGERGKKGST  GPKGDKGDQG
601   APGLDAPCPL  GEDGLPVQGC  WNK
```

Structural and functional sites

Signal peptide: 1–21 (no signal peptide if transmembrane collagen)
NC4 domain: 22–38
COL3 domain: 39–142
NC3 domain: 143–176
COL2 domain: 177–348
NC2 domain: 349–370
COL1 domain: 371–605
NC1 domain: 606–623
Alternatively spliced domains: 42–88, 146–167 (replaced by ECLSSMPAALRSSQIIALK; this gives a sequence that is three residues shorter), 457–471, 522–533, 582–610

Gene structure

The gene encoding the α1(XIII) chain spans at least 140 kb and comprises 40/41 exons. Twenty-one exons are multiples of 9 bp, the remaining exons vary between 24 and 153 bp. The gene has been localized to chromosome 10 at locus q22 [2,6].

References

1. Pihlajaniemi, T. and Tamminen, M. (1990) The α1 chain of type XIII collagen consists of three collagenous and four noncollagenous domains, and its primary transcript undergoes complex alternative splicing. J. Biol. Chem. 265: 16922–16928.
2. Tikka, L. et al. (1991) Human α1(XIII) collagen gene. Multiple forms of the gene transcripts are generated through complex alternative splicing of several short exons. J. Biol. Chem. 266: 17713–17719.
3. Juvonen, M. and Pihlajaniemi, T. (1992) Characterization of the spectrum of alternative splicing of α1(XIII) collagen transcripts in HT1080 cells and calvarial tissue resulted in identification of two previously unidentified alternatively spliced sequences, one previously unidentified exon, and nine new mRNA variants. J. Biol. Chem. 267: 24693–24699.

[4] Juvonen, M. et al. (1992) Patterns of expression of the six alternatively spliced exons affecting the structures of the COL1 and NC2 domains of the α1(XIII) collagen chain in human tissues and cell lines. J. Biol. Chem. 267: 24700–24707.
[5] Peltonen, S. et al. (1997) Alternative splicing of mouse α1(XIII) collagen RNAs results in at least 17 different transcripts, predicting α1(XIII) collagen chains with lengths varying between 651 and 710 amino acid residues. DNA Cell Biol. 16: 227–234.
[6] Shows, T.B. et al. (1989) Assignment of the human collagen α1(XIII) chain gene (COL13A1) to the q22 region of chromosome 10. Genomics 5: 128–133.

Collagen type XIV

Type XIV collagen is a member of the FACIT collagens that include type IX, XII, XVI and XIX collagens. Type XIV collagen is found mainly in tissues rich in collagen type I, particularly in regions where high mechanical stresses are involved, such as bone-ligament junctions. It is also present in cartilaginous tissues that contain type II collagen.

Molecular structure

Type XIV collagen is synthesized as a disulphide-bonded homotrimeric molecule comprising three α1(XIV) chains and is not processed prior to deposition in the extracellular matrix. Each chain consists of two collagenous triple-helical domains (COL1–2) and three non-collagenous domains (NC1–3). The COL1 domain of α1(XIV) is homologous to the COL1 domain of type IX, XII and XVI collagens. The NC3 domain comprises eight fibronectin type III repeats, two von Willebrand factor A repeats and a PARP repeat. Type XIV collagen contains two cysteine residues that participate in inter-chain disulphide bonding (residues 1769 and 1774) and that are conserved in type IX, XII, XVI and XIX collagens. The triple helical COL1 domain also contains two imperfections in Gly-X-Y triplets (residues 1715–1716 and 1735–1736) that are conserved in the COL1 domains of type IX, XII and XVI collagens. The seven fibronectin type III repeats (352–712, 741–1010) and the connecting region show 71% sequence identity with a non-collagenous protein, undulin. Undulin may arise either by alternative splicing of a single pre-mRNA encoding the α1(XIV) chain or by proteolysis of the type XIV collagen protein. In addition, two variants of type XIV collagen exist which differ in their NC1 domains and which may arise by alternative splicing. As for type IX and XII collagens, type XIV collagen can also exist in a proteoglycan form in some tissues (e.g. fetal bovine cartilage)[1–5]. It interacts *in vitro* with the glycosaminoglycan chains of decorin and a variant of the CD44 cell surface receptor as well as with type VI collagen. It also binds to procollagen I N-proteinase which may be important during fibrillogenesis[6,7].

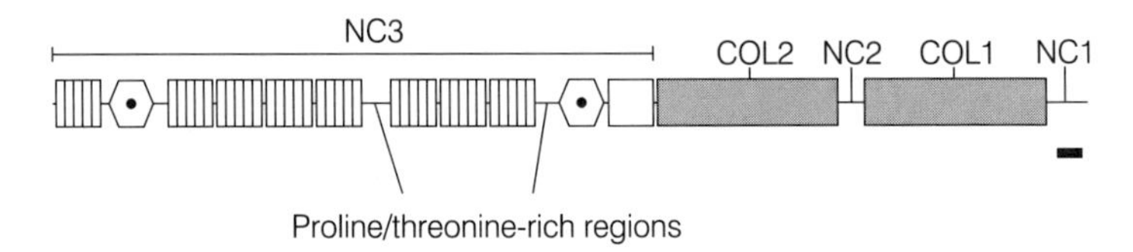

Isolation

The intact, native form of type XIV collagen can be isolated by extraction of tissues with low-molarity NaCl solutions since it is not covalently cross-linked within the extracellular matrix[1,3].

Accession number

X70792–X70793

Primary structure: α1(XIV) chain (chicken)

Ala	A	130	Cys	C	20	Asp	D	100	Glu	E	120
Phe	F	65	Gly	G	201	His	H	26	Ile	I	99
Lys	K	96	Leu	L	143	Met	M	27	Asn	N	61
Pro	P	166	Gln	Q	73	Arg	R	80	Ser	S	137
Thr	T	136	Val	V	131	Trp	W	16	Tyr	Y	61

Mol. wt (calc.) = 202 456 Residues = 1888

```
1     MLCWNEVQSC FLLAFLAVAA YSVSDAQGQV SPPTRLRYNV VNPDSVQISW
51    KAPKGQFSGY KLLVTPSSGG KTNQLILQNT ATKAIIQGLI PDQNYALQII
101   AFSDDKESKP AQGQFRIKDI ERRKETSKSK VKDPEKTNAS KPTPEGNLFT
151   CKTPAIADIV ILVDGSWSIG RFNFRLVRLF LENLVSAFNV GSEKTRVGLA
201   QYSGDPRIEW HLNAYGTKDA VLDAVRNLPY KGGNTLTGLA LTYILENSFK
251   PEAGARPGVS KIGILITDGK SQDDVIPPAK NLRDAGIELF AIGVKNADIN
301   ELKEIASEPD STHVYNVADF NFMNSIVEGL TRTVCSRVEE QEKEIKGTIA
351   ASLGAPTDLV TSDITARGFR VSWTHSPGKV EKYRVVYYPT RGGQPEEVVV
401   DGSSSTAVLK NLMSLTEYQI AVFAIYSNAA SEGLRGTETT LALPMASDLK
451   LYDVSHSSMR AKWNGVAGAT GYMILYAPLT EGLAADEKEI KIGEASTELE
501   LDGLLPNTEY TVTVYAMFGE EASDPLTGQE TTLPLSPPSN LKFSDVGHNS
551   AKLTWDPASK NVKGYRIMYV KTDGTETNEV EVGPVSTHTL KSLTALTEYT
601   VAIFSLYDEG QSEPLTGSFT TRKVPPPQHL EVDEASTDSF RVSWKPTSSD
651   IAFYRLAWIP LDGGESEEVV LSGDADSYVI EGLLPNTEYE VSLLAVFDDE
701   TESEVVAVLG ATIVGTTAIP TTVTTTTTTT ATTPKPTIAV FRTGVRNLVI
751   DDETTSSLRV VWDISDHNAQ QFRVTYLTAK GDRAEEAIMV PGRQNTLLLQ
801   PLLPDTEYKV TITPIYADGE GVSVSAPGKT LPLSAPRNLR VSDEWYNRLR
851   ISWDAPPSPT MGYRIVYKSI NVPGPALETF VGDDINTILI LNLFSGTEYS
901   VKVFASYSTG FSDALTGVAK TLYLGVTNLD TYQVRMTSLC AQWQLHRHAT
951   AYRVVIESLV DGKKQEVNLG GGVPRHCFFE LMPGTEYKIS VHAQLQEIEG
1001  PAVSIMETTL PFPTQPPTSP STTLPPPTIP PAKEVCKAAK ADLVFLVDGS
1051  WSIGDDNFNK IISFLYSTVG ALDKIGPDGT QVAIIQFSDD PRTEFKLNAY
1101  KTKETLLEAI QQIAYKGGNT KTGKAIKHAR EVLFTGEAGM RKGIPKVLVV
1151  ITDGRSQDDV NKVSREMQLD GFSFFAIGVA DADYSELVNI GSKPSERHVF
1201  FVDDFDAFTK IEDELITFVC ETASATCPLV FKDGDKLAGF KMMEMFGLVE
1251  KEFSAIDGVS MEPGTFNVYP CYRLHKDALV SQPTKYLHPE GLPSDYTITF
1301  LFRILPDTPQ EPFALWEILN EQYEPLVGVI LDNGGKTLTF FNYDYKGDFQ
1351  TVTFEGPEIR KIFYGSFHKL HVVISKTTAK IIIDCKEAGE KTINAAGNIS
1401  SDGIEVLGRM VRSRGPRDNS APLQLQMFDI VCATSWANRD KCCELPGLRD
1451  EENCPALPHA CSCSEANKGP LGPPGPPGGP GVRGAKGHRG DPGPKGPDGP
1501  RGEIGVPGPQ GPPGPQGPSG LSIQGLPGPP GEKGEKGDLG FPGLQGVPGA
1551  SGSPGRDGAQ GQRGLPGKDG PTGPQGPPGP VGIPGAPGVP GITGSQGPQG
1601  DVGAPGAPGP KGERGERGDL QSQAMVRAVA RQVCEQLIQG HMARYNSILN
1651  QIPSQSVSTR TIAGPPGEPG RPGAPGPQGE QGSPGMQGFP GNPGQPGRPG
1701  ERGLPGEKGD RGNPGVGTQG PRGPPGSTGP PGESRTGSPG PPGSPGPRGP
1751  AGHTGPPGSQ GPAGPPGYCD PSSCAGYGMG GGYGEPTDQD IPVVQLPHNS
1801  YQIYDPEDLY DGEQQPYVVH GSYPLPSPYS QSSYPSPHLA QPEFTPVREE
1851  MEAVELRSPG ISRFRRKIAK RSIKTLEHKR EMAKEPSQ
```

Structural and functional sites

Signal peptide: 1–28
NC3 domain: 29–1468
COL2 domain: 1469–1620

NC2 domain: 1621–1663
COL1 domain: 1664–1769
NC1 domain: 1770–1888
Alternatively spliced domain: 1813–1843 (residue 1812G becomes E)
Fibronectin type III repeats: 29–118, 352–441, 442–533, 534–622, 623–712, 741–831, 832–922, 923–1010
von Willebrand factor A repeats: 148–336, 1032–1221
PARP repeat: 1222–1468
Potential N-linked glycosylation sites: 138, 1398
Imperfections in Gly-X-Y triplets: 1715–1716, 1735–1736 (conserved in the COL1 domains of type IX, XII and XVI collagens)
Cysteines 1769, 1774 conserved in type IX, XII, XVI and XIX collagens

Gene structure

The size of the gene for the α1(XIV) chain has not been determined. The gene for undulin has been localized to human chromosome 8q23 [8].

References

1. Aubert-Foucher, E. et al. (1992) Purification and characterization of native type XIV collagen. J. Biol. Chem. 266: 15759–15764.
2. Trueb, J. and Trueb, B. (1992) Type XIV collagen is a variant of undulin. Eur. J. Biochem. 207: 549–557.
3. Watt, S.L. et al. (1992) Characterization of collagen types XII and XIV from fetal bovine cartilage. J. Biol. Chem. 267: 20093–20099.
4. Walchli, C. et al. (1993) Complete primary structure of chicken collagen XIV. Eur. J. Biochem. 212: 483–490.
5. Gerecke, D.R. et al. (1993) Type XIV collagen is encoded by alternative transcripts with distinct 5′ regions and is a multidomain protein with homologies to von Willebrand's factor, fibronectin, and other matrix proteins. J. Biol. Chem. 268: 12177–12184.
6. Colige, A. et al. (1995) Characterization and partial amino acid sequencing of a 107-kDa procollagen I N-proteinase purified by affinity chromatography on immobilized type XIV collagen. J. Biol. Chem. 270: 16724–16730.
7. van der Rest, M. and Dublet, B. (1996) Type XII and type XIV collagens: Interfibrillar constituents of dense connective tissues. Semin. Cell Develop. Biol. 7: 639–648.
8. Schnittger, S. et al. (1995) Localization of the undulin gene (UND) to human chromosome band 8q23. Cytogent. Cell Genet. 68: 233–234.

Collagen type XV

Type XV collagen (like type XVIII collagen) is a member of the sub-family of collagens designated MULTIPLEXINs (*multiple* triple-helix domains and *in*terruptions). It has been characterized at the cDNA and genomic level only. The molecule was identified from clones and transcripts of human placenta. It is expressed widely in the basement membrane zones of many internal organs, including placenta, adrenal gland, kidney, heart, skeletal muscle, ovary and testis.

Molecular structure

The α1(XV) chain comprises nine collagenous and ten non-collagenous domains [1–3]. The NC1, COL1–6 and COL8 domains, and the PARP motif are homologous to those of mouse type XVIII collagen. The homologous COL domains have imperfections in the Gly-X-Y sequences in identical positions in the two collagens. The C-terminal end of the NC11 domain is homologous to rat cartilage proteoglycan core protein. Differences between the molecular mass of the α1(XV) chain calculated from sequence data and that estimated from Western blots suggests that the amino-terminal domain may be processed [4]. The molecular structure of type XV collagen is unknown. (N.B. While these COL and NC domain designations follow the usual convention and are numbered from C to N, it should be noted that in reference 3 the domains are numbered from N to C.)

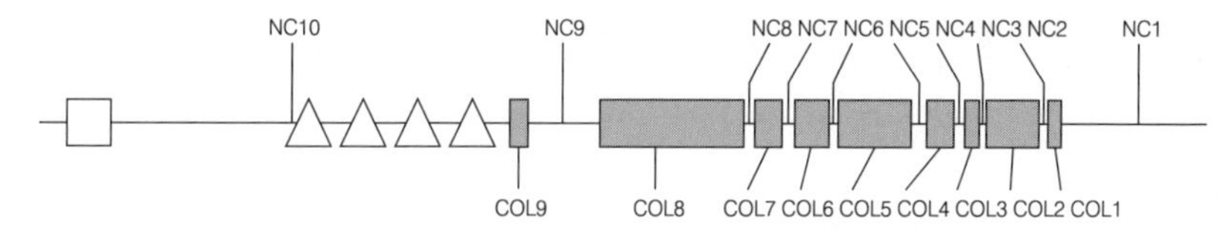

Isolation

The type XV collagen protein has not been isolated. However, collagenous polypeptides of approximately 125 kDa and 116 kDa have been identified by Western blotting HeLa cell lysates and placenta homogenates, respectively, using antisera raised to regions of the non-collagenous COOH-terminal domain [1,4].

Accession number

P39059

Primary structure: α1(XV) chain

Ala	A	106	Cys	C	10	Asp	D	59	Glu	E	97
Phe	F	43	Gly	G	221	His	H	25	Ile	I	54
Lys	K	52	Leu	L	113	Met	M	30	Asn	N	38
Pro	P	192	Gln	Q	40	Arg	R	49	Ser	S	98
Thr	T	74	Val	V	63	Trp	W	10	Tyr	Y	14

Mol. wt (calc.) = 141 930   Residues = 1388

```
1     MAPRRNNGQC WCLLMLLSVS TPLPAVTQTR GATETASQGH LDLTQLIGDP
51    LPSSVSFVTG YGGFPAYSFG PGANVGRPAR TLIPSTFFRD FAIRLVVKPS
101   STRGGVLFAI TDAFQKVIYL GLRLSGVEDG HQRIILYYTE PGSHVSQEAP
151   AFSVPVMTHR WNRFAMIVQG EEVTLLVNCE EHSRIPFQRS SQALAFESSA
201   GIFMGNAGAT GLERFTGSLQ QLTVHPDPRT PEELCDPEES SASGETSGLQ
251   EADGVAEILE AVTYTQASPK EAKVEPINTP PTPSSPFEDM ELSGEPVPEG
301   TLETTNMSII QHSSPKQGSG EILNDTLEGV HSVDGDPITD SGSGAGAFLD
351   IAEEKNLAAT AAGLAEVPIS TAGEAEASSV PTGGPTLSMS TENPEEGVTP
401   GPDNEERLRA TAAGEAEALA SMPGEVEASG VAPGELDLSM SAQSLGEEAT
451   VGPSSEDSLT TAAAATEVSL STFEDEEASG VPTDGLAPLT ATMAPERAVT
501   SGPGDEEDLA AATTEEPLIT AGGEESGSPP PDGPPLPLPT VAPERWITPA
551   QREHVGMKGQ AGPKGEKGDA GEELPGPPEP SGPVGPTAGA EAEGSGLGWG
601   SDVGSGSGDL VGSEQLLRGP PGPPGPPGLP GIPGKPGTDV FMGPPGSPGE
651   DGPAGEPGPP GPEGQPGVDG ATGLPGMKGE KGARGPNGSV GEKGDPGNRG
701   LPGPPGKKGQ AGPPGVMGPP GPPGPPGPPG PGCTMGLGFE DTEGSGSTQL
751   LNEPKLSRPT AAIGLKGEKG DRGPKGERGM DGASIVGPPG PRGPPGHIKV
801   LSNSLINITH GFMNFSDIPE LVGPPGPDGL PGLPGFPGPR GPKGDTGLPG
851   FPGLKGEQGE KGEPGAILTE DIPLERLMGK KGEPGMHGAP GPMGPKGPPG
901   HKGEFGLPGR PGRPGLNGLK GTKGDPGVIM QGPPGLPGPP GPPGPPGAVI
951   NIKGAIFPIP VRPHCKMPVD TAHPGSPELI TFHGVKGEKG SWGLPGSKGE
1001  KGDQGAQGPP GPPLDLAYLR HFLNNLKGEN GDKGFKGEKG EKGDINGSFL
1051  MSGPPGLPGN PGPAGQKGET VVGPQGPPGA PGLPGPPGFG RPGDPGPPGP
1101  PGPPGPPAIL GAAVALPGPP GPPGQPGLPG SRNLVTAFSN MDDMLQKAHL
1151  VIEGTFIYLR DSTEFFIRVR DGWKKLQLGE LIPIPADSPP PPALSSNPHQ
1201  LLPPPNPISS ANYEKPALHL AALNMPFSGD IRADFQCFKQ ARAAGLLSTY
1251  RAFLSSHLQD LSTIVRKAER YSLPIVNLKG QVLFNNWDSI FSGHGGQFNM
1301  HIPIYSFDGR DIMTDPSWPQ KVIWHGSSPH GVRLVDNYCE AWRTADTAVT
1351  GLASPLSTGK ILDQKAYSCA NRLIVLCIEN SFMTDARK
```

Structural and functional sites

Signal peptide: 1–25
NC10 domain: 26–555
COL9 domain: 556–573
NC9 domain: 574–618
COL8 domain: 619–732
NC8 domain: 733–763
COL7 domain: 764–798
NC7 domain: 799–822
COL6 domain: 823–867
NC6 domain: 868–878
COL5 domain: 879–949
NC5 domain: 950–983
COL4 domain: 984–1013
NC4 domain: 1014–1027
COL3 domain: 1028–1045
NC3 domain: 1046–1052
COL2 domain: 1053–1107
NC2 domain: 1108–1117
COL1 domain: 1118–1132
NC1 domain: 1133–1388

PARP repeat: 40–239
Tandem repeat structure homologous to rat aggrecan core protein (link protein repeats): 358–555 [358–408, 409–459, 460–509, 510–555]
Potential N-linked glycosylation sites: 306, 324, 687, 807, 814, 1046
Potential O-linked glycosaminoglycans: 319, 341, 343, 526, 595, 605, 607, 745, 1292
Imperfections in Gly-X-Y: 640–642, 785–786, 930–931, 1071–1072, 1091–1092
Cysteines 1237, 1339, 1369, 1377 are conserved in type XVIII collagen

Gene structure

The gene encoding the α1(XV) chain has been localized to chromosome 9 at locus q21–22[5]. Partial analysis of the gene indicates that the sizes of the exons are highly variable (varying between 36 and 140 bp)[3].

References

1. Myers, J. C. et al. (1992) Identification of a previously unknown collagen chain, α1(XV), characterised by extensive interruptions in the triple-helical region. Proc. Natl Acad. Sci. USA 89: 10144–10148.
2. Muragaki, Y. et al. (1994) The human α1(XV) collagen chain contains a large amino-terminal non-triple helical domain with a tandem repeat structure and homology to α1(XVIII) collagen. J. Biol. Chem. 269: 4042–4046.
3. Kivirikko, S. et al. (1994) Primary structure of the α1 chain of human type XV collagen and exon–intron organization in the 3′ region of the corresponding gene. J. Biol. Chem. 269: 4773–4779.
4. Myers, J.C. et al. (1996) Type XV collagen exhibits a widespread distribution in human tissues but a distinct localization in basement membrane zones. Cell Tissue Res. 286: 493–505.
5. Huebner, K. et al. (1992) Chromosomal assignment of a gene encoding a new collagen type (COL15A1) to 9q21–q22. Genomics 14: 220–224.

Collagen type XVI

Type XVI collagen is a member of the FACIT collagens that include type IX, XII, XIV and XIX collagens. It has a broad tissue distribution, being localized predominantly in heart, kidney, intestine, ovary, testis, eye, arterial walls and smooth muscles.

Molecular structure

The predicted α1(XVI) chain comprises ten collagenous domains (COL1–10) which range in size from 15 to 422 amino acids, and 11 non-collagenous domains (NC1–11), ranging from 11 to 39 amino acids long, except for the large amino-terminal domain NC11 which has 312 residues. Thirty-two cysteine residues are present, mainly in the NC domains. The two cysteines present four amino acids apart at the COL1/NC1 junction and the size of the COL1 domain are characteristic of FACIT collagens [1,2]. Biosynthetic studies on stably transfected kidney cell clones [3], and on human fibroblasts and smooth muscle cells [4,5], indicate that type XVI collagen is a disulphide-bonded homotrimer comprising 200–220 kDa α1(XVI) chains. Type XVI collagen may be processed with the release of the PARP-like domain, as occurs for the α2(XI) chain, resulting in a protein comprising chains of 150–160 kDa [2,3,5].

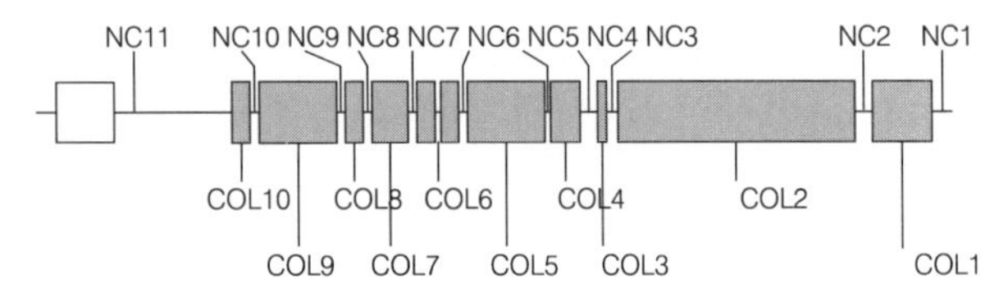

Isolation

The type XVI collagen protein has been isolated from the medium of stably transfected kidney cell clones [3] and immunoprecipitated from human fibroblast and smooth muscle cell extracts [4,5].

Accession number

M92642

Primary structure: α1(XVI) chain

Ala	A	86	Cys	C	32	Asp	D	52	Glu	E	92
Phe	F	30	Gly	G	390	His	H	19	Ile	I	41
Lys	K	88	Leu	L	99	Met	M	26	Asn	N	23
Pro	P	278	Gln	Q	79	Arg	R	63	Ser	S	77
Thr	T	44	Val	V	63	Trp	W	8	Tyr	Y	13

Mol. wt (calc.) = 157 516 Residues = 1603

```
1     MWVSWAPGLW  LLGLWATFGH  GANTGAQCPP  SQQEGLKLEH  SSSLPANVTG
51    FNLIHRLSLM  KKSAIKKIRN  PKGPLILRLG  AAPVTQPTRR  VFPRGLPEEF
101   ALVLTLLLKK  HTHQKTWYLF  QVTDANGYPQ  ISLEVNSQER  SLELRAQGQD
151   GDFVSCIFPV  PQLFDLRWHK  LMLSVAGRVA  SVHVDCSSAS  SQPLGPRRPM
```

```
201   RPVGHVFLGL DAEQGKPVSF DLQQVHIYCD PELVLEEGCC EILPAGCPPE
251   TSKARRDTQS NELIEINPQS EGKVYTRCFC LEEPQNSEVD AQLTGRISQK
301   AERGAKVHQE TAADECPPCV HGARDSNVTL APSGPKGGKG ERGLPGPPGS
351   KGEKGARGND CVRISPDAPL QCAEGPKGEK GESGALGPSG LPGSTGEKGQ
401   KGEKGDGGIK GVPGKPGRDA PGEICVIGPK GQKGDPGFVG PEGLAGEPGP
451   PGLPGPPGIG LPGTPGDPGG PPGPKGDKGS SGIPGKEGPG GKPGKPGVKG
501   EKGDPCEVCP TLPEGFQNFV GLPGKPGPKG EPGDPVRARG DPGIQGIKGE
551   KGEPCLSCSS VVGAQHLVSS TGASGDVGSP GFGLPGLPGR AGVPGLKGEK
601   GNFGEAGPAG SPGPPGPVGP AGIKGAKGEP CEPCPALSNL QDGDVRVVAL
651   PGPSGEKGEP GPPGFGLPGK QGKAGERGLK GQKGDAGNPG DPGTPGTTGR
701   PGLSGEPGVQ GPAGPKGEKG DGCTACPSLQ GTVTDMAGRP GQPGPKGEQG
751   PEGVGRPGKP GQPGLPGVQG PPGLKGVQGE PGPPGRGVQG PQGEPGAPGL
801   PGIQGLPGPR GPPGPTGEKG AQGSPGVKGA TGPVGPPGAS VSGPPGRDGQ
851   QGQTGLRGTP GEKGPRGEKG EPGECSCPSQ GDLIFSGMPG APGLWMGSSW
901   QPGPQGPPGI PGPPGPPGVP GLQGVPGNNG LPGQPGLTAE LGSLPIEQHL
951   LKSICGDCVQ GQRAHPGYLV EKGEKGDQGI PGVPGLDNCA QCFLSLERPR
1001  AEEARGDNSE GDPGCVGSPG LPGPPGLPGQ RGEEGPPGMR GSPGPPGPIG
1051  PPGFPGAVGS PGLPGLQGER GLTGLTGDKG EPGPPGQPGY PGATGPPGLP
1101  GIKGERGYTG SAGEKGEPGP PGSEGLPGPP GPAGPRGERG PQGNSGEKGD
1151  QGFQGQPGFT GPTGSPGFPG KVGSPGPPGP QAEKGSEGIR GPSGLPGSPG
1201  PPGPPGIQGP AGLDGLDGKD GKPGLRGDPG PAGPPGLMGP PGFKGKTGHP
1251  GLPGPKGDCG KPGPPGSTGR PGAEGEPGAM GPQGRPGPPG HVGPPGPPGQ
1301  PGPAGISAVG LKGDRGATGE RGLAGLPGQP GPPGHPGPPG EPGTDGAAGK
1351  EGPPGKQGFY GPPGPKGDPG AAGQKGQAGE KGRAGMPGGP GKSGSMGPVG
1401  PPGPAGERGH PGAPGPSGSP GLPGVPGSMG DMVNYDEIKR FIRQEIIKMF
1451  DERMAYYTSR MQFPMEMAAA PGRPGPPGKD GAPGRPGAPG SPGLPGQIGR
1501  EGRQGLPGVR GLPGTKGEKG DIGIGIAGEN GLPGPPGPQG PPGYGKMGAT
1551  GPMGQQGIPG IPGPPGPMGQ PGKAGHCNPS DCFGAMPMEQ QYPPMKTMKG
1601  PFG
```

Structural and functional sites

Signal peptide: 1–21
NC11 domain: 22–333
COL10 domain: 334–360
NC10 domain: 361–374
COL9 domain: 375–505
NC9 domain: 506–520
COL8 domain: 521–554
NC8 domain: 555–571
COL7 domain: 572–630
NC7 domain: 631–651
COL6 domain: 652–722
NC6 domain: 723–737
COL5 domain: 738–875
NC5 domain: 876–886
COL4 domain: 887–838
NC4 domain: 839–972
COL3 domain: 973–987
NC3 domain: 988–1010
COL2 domain: 1011–1432
NC2 domain: 1433–1471

COL1 domain: 1472–1577
NC1 domain: 1578–1603
PARP repeat 22–255
Possible proteolytic cleavage of NC11: 256–257
Potential N-linked glycosylation sites: 47, 327, 1578
Imperfections in Gly-X-Y: 420–421, 425–427, 458–459, 536–539, 581–582, 664–665, 753–754, 785–786, 841–842, 896–902, 1182–1184, 1308–1309, 1523–1524, 1543–1544
Cysteines 1577, 1582 are conserved in type IX, XII, XIV and XIX collagens

Gene structure

The gene for the α1(XVI) chain has been localized to human chromosome 1 at locus p34–35 [1].

References

1. Pan, T.-C. et al. (1992) Cloning and chromosomal location of human α1(XVI) collagen. Proc. Natl Acad. Sci. USA 89: 6565–6569.
2. Yamaguchi, N. et al. (1992) Molecular cloning and partial characterisation of a novel collagen chain, α1 (XVI), consisting of repetitive collagenous domains and cysteine-containing non-collagenous segments. J. Biochem. 112: 856–863.
3. Tillet, E. et al. (1995) Recombinant analysis of human α1(XVI) collagen. Evidence for processing of the N-terminal globular domain. Eur. J. Biochem. 228: 160–168.
4. Lai, C.-H. and Chu, M.-L. (1996) Tissue distribution and developmental expression of type XVI collagen in the mouse. Tissue Cell 28: 155–164.
5. Grassel, S. et al. (1996) Biosynthesis and processing of type XVI collagen in human fibroblasts and smooth muscle cells. Eur. J. Biochem. 242: 576–584.

Collagen type XVII

Type XVII collagen is identified as the 180 kDa autoantigen associated with the blistering skin disease bullous pemphigoid as well as with bullous diseases of other epithelia including cornea and mucous membranes. It is a hemidesmosomal transmembrane glycoprotein comprising an interrupted collagenous domain in its extracellular part. Several mutations in the COL17A1 gene (all in regions encoding the extracellular domain) have been identified which result in premature chain termination and absence of the protein. These mutations have been linked to the non-lethal junctional blistering disease of generalized atrophic benign epidermolysis bullosa in which the tissue separates within the cutaneous basement membrane zone at, or just below, the level of hemidesmosomes [1–5].

Molecular structure

The α1(XVII) chain comprises 15 collagenous (COL1–15) domains ranging from 15 to 242 amino acids long and 16 non-collagenous (NC1–16) domains, the large amino-terminal NC16 domain containing 566 amino acids. Rotary shadowing images show a quaver-like molecule consisting of a globular head (cytoplasmic NC16 domain), a central 60–70 nm rod (large COL15 domain) and a flexible tail (interrupted collagenous domain) [1–3]. Molecular genetic studies in an epithelial cell line indicate that the amino-terminus plays a role in targeting and polarizing type XVII collagen within the hemidesmosome while a short non-collagenous extracellular region adjacent to the cell membrane is important for interaction with other hemidesmosomal components, including the α6 subunit of the α6β4 integrin [4]. This region also contains the major epitope recognized by autoantibodies in the sera of patients with both bullous pemphigoid and herpes gestationis [5]. Recent studies indicate that type XVII collagen is a homotrimer [3,6].

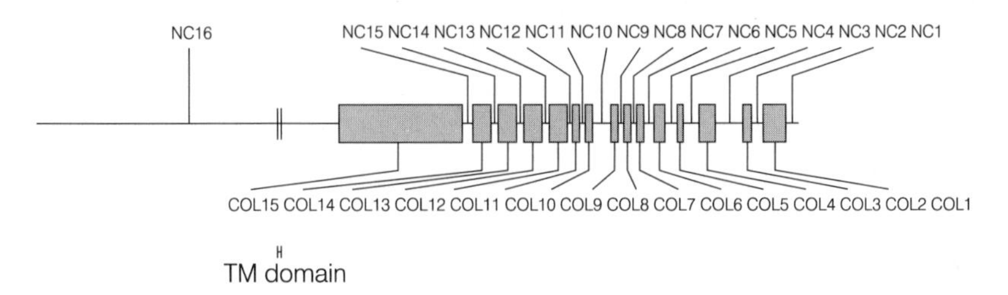

Isolation

Type XVII collagen has been solubilized from a bovine mammary gland epithelial cell line using 0.5% Triton X-100 and purified by immunoaffinity chromatography using a monoclonal antibody to the bullous antigen [3].

Accession number

U76604

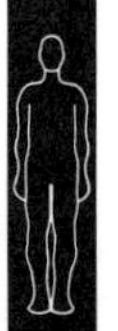

Primary structure: α1(XVII) chain

Ala	A	85	Cys	C	7	Asp	D	54	Glu	E	63
Phe	F	26	Gly	G	281	His	H	23	Ile	I	42
Lys	K	60	Leu	L	107	Met	M	31	Asn	N	31
Pro	P	204	Gln	Q	48	Arg	R	68	Ser	S	178
Thr	T	84	Val	V	59	Trp	W	10	Tyr	Y	36

Mol. wt (calc.) = 150 459 Residues = 1497

```
1     MDVTKKNKRD GTEVTERIVT ETVTTRLTSL PPKGGTSNGY AKTASLGGGS
51    RLEKQSLTHG SSGYINSTGS TRGHASTSSY RRAHSPASTL PNSPGSTFER
101   KTHVTRHAYE GSSSGNSSPE YPRKEFASSS TRGRSQTRES EIRVRLQSAS
151   PSTRWTELDD VKRLLKGSRS ASVSPTRNSS NTLPIPKKGT VETKIVTASS
201   QSVSGTYDAT ILDANLPSHV WSSTLPAGSS IGTYHNNMTT QSSSLLNTNA
251   YSAGSVFGVP NNMASCSPTL HPGLSTSSSV FGMQNNLAPS LTTLSHGTTT
301   TSTAYGVKKN MPQSPAAVNT GVSTSAACTT SVQSDDLLHK DCKFLILEKD
351   NTPAKKEMEL LIMTKDSGKV FTASPASIAA TSFSEDTLKK EKQAAYNADS
401   GLKAEANGDL KTVSTKGKTT TADIHSYSSS GGGGSGGGGG VGGAGGGPWG
451   PAPAWCPCGS CCSWWKWLLG LLLTWLLLLG LLFGLIALAE EVRKLKARVD
501   ELERIRRSIL PYGDSMDRIE KDRLQGMAPA AGADLDKIGL HSDSQEELWM
551   FVRKKLMMEQ ENGNLRGSPG PKGDMGSPGP KGDRGFPGTP GIPGPLGHPG
601   PQGPKGQKGS VGDPGMEGPM GQRGREGPMG PRGEAGPPGS GEKGERGAAG
651   EPGPHGPPGV PGSVGPKGSS GSPGPQGPPG PVGLQGLRGE VGLPGVKGDK
701   GPVGPPGPKG DQGEKGPRGL TGEPGMRGLP GAVGEPGAKG AMGPAGPDGH
751   QGPRGEQGLT GMPGIRGPPG PSGDPGKPGL TGPQGPQGLP GTPGRPGIKG
801   EPGAPGKIVT SEGSSMLTVP GPPGPPGAMG PPGPPGAPGP AGPAGLPGHQ
851   EVLNLQGPPG PPGPRGPPGP SIPGPPGPRG PPGEGLPGPP GPPGSFLSNS
901   ETFLFGPPGP PGPPGPKGDQ GPPGPRGHQG EQGLPGFSTS GSSSFGLNLQ
951   GPPGPPGPQG PKGDKGDPGV PGALGIPSGP SEGGSSSTMY VSGPPGPPGP
1001  PGPPGSISSS GQEIQQYISE YMQSDSIRSY LSGVQGPPGP PGPPGPVTTI
1051  TGETFDYSEL ASHVVSYLRT SGYGVSLFSS SISSEDILAV LQRDDVRQYL
1101  RQYLMGPRGP PGPPGASGDG SLLSLDYAEL SSRILSYMSS SGISIGLPGP
1151  PGPPGLPGTS YEELLSLLRG SEFRGIVGPP GPPGPPGIPG NVWSSISVED
1201  LSSYLHTAGL SFIPGPPGPP GPPGPRGPPG VSGALATYAA ENSDSFRSEL
1251  ISYLTSPDVR SFIVGPPGPP GPQGPPGDSR LLSTDASHSR GSSSSSHSSS
1301  VRRGSSYSSS MSTGGGGAGS LGAGGAFGEA AGDRGPYGTD IGPGGGYGAA
1351  AEGGMYAGNG GLLGADFAGD LDYNELAVRV SESMQRQGLL QGMAYTVQGP
1401  PGQPGPQGPP GISKVFSAYS NVTADLMDFF QTYGAIQGPP GQKGEMGTPG
1451  PKGDRGPAGP PGHPGPPGPR GHKGEKGDKG DQVYAGRRRR RSIAVKP
```

Structural and functional sites

No signal sequence
NC16 domain: 1–566 [intracellular domain with four tandem repeats (24–26 residues, 227–324) 1–466; transmembrane domain (type II orientation) 467–488; extracellular domain 489–566]
COL15 domain: 567–808
NC15 domain: 809–820
COL14 domain: 821–850
NC14 domain: 851–856
COL13 domain: 857–896
NC13 domain: 897–905

COL12 domain: 906–938
NC12 domain: 939–950
COL11 domain: 951–977
NC11 domain: 978–992
COL10 domain: 993–1007
NC10 domain: 1008–1032
COL9 domain: 1033–1047
NC9 domain: 1048–1105
COL8 domain: 1106–1120
NC8 domain: 1121–1145
COL7 domain: 1146–1160
NC7 domain: 1161–1174
COL6 domain: 1175–1192
NC6 domain: 1193–1214
COL5 domain: 1215–1235
NC5 domain: 1236–1264
COL4 domain: 1265–1279
NC4 domain: 1280–1313
COL3 domain: 1314–1350
NC3 domain: 1351–1398
COL2 domain: 1399–1413
NC2 domain: 1414–1437
COL1 domain: 1438–1482
NC1 domain: 1483–1497
Potential N-linked glycosylation sites: 66, 116, 178, 237, 1421
Imperfections in Gly-X-Y triplets: 639–640, 872–873, 883–884, 1320–1321, 1331, 1341
NC16 residues 1–36 required for correct basal polarization; 493–519 important for extracellular interaction with other hemidesmosomal components, e.g. $\alpha6\beta4$ integrin
Potential antigenic sequences: 5–10, 517–523, 1486–1492 [2]
Immunodominant site: 507–520 [5]

Gene structure

The gene for the α1(XVII) chain comprises 56 exons which span approximately 52 kb of the genome on the long arm of human chromosome 10, locus 10q24.3 [2,7]. The size of the exons vary from 27 bp (exon 20) to 390 bp (exon 52). Several of the introns are larger than 1 kb and intron 1 is approximately 5 kb long. The intracellular globular domain is encoded by exons 2–17, the transmembrane domain by the 3′-end of exon 17 and the extracellular domain by exons 18–56.

References

1 Giudice, G.J. et al. (1992) Cloning and primary structural analysis of the bullous pemphigoid autoantigen BP180. J. Invest. Dermatol. 99: 243–250.
2 Gatalica, B. et al. (1997) Cloning of the human type XVII collagen gene (COL17A1) and detection of novel mutations in generalised atrophic benign epidermolysis bullosa. Am. J. Hum. Genet. 60: 352–365.

[3] Hirako, Y. et al. (1996) Demonstration of the molecular shape of BP180, a 180-kDa bullous pemphigoid antigen and its potential for trimer formation. J. Biol. Chem. 271: 13739–13745.
[4] Hopkinson, S.B. et al. (1995) Molecular genetic studies of a human epidermal autoantigen (the 180-kD bullous pemphigoid antigen/BP180): Identification of functionally important sequences within the BP180 molecule and evidence for an interaction between BP180 and $\alpha 6$ integrin. J. Cell Biol. 130: 117–125.
[5] Giudice, G.J. et al. (1993) Bullous pemphigoid and herpes gestationis autoantibodies recognize a common non-collagenous site on the BP180 ectodomain. J. Immunol. 151: 5742–5750.
[6] Limardo, M. et al. (1996) Evidence that the 180-kDa bullous pemphigoid antigen is a transmembrane collagen, type XVII, in a triple helical conformation and in type II transmembrane topography. J. Invest. Dermatol. 106: 860.
[7] Li, K. et al. (1991) Genomic organization of collagenous domains and chromosomal assignment of human 180-kDa bullous pemphigoid antigen-2, a novel collagen of stratified squamous epithelium. J. Biol. Chem. 266: 24064–24069.

Collagen type XVIII

Type XVIII collagen (like type XV collagen) is a member of the sub-family of collagens designated MULTIPLEXINs (*multiple* triple-helix domains and *in*terruptions). It exists in three variant forms which are expressed in the basement membrane zones of vessels in a tissue-specific manner. The long unspliced form is expressed in liver, lung and kidney, and to a lesser extent in heart, brain and skeletal muscle. The short variant and the spliced variant of the long form are predominantly expressed in kidney and liver, respectively.

Molecular structure

The three variants of the α1(XVIII) chain each consists of 10 collagenous and 11 non-collagenous domains and differ only in their amino-terminal NC11 domain. They originate from the use of two alternative promoters of the α1(XVIII) gene. The upstream promoter directs the synthesis of the short form and the downstream promoter directs that of the long form, which is further subject to alternative splicing. The short and long forms have different signal sequences. All three forms share a common 299 amino acid residues at the carboxy terminal end of the NC11 domain. The longest form contains a segment of 110 residues with 10 cysteines designated a frizzled motif (fz). This motif is found in frizzled proteins which resemble the G-protein-coupled membrane receptors. The NC1, COL1–6 and COL8 domains and the PARP motif are homologous to those of human type XV collagen [1–5]. The most C-terminal 184 amino acid residues of the NC1 domain (1591–1774) have been isolated from a murine haemangioendothelioma cell line as a 20 kDa inhibitor of angiogenesis called endostatin [6]. The molecular structure of type XVIII collagen is unknown. (N.B. While these COL and NC domain designations follow the usual convention and are numbered from C to N, it should be noted that in references 2, 3 and 5, the domains are numbered from N to C.)

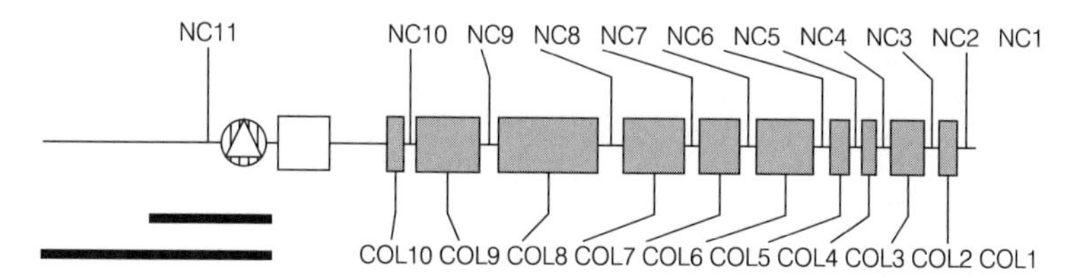

Isolation

The type XVIII collagen protein has been detected in Western blots of extracts from mouse embryonic stem cells as a 200 kDa band, corresponding to the short form [4].

Accession number

U03713–6; U03718; U11636–7; L16898

Primary structure: α1(XVIII) chain (mouse)

Ala	A	118	Cys	C	18	Asp	D	83	Glu	E	95
Phe	F	61	Gly	G	285	His	H	37	Ile	I	45
Lys	K	57	Leu	L	139	Met	M	17	Asn	N	29
Pro	P	275	Gln	Q	80	Arg	R	79	Ser	S	147
Thr	T	72	Val	V	94	Trp	W	20	Tyr	Y	23

Mol. wt (calc.) = 182 230 Residues = 1774

```
1     MAPDPSRRLC LLLLLLLSCR LVPASADGNS LSPLNPLVWL WPPKTSDSLE
51    GPVSKPQNSS PVQSTENPTT HVVPQDGLTE QQTTPASSEL PPEEEEEEDQ
101   KAGQGGSPAT PAVPIPLVAP AASPDMKEEN VAGVGAKILN VAQGIRSFVQ
151   LWDEDSTIGH SAGTEVPDSS IPTVLPSPAE LSSAPQGSKT TLWLSSAIPS
201   SPDAQTTEAG TLAVPTQLPP FQSNLQAPLG RPSAPPDFPG RAFLSSSTDQ
251   GSSWGNQEPP RQPQHLEGKG FLPMTARSSQ QHRHSDVHSD IHGHVPLLPL
301   VTGPLVTASL SVHGLLSVPS SDPSGQLSQV AALPGFPGTW VSHVAPSSGT
351   GLSNDSALAG NGSLTSTSRC LPLPPTLTLC SRLGIGHFWL PNHLHHTDSV
401   EVEATVQAWG RFLHTNCHPF LAWFFCLLLA PSCGPGPPPP LPPCRQFCEA
451   LEDECWNYLA GDRLPVVCAS LPSQEDGYCV FIGPAAENVA EEVGLLQLLG
501   DPLPEKISQI DDPHVGPAYI FGPDSNSGQV AQYHFPKLFF RDFSLLFHVR
551   PATEAAGVLF AITDAAQVVV SLGVKLSEVR DGQQNISLLY TEPGASQTQT
601   GASFRLPAFV GQWTHFALSV DGGSVALYVD CEEFQRVPFA RASQGLELER
651   GAGLFVGQAG TADPDKFQGM ISELKVRKTP RVSPVHCLDE EDDDEDRASG
701   DFGSGFEESS KSHKEDTSLL PGLPQPPPVT SPPLAGGSTT EDPRTEETEE
751   DAAVDSIGAE TLPGTGSSGA WDEAIQNPGR GLIKGGMKGQ KGEPGAQGPP
801   GPAGPQGPAG PVVQSPNSQP VPGAQGPPGP QGPPGKDGTP GRDGEPGDPG
851   EDGRPGDTGP QGFPGTPGDV GPKGEKGDPG IGPRGPPGPP GPPGPSFRQD
901   KLTFIDMEGS GFSGDIESLR GPRGFPGPPG PPGVPGLPGE PGRFGINGSY
951   APGPAGLPGV PGKEGPPGFP GPPGPPGPPG KEGPPGVAGQ KGSVGDVGIP
1001  GPKGSKGDLG PIGMPGKSGL AGSPGPVGPP GPPGPPGPPG PGFAAGFDDM
1051  EGSGIPLWTT ARSSDGLQGP PGSPGLKGDP GVAGLPGAKG EVGADGAQGI
1101  PGPPGREGAA GSPGPKGEKG MPGEKGNPGK DGVGRPGLPG PPGPPGPVIY
1151  VSSEDKAIVS TPGPEGKPGY AGFPGPAGPK GDLGSKGEQG LPGPKGEKGE
1201  PGTIFSPDGR ALGHPQKGAK GEPGFRGPPG PYGRPGHKGE IGFPGRPGRP
1251  GTNGLKGEKG EPGDASLGFS MRGLPGPPGP PGPPGPPGMP IYDSNAFVES
1301  GRPGLPGQQG VQGPSGPKGD KGEVGPPGPP GQFPIDLFHL EAEMKGDKGD
1351  RGDAGQKGER GEPGAPGGGF FSSSVPGPPG PPGYPGIPGP KGESIRGPPG
1401  PPGRQGPPGI GYEGRQGPPG PPGPPGPPSF PGPHRQTVSV PGPPGPPGPP
1451  GPPGAMGASA GQVRIWATYQ TMLDKIREVP EGWLIFVAER EELYVRVRNG
1501  RKVLLEART  ALPRGTGNEV AALQPPLVQL HEGSPYTRRE YSYSTARPWR
1551  ADDILANPPR LPDRQPYPGV PHHHSSYVHL PPARPTLSLA HTHQDFQPVL
1601  HLVALNTPLS GGMRGIRGAD FQCFQQARAV GLSGTFRAFL SSRLQDLYSI
1651  VRRADRGSVP IVNLKDEVLS PSWDSLFSGS QGQLQPGARI FSFDGRDVLR
1701  HPAWPQKSVW HGSDPSGRRL MESYCETWRT ETTGATGQAS SLLSGRLLEQ
1751  KAASCHNSYI VLCIENSFMT SFSK
```

Structural and functional sites

Signal peptide: 1–21/24
NC11 domain: 22/25–785 [Residues 240–486 missing in alternatively spliced form]
COL10 domain: 786–812

NC10 domain: 813–822
COL9 domain: 823–896
NC9 domain: 897–920
COL8 domain: 921–1042
NC8 domain: 1043–1065
COL7 domain: 1066–1148
NC7 domain: 1149–1162
COL6 domain: 1163–1204
NC6 domain: 1205–1217
COL5 domain: 1218–1290
NC5 domain: 1291–1300
COL4 domain: 1301–1333
NC4 domain: 1334–1345
COL3 domain: 1346–1366
NC3 domain: 1367–1376
COL2 domain: 1377–1428
NC2 domain: 1429–1441
COL1 domain: 1442–1459
NC1 domain: 1460–1774
Frizzle (fz) repeat: 370–479
PARP repeat: 487–694
Potential N-linked glycosylation sites: 58, 354, 361, 585, 947
Potential O-linked glycosaminoglycans: 348, 910, 1053
Putative RGD cell adhesion site: 1351–1353
Imperfections in Gly-X-Y: 880–881, 951–952, 1132–1133, 1266–1267, 1271–1272, 1396–1397, 1409–1410
Cysteines 1623, 1725, 1755 and 1763 are conserved in type XV collagen

Primary structure: α1(XVIII) chain (mouse) (alternative short form)

```
1      MAPRWHLLDV  LTSLVLLLVA  RVSWAEP
```

Structural and functional sites

Signal peptide: 1–25
NC11 domain: 26–326 [Residues 28–326 are identical to residues 487–785 of the long form]
COL11 to NC1 domains: identical in sequence to long form

Gene structure

The mouse gene for the α1(XVIII) chain is greater than 102 kb and consists of 43 exons ranging between 26 and 1770 bp. Exons 4–9 encode a region in the NC11 domain that is common to all three variants and exons 9–43 encode the common collagenous and non-collagenous domains COL1–10 and NC1–10. Most of the exons encoding the collagenous regions are characteristic of non-fibrillar collagens being multiples of 9 bp with few 54 bp. The two promoters are separated by 50 kb. The upstream promoter and exons 1 and 2 give rise to the short form. The downstream promoter and exon 3 result in the longer forms. The mouse gene has been mapped to chromosome 10 and the human gene to chromosome 21 at locus q22.3, close to the locus for the genes encoding the α1(VI) and α2(VI) chains[5,7].

References

[1] Rehn, M. and Pihlajaniemi, T. (1994) α1(XVIII), a collagen chain with frequent interruptions in the collagenous sequence, a distinct tissue distribution, and homology with type XV collagen. Proc. Natl Acad. Sci. USA 91: 4234–4238.

[2] Rehn, M. et al. (1994) Primary structure of the α1 chain of mouse type XVIII collagen, partial structure of the corresponding gene, and comparison of the α1(XVIII) chain with its homologue, the α1(XV) collagen chain. J. Biol. Chem. 269: 13929–13935.

[3] Rehn, M. and Pihlajaniemi, T. (1995) Identification of three N-terminal ends of type XVIII collagen chains and tissue-specific differences in the expression of the corresponding transcripts. The longest form contains a novel motif homologous to rat and *drosophila* frizzled proteins. J. Biol. Chem. 270: 4705–4711.

[4] Muragaki, Y. et al. (1995) Mouse *Col18α1* is expressed in a tissue-specific manner as three alternative variants and is localized in basement membrane zones. Proc. Natl Acad. Sci. USA 92: 8763–8767.

[5] Rehn, M. et al. (1996) Characterization of the mouse gene for the α1 chain of type XVIII collagen (Col18a1) reveals that the three variant N-terminal polypeptide forms are transcribed from two widely separated promoters. Genomics 32: 436–446.

[6] O'Reilly, M.S. et al. (1997) Endostatin: An endogenous inhibitor of angiogenesis and tumor growth. Cell 88: 277–285.

[7] Oh, S.P. et al. (1994) Cloning of cDNA and genomic DNA encoding human type XVIII collagen and localization of the α1(XVIII) collagen gene to mouse chromosome 10 and human chromosome 21. Genomics 19: 494–499.

Collagen type XIX

Type XIX collagen has been characterized at the cDNA and genomic levels only and is a member of the FACIT collagens, which include type IX, XII, XIV and XVI collagens. It is expressed in human rhabdomyosarcoma and fibroblast cell lines[1–3].

Molecular structure

The predicted α1(XIX) chain comprises five collagenous domains (COL1–5), which range in size from 70 to 224 amino acids, and six non-collagenous domains (NC1–6), varying between 19 and 44 amino acids long. The COL1 domain is smaller than the COL1 domains of type IX, XII, XIV and XVI collagens. The amino-terminal domain contains only the PARP motif as in the case of type IX and XVI collagens, and differs from that of types XII and XIV collagens. The molecular structure of type XIX collagen is unknown[3].

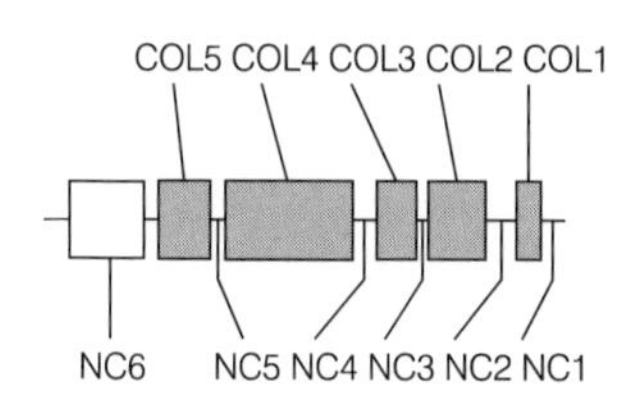

Isolation

Type XIX collagen protein has not been isolated.

Accession number

JX0369

Primary structure: α1(XIX) chain

Ala	A	52	Cys	C	14	Asp	D	49	Glu	E	67
Phe	F	25	Gly	G	261	His	H	16	Ile	I	57
Lys	K	77	Leu	L	71	Met	M	18	Asn	N	30
Pro	P	173	Gln	Q	46	Arg	R	47	Ser	S	53
Thr	T	30	Val	V	33	Trp	W	7	Tyr	Y	16

Mol. wt (calc.) = 115 220 Residues = 1142

```
1     MRLTGPWKLW  LWMSIFLLPA  STSVTVRDKT  EESCPILRIE  GHQLTYDNIN
51    KLEVSGFDLG  DSFSLRRAFC  ESDKTCFKLG  SALLIRDTIK  IFPKGLPEEY
101   SVAAMFRVRR  NAKKERWFLW  QVLNQQNIPQ  ISIVVDGGKK  VVEFMFQATE
151   GDVLNYIFRN  RELRPLFDRQ  WHKLGISIQS  QVISLYMDCN  LIARRQTDEK
201   DTVDFHGRTV  IATRASDGKP  VDIELHQLKI  YCSANLIAQE  TCCEISDTKC
251   PEQDGFGNIA  SSWVTAHASK  MSSYLPAKQE  LKDQCQCIPN  KGEAGLPGAP
301   GSPGQKGHKG  EPGENGLHGA  PGFPGQKGEQ  GFEGSKGETG  EKGEQGEKGD
351   PALAGLNGEN  GLKGDLGPHG  PPGPKGEKGD  TGPPGPPALP  GSLGIQGPQG
401   PPGKEGQRGR  RGKTGPPGKP  GPPGPPGPPG  IQGIHQTLGG  YYNKDNKGND
451   EHEAGGLKGD  KGETGLPGFP  GSVGPKGQKG  EPGEPFTKGE  KGDRGEPGVI
501   GSQGVKGEPG  DPGPPGLIGS  PGLKGQQGSA  GSMGPRGPPG  DVGLPGEHGI
```

```
551  PGKQGIKGEK GDPGGIIGPP GLPGPKGEAG PPGKSLPGEP GLDGNPGAPG
601  PRGPKGERGL PGVHGSPGDI GPQGIGIPGR TGAQGPAGEP GIQGPRGLPG
651  LPGTPGTPGN DGVPGRDGKP GLPGPPGDPI ALPLLGDIGA LLKNFCGNCQ
701  ASVPGLKSNK GEEGGAGEPG KYDSMARKGD IGPRGPPGIP GREGPKGSKG
751  ERGYPGIPGE KGDEGLQGIP GIPGAPGPTG PPGLMGRTGH PGPTGAKGEK
801  GSDGPPGKPG PPGPPGIPFN ERNGMSSLYK IKGGVNVPSY PGPPGPPGPK
851  GDPGPVGEPG AMGLPGLEGF PGVKGDRGPA GPPGIAGMSG KPGAPGPPGV
901  PGEPGERGPV GDIGFPGPEG PSGKPGINGK DGIPGAQGIM GKPGDRGPKG
951  ERGDQGIPGD RGSQGERGKP GLTGMKGAIG PMGPPGNKGS MGSPGHQGPP
1001 GSPGIPGIPA DAVSFEEIKK YINQEVLRIF EERMAVFLSQ LKLPAAMLAA
1051 QAYGRPGPPG KDGLPGPPGD PGPQGYRGQK GERGEPGIGL PGSPGLPGTS
1101 ALGLPGSPGA PGPQGPPGPS GRCNPEDCLY PVSHAHQRTG GN
```

Structural and functional sites

Signal peptide: 1–23
NC6 domain: 24–291
COL5 domain: 292–435
NC5 domain: 436–455
COL4 domain: 456–679
NC4 domain: 680–710
COL3 domain: 711–818
NC3 domain: 819–841
COL2 domain: 842–1009
NC2 domain: 1010–1053
COL1 domain: 1054–1123
NC1 domain: 1124–1142
PARP repeat: 26–255
Potential N-linked glycosylation sites: none
Imperfections in Gly-X-Y: 352–354, 388–390, 486–488, 567, 586–587, 624–625, 723–728, 1087–1088, 1101–1102
Cysteines 34, 189, 232 and 242 in the NC6 domain are conserved in the NC4 domain of α1(IX), NC3 domains of α1(XII) and α1(XIV), and NC11 of α1(XVI)
Cysteines 1123 and 1128 at the COL1/NC1 junction are conserved at this junction in collagens IX, XII, XIV and XVI

Gene structure

The gene for the α1(XIX) chain is greater than 10 kb with an unusually long 3′-untranslated region (>5 kb). It is located on chromosome 6 at position 6q12–q14, the same region in which the genes for α1(IX) and α1(XII) are located[4]. Sequence comparison between cDNA and genomic clones indicates unusual alternative splicing events involving incomplete recognition at acceptor sites. However, it is not known whether the resulting truncated variants are functional[3].

References

[1] Myers, J.C. et al. (1993) Human cDNA clones transcribed from an unusually high-molecular weight RNA encode a new collagen chain. Gene 123: 211–217.

[2] Myers, J.C. et al. (1994) The triple-helical region of human type XIX collagen consists of multiple subdomains and exhibits limited sequence homology to α1(XVI). J. Biol. Chem. 269: 18549–18557.
[3] Inoguchi, K et al. (1995) The mRNA for α1(XIX) collagen chain, a new member of FACITs, contains a long unusual 3′ untranslated region and displays many unique splicing events. J. Biochem. 117: 137–146.
[4] Yoshioka, H. et al. (1992) Synteny between the loci for a novel FACIT-like collagen locus (D6S228E) and α1(IX) collagen (COL9A1) on 6q12–q14 in humans. Genomics 13: 884–886.

Decorin PG-S2, PGII, DCN

Decorin is a member of the family of small chondroitin sulphate/dermatan sulphate proteoglycans and characterized protein cores containing a number of leucine-rich repeats. Decorin is relatively abundant in bone, tendon, sclera and cornea. Rotary shadowing electron microscopy shows decorin containing one glycosaminoglycan (GAG) chain. Decorin binds type I collagen and has been observed associated with collagen fibrils. Furthermore, decorin can inhibit collagen fibrillogenesis of both type I and II collagens *in vitro* in a manner that is not affected by removal of the GAG chain. Decorin can bind transforming growth factor β (TGFβ) and neutralize its biological activity. The decorin core protein binds with relatively high affinity ($K_d = 3 \times 10^{-7}$ M) to type VI collagen. Decorin knockout leads to abnormal collagen fibril morphology and skin fragility [1–13].

Molecular structure

Most of the decorin protein consists of 11 repeats of a 23-residue leucine-rich sequence. These repeats are homologous to sequences in other proteins including biglycan, fibromodulin, the serum protein LRG, platelet surface protein GPIb, ribonuclease/angiotensin inhibitor, chaoptin, toll protein and adenylate cyclase. Decorin, biglycan and fibromodulin are similar in size and their cysteine residues are located in conserved positions. There is preliminary evidence that decorin is synthesized as a procore protein [1–11].

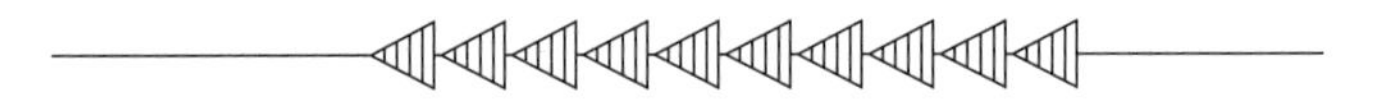

Isolation

Decorin can be isolated from a variety of tissues including articular cartilage by guanidine–HCl extraction, density gradient centrifugation, ion exchange chromatography and octyl-Sepharose chromatography [12].

Accession number

P07585

Primary structure

Ala	A	17	Cys	C	6	Asp	D	18	Glu	E	17
Phe	F	13	Gly	G	24	His	H	9	Ile	I	21
Lys	K	27	Leu	L	48	Met	M	5	Asn	N	27
Pro	P	23	Gln	Q	15	Arg	R	13	Ser	S	27
Thr	T	16	Val	V	23	Trp	W	2	Tyr	Y	8

Mol. wt (calc.) = 39 701 Residues = 359

```
1     MKATIILLLL  AQVSWAGPFQ  QRGLFDFMLE  DEASGIGPEV  PDDRDFEPSL
51    GPVCPFRCQC  HLRVVQCSDL  GLDKVPKDLP  PDTTLLDLQN  NKITEIKDGD
101   FKNLKNLHAL  ILVNNKISKV  SPGAFTPLVK  LERLYLSKNQ  LKELPEKMPK
151   TLQELRAHEN  EITKVRKVTF  NGLNQMIVIE  LGTNPLKSSG  IENGAFQGMK
201   KLSYIRIADT  NITSIPQGLP  PSLTELHLDG  NKISRVDAAS  LKGLNNLAKL
```

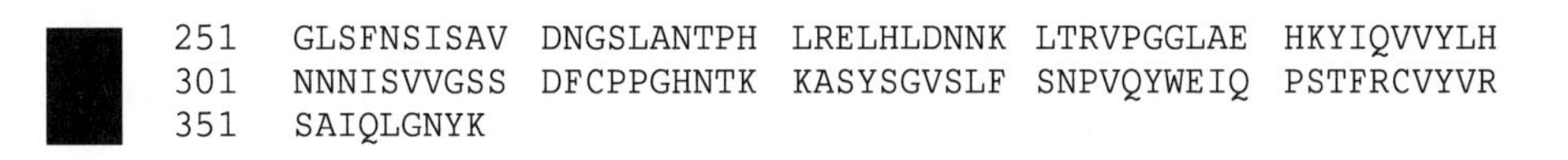

```
251   GLSFNSISAV  DNGSLANTPH  LRELHLDNNK  LTRVPGGLAE  HKYIQVVYLH
301   NNNISVVGSS  DFCPPGHNTK  KASYSGVSLF  SNPVQYWEIQ  PSTFRCVYVR
351   SAIQLGNYK
```

Structural and functional sites

Signal peptide: 1–16
Propeptide: 17–30
Leucine-rich repeats: 77–98, 99–122, 123–145, 146–167, 168–193, 194–217, 218–238, 239–262, 263–285, 286–308
Glycosaminoglycan attachment sites: 34
Potential N-linked glycosylation sites: 211, 262, 303

Gene structure

The human decorin gene is present in two copies on chromosome 12q21–q22 [11].

References

1. Scott, J.E. and Orford, C.R. (1981) Dermatan sulfate-rich proteoglycan associates with rat tail tendon collagen at the d-band in the gap region. Biochem. J. 197: 213–216.
2. Vogel, K.G. and Heinegård, D. (1985) Characterization of proteoglycans from adult bovine tendon. J. Biol. Chem. 260: 9298–9306.
3. Krusius, T. and Ruoslahti, E. (1986) Primary structure of an extracellular matrix proteoglycan core protein deduced from cloned cDNA. Proc. Natl Acad. Sci. USA 83: 7683–7687.
4. Scott, P.G. et al. (1986) A role for disulfide bridges in the protein core in the interaction of proteodermatan sulfate and collagen. Biochem. Biophys. Res. Comm. 138: 1348–1354.
5. Roughley, P.J. and White, R.J. (1989) Dermatan sulphate proteoglycans of human articular cartilage. The properties of dermatan sulphate proteoglycans I and II. Biochem. J. 262: 823–827.
6. Vogel, K.G. and Brown, D.C. (1990) Characteristics of the in vitro interaction of a small proteoglycan (PG-II) of bovine tendon with type I collagen. Matrix 9: 468–478.
7. Yamaguchi, Y. et al. (1990) Negative regulation of transforming growth factor β by the proteoglycan decorin. Nature 346: 281–284.
8. Sawhney, R.S. et al. (1991) Biosynthesis of small proteoglycan II (decorin) by chondrocytes and evidence for a procore protein. J. Biol. Chem. 266: 9231–9240.
9. Bidanset, D.J. et al. (1992) Binding of the proteoglycan decorin to collagen type VI. J. Biol. Chem. 267: 5250–5256.
10. Border, W.A. et al. (1992) Natural inhibition of transforming growth factor β protects against scarring in experimental kidney disease. Nature 360: 361–364.
11. Pulkkinen, L. et al. (1992) Expression of decorin in human tissues and cell lines and defined chromosomal assignment of the gene locus (DCN). Cytogenet. Cell Genet. 60: 107–111.

[12] Choi, H.U. et al. (1989) Characterization of the dermatan sulfate proteoglycans, DS-PGI and DS-PGII, from bovine articular cartilage and skin isolated by octyl-Sepharose chromatography. J. Biol. Chem. 264: 2876–2884.

[13] Danielson, K.G. et al. (1997) Targeted disruption of decorin leads to abnormal fibril morphology and skin fragility. J. Cell Biol. 136: 729–743.

Dentine matrix protein

dentin matrix phosphoprotein, DMP1, AG1

Dentine extracellular matrix contains a variety of non-collagenous proteins, including phosphophoryns, dentine sialoprotein and dentine matrix protein, which appear to be distinct from those found in bone. Dentine matrix protein is an acidic phosphoprotein initially identified from screening a rat cDNA library. Initially thought to be unique to dentine, it has subsequently been shown to be expressed in calvaria, but not other bone, and preameloblasts. It is considered to play a role in mineralization.

Molecular structure

Sequencing predicts a highly acidic, serine-rich protein of 513 amino acids. It contains 12.1% aspartic acid and 15.6% glutamic acid, and has a composition intermediate between the glutamic acid-rich phosphoproteins of bone and the aspartic acid proteins of dentine. Approximately half the glutamic acid residues appear as consecutive EE sequences, while about 50% of the serine residues can be readily phosphorylated[1–6].

Isolation

Dentine matrix protein has not been isolated.

Accession number

L11354

Primary structure (rat)

Ala	A	20	Cys	C	2	Asp	D	62	Glu	E	72
Phe	F	7	Gly	G	36	His	H	8	Ile	I	5
Lys	K	11	Leu	L	15	Met	M	5	Asn	N	17
Pro	P	19	Gln	Q	33	Arg	R	29	Ser	S	107
Thr	T	26	Val	V	8	Trp	W	2	Tyr	Y	5

Mol. wt (calc.) = 53 000 Residues = 489

```
1     MKTVILLTFL  WGLSCALPVA  RYQNTESESS  EGRTGNLAQS  PFPFMANSDH
51    TDSSESGEEL  GSDRSQYRFA  GGLSKSAGMD  ADKEEDEDDS  GDDTFGDEDN
101   GPGPEERQWG  GPSRLDSDED  SADTTQSSED  STSQENSAQD  TPSDSKDHHS
151   DEADSRPEAG  DSTQDSESEG  YRVGGGSEGE  SSHGDGSEFD  DEGMQSDDPG
201   STRSDRGHTR  MSSADISSGE  SKGDHEPTST  QDSDDSQDVE  FSSRKSFRRS
251   RVSEEDDRGE  LADSNSRGTQ  SVSTEDFRSK  EESDSETQGD  TAETQSQEDS
301   PEGQDPSSES  SEEAGEPSQE  SSSGSQEGVA  SESRGDNPDN  TSQTGDQRDS
351   ESSEEDRLNT  FSSSESQSTE  EQGDSESNES  LSLSEESQES  AQDEDSSSQE
401   GLQSQSASRE  SRSQESQSEG  RSRSEENRDS  DSQDSSRSKE  ESNSTGSTSS
451   SEEDNIFKNI  EADNRKLIVD  AYHNKPIGDE  DDNDCQDGY
```

Structural and functional sites

Signal peptide: 1–16
Potential N-linked glycosylation site: 340
Acidic patches: 84–89, 254–257
Putative RGD cell adhesion site: 334–336

Gene structure

DMP1 has been mapped to human chromosome 4q21 and contains six exons [6].

References

[1] Robey, P.G. (1996) Vertebrate mineralized matrix proteins: structure and function. Connect. Tiss. Res. 34: 269–276.
[2] Butler, W.T. et al. (1983) Multiple forms of rat dentin phosphoproteins. Arch. Biochem. Biophys. 225: 178–186.
[3] Butler, W.T. et al. (1997) Extracellular matrix proteins of dentine. CIBA Foundation Symp. 205: 107–117.
[4] George, A. et al. (1993) Characterisation of a novel dentin matrix acidic phosphoprotein. Implications for induction of biomineralization. J. Biol. Chem. 268: 12624–12630.
[5] George, A. et al. (1994) In situ localisation and chromosomal mapping of the AG1 (Dmp1) gene. J. Histochem. Cytochem. 42: 1527–1531.
[6] Hirst, K.L. et al. (1997) Elucidation of the sequence and the genomic organisation of the human dentin matrix acidic phosphoprotein 1 (DMP1) gene: Exclusion of the locus from a causative role in the pathogenesis of dentinogenesis imperfecta type II. Genomics 42: 38–45.

Dentine sialoprotein

DMP3

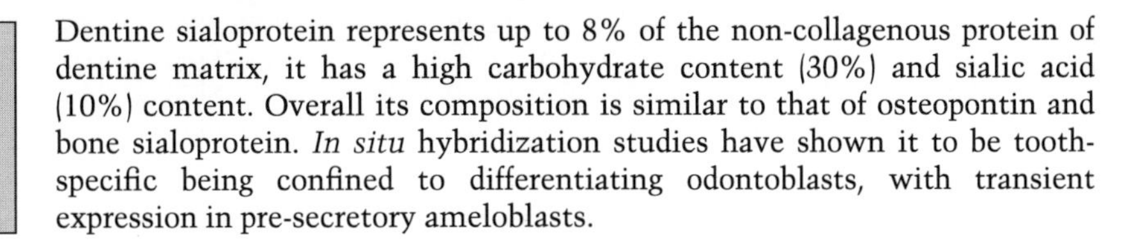

Dentine sialoprotein represents up to 8% of the non-collagenous protein of dentine matrix, it has a high carbohydrate content (30%) and sialic acid (10%) content. Overall its composition is similar to that of osteopontin and bone sialoprotein. *In situ* hybridization studies have shown it to be tooth-specific being confined to differentiating odontoblasts, with transient expression in pre-secretory ameloblasts.

Molecular structure

Rat dentine sialoprotein contains 366 amino acids, predominantly aspartic acid, serine, glycine and glutamic acid. Carbohydrate (29.6%) increases the molecular weight from 37 344 to 53 045. There are six potential N-linked glycosylation sites[1–4].

Isolation

Dentine was extracted in 0.5 M EDTA in 4 M guanidinium chloride containing protease inhibitors for 48 h, with the extracts concentrated and passed over Sephacryl S-200 column. The high molecular weight fraction was passed over DEAE-Sephacel in 6 M urea buffers and eluted with 0.2 M NaCl. This fraction was further purified on a Bio-Gel A1-5m column[2].

Accession number

U67916

Primary structure (mouse)

Ala	A	30	Cys	C	4	Asp	D	174	Glu	E	63
Phe	F	3	Gly	G	78	His	H	20	Ile	I	24
Lys	K	26	Leu	L	14	Met	M	3	Asn	N	48
Pro	P	21	Gln	Q	25	Arg	R	12	Ser	S	336
Thr	T	32	Val	V	20	Trp	W	2	Tyr	Y	5

Mol. wt (calc.) = 94 642 Residues = 940

```
1    MKMKIIIYIC IWATAWAIPV PQLVPLERDI VENSVAVPLL THPGTAAQNE
51   LSINSTTSNS NDSPDGSEIG EQVLSEDGYK RDGNGSESIH VGGKDFPTQP
101  ILVNEQGNTA EEHNDIETYG HDGVHARGEN STANGIRSQV GIVENAEEAE
151  SSVHGQAGQN TKSGGASDVS QNGDATLVQE NEPPEASIKN STNHEAGIHG
201  SGVATHETTP QREGLGSENQ GTEVTPSIGE DAGLDDTDGS PSGNGVEEDE
251  DTGSGDGEGA EAGDGRESHD GTKGQGGQSH GGNTDHRGQS SVSTEDDDSK
301  EQEGFPNGHN GDNSSEENGV EEGDSTQATQ DKEKLSPKDT RDAEGGIISQ
351  SEACPSGKSQ DQGIETEGPN KGNKSIITKE SGKLSGSKDS NGHQGVELDK
401  RNSPKQGESD KPQGTAEKSA AHSNLGHSRI GSSSNSDGHD SYEFDDESMQ
451  GDDPKSSDES NGSDESDTNS ESANESGSRG DASYTSDESS DDDNDSDSHA
501  GEDDSSDDSS GDGDSDSNGD GDSESEDKDE SDSSDHDNSS DSESKSDSSD
551  SSDDSSDSSD SSDSSDSSDS SDSSDSSDSS DSSDSNSSSD SSDSSGSSDS
601  SDSSDTCDSS DSSDSSDSSD SSDSSDSSDS SDSSDSSDSS DSSSSSDSSD
651  SSSCSDSSDS SDSSDSSDSS DSSDSSSSDS SSSSNSSDSS DSSDSSSSSD
701  SSDSSDSSDS SDSSGSSDSS DSSASSDSSS SSDSSDSSSS SDSSDSSDSS
```

```
751   DSSDSSESSD  SSNSSDSSDS  SDSSDSSDSS  DSSDSSDSSD  SSNSSDSSDS
801   SDSSDSSDSS  NSSDSSDSSD  SSDSSDSSDS  SDSSDSSDSS  DSSDSSDSSD
851   SSDSSDSSDS  SDSSDSSDSS  DSSDSSDSSN  SSDSSDSDSK  DSSSDSSDGD
901   SKSGNGNSDS  NSDSNSDSDS  DSEGSDSNHS  TSDDEIRENP
```

Structural and functional sites

Signal peptide: 1–17
N-linked glycosylation: 54, 61, 84, 130, 190, 313, 373
The dentine sialoprotein is represented by residues 18–387; phosphophoryn by residues 452–940

Gene structure

Evidence has been presented which localizes the gene to human chromosome 4, and that this gene codes for both dentine sialoprotein and phosphophoryn. Northern blot analysis of RNA from mouse odontoblast cell lines showed two transcripts 4.4 and 2.2 kb, which suggests alternative splicing of the gene[5,6].

References

1 Butler, W.T. (1995) Dentin matrix proteins and dentinogenesis. Connect. Tissue Res. 32: 381–387.
2 Butler, W.T. et al. (1992) Isolation, characterisation and immunolocalisation of a 53 kDa dentin sialoprotein (DSP). Matrix 12: 343–351.
3 Ritchie, H.H. et al. (1994) Cloning and sequence determination of rat dentin sialoprotein, a novel dentin protein. J. Biol. Chem. 269: 3698–3702.
4 Ritchie, H.H. et al. (1996) Partial cDNA sequence of mouse dentin sialoprotein and detection of its specific expression by odontoblasts. Arch. Oral Biol. 41: 571–575.
5 MacDougall, M. et al. (1997) Dentin phosphoprotein and dentin sialoprotein are cleavage products expressed from a single transcript coded by a gene on human chromosome 4. J. Biol. Chem. 272: 835–842.
6 Ritchie, H. and Wang, L.H. (1997) A mammalian bicistronic transcript encoding two dentin-specific proteins. Biochem. Biophys. Res. Commun. 231: 425–428.

Elastin

Elastin is the major protein of the elastic fibres that form a randomly oriented, inter-connected network in many tissues. Elastin content may vary from 2% of dry weight in skin to over 70% in the nuchal ligament of grazing animals. A high content of hydrophobic amino acids makes elastin one of the most chemically and proteinase-resistant proteins in the body. Its principal function is to provide elasticity and resilience to tissues; however, elastin promotes cell adhesion and elastin peptides have been shown to be chemotactic.

Molecular structure

Tropoelastin, the soluble precursor of elastin, is a single polypeptide chain; however, alternative splicing produces a number of different isoforms. Splicing is regulated in a developmental and/or tissue-specific manner. Individual chains are secreted into the extracellular space and, in association with microfibrillar components, assemble to form elastic fibres. The elastin polypeptide is made up of a number of alternating hydrophobic and cross-link repeats. Deamination of specific lysines by lysyl oxidase allows the introduction of covalent cross-links to stabilize the elastic fibres. These cross-links (desmosine and isodesmosine) are specific to elastic fibres. Lysines in cross-link repeats usually occur in pairs, but in two instances three lysines are found near one another (residue numbers 375–382 and 558–567). Strong homology exists between the human, bovine and porcine sequences, but these differ considerably from the chicken. Several VGVAPG sequences and their homologues are reported to mediate cell adhesion via a 67-kDa binding protein[1–6].

Isolation

Early methods for preparing elastin from tissues relied on its chemical resistance to reflux in 0.1 N NaOH. The absence of methionine also allows CNBr digestion of the tissue under denaturing conditions to remove other proteins[7].

Accession number

A32707, P15502

Primary structure

Ala	A	167	Cys	C	2	Asp	D	3	Glu	E	5
Phe	F	16	Gly	G	223	His	H	2	Ile	I	17
Lys	K	35	Leu	L	52	Met	M	1	Asn	N	0
Pro	P	100	Gln	Q	10	Arg	R	12	Ser	S	16
Thr	T	12	Val	V	98	Trp	W	0	Tyr	Y	15

Mol. wt (calc.) = 68 419 Residues = 786

```
1     MAGLTAAAPR  PGVLLLLLSI  LHPSRPGGVP  GAIPGGVPGG  VFYPGAGLGA
51    LGGGALGPGG  KPLKPVPGGL  AGAGLGAGLG  AFPAVTFPGA  LVPGGVADAA
101   AAYKAAKAGA  GLGGVPGVGG  LGVSAGAVVP  QPGAGVKPGK  VPGVGLPGVY
151   PGGVLPGARF  PGVGVLPGVP  TGAGVKPKAP  GVGGAFAGIP  GVGPFGGPQP
201   GVPLGYPIKA  PKLPGGYGLP  YTTGKLPYGY  GPGGVAGAAG  KAGYPTGTGV
251   GPQAAAAAAA  KAAAKFGAGA  AGVLPGVGGA  GVPGVPGAIP  GIGGIAGVGT
301   PAAAAAAAAA  AKAAKYGAAA  GLVPGGPGFG  PGVVGVPGAG  VPGVGVPGAG
351   IPVVPGAGIP  GAAVPGVVSP  EAAAKAAAKA  AKYGARPGVG  VGGIPTYGVG
401   AGGFPGFGVG  VGGIPGVAGV  PSVGGVPGVG  GVPGVGISPE  AQAAAAAKAA
451   KYGAAGAGVL  GGLVPGPQAA  VPGVPGTGGV  PGVGTPAAAA  AKAAAKAAQF
501   GLVPGVGVAP  GVGVAPGVGV  APGVGLAPGV  GVAPGVGVAP  GVGVAPGIGP
551   GGVAAAAKSA  AKVAAKAQLR  AAAGLGAGIP  GLGVGVGVPG  LGVGAGVPGL
601   GVGAGVPGFG  AGADEGVRRS  LSPELREGDP  SSSQHLPSTP  SSPRVPGALA
651   AAKAAKYGAA  VPGVLGGLGA  LGGVGIPGGV  VGAGPAAAAA  AAKAAAKAAQ
701   FGLVGAAGLG  GLGVGGLGVP  GVGGLGGIPP  AAAAKAAKYG  AAGLGGVLGG
751   AGQFPLGGVA  ARPGFGLSPI  FPGGACLGKA  CGRKRK
```

Structural and functional sites

Signal peptide: 1–26
Hydrophobic repeats: 28–53, 66–77, 109–124, 143–158, 181–190, 215–228, 229–248, 267–296, 317–365, 384–438, 453–481, 501–548, 570–644, 658–681, 702–726, 740–757, 758–772
Cross-link repeats: 54–65, 78–108, 125–142, 159–180, 191–214, 249–266, 297–316, 366–383, 439–452, 482–500, 549–569, 645–657, 682–701, 727–739
Alternatively spliced repeats: 453–481, 482–500, 501–548, 570–644, 740–757, 758–772
Free cysteine residues: 781, 776
β spiral motif: 505–546

Gene structure

Human elastin is encoded by a single copy gene found on chromosome 7 (at locus q11.2). The gene is approximately 45 kb long and contains 34 exons separated by large introns. The hydrophobic and cross-link repeats of the protein are encoded by distinct exons. In most cases, alternative splicing either includes or deletes an exon in a cassette-like fashion. In two cases, splicing occurs within an exon (501–548 and 612–644). At the exon–intron border, codons are split in the same way; the 5′-border provides the second and third nucleotides while the first nucleotide is found at the 3′-border [1,3,4,6].

References

[1] Sandberg, L.B. and Davidson, J.M. (1984) Elastin and its gene. Peptide Protein Rev. 3: 169–193.

[2] Senior, R.M. et al. (1984) Val-Gly-Val-Ala-Pro-Gly, a repeating peptide in elastin, is chemotactic for fibroblasts and monocytes. J. Cell Biol. 99: 870–874.

[3] Yeh, H. et al. (1989) Structure of the bovine elastin gene and S1 nuclease analysis of the alternative splicing of elastin mRNA in the bovine nuchal ligament. Biochemistry 28: 2365–2370.

[4] Indik, Z. et al. (1990) Structure of the elastin gene and alternative splicing of elastin mRNA. In: Extracellular Matrix Genes, ed. Sandell, L.J. and Boyd C.D., Academic Press, New York, pp. 221–250.
[5] Pollock, J. et al. (1990) Chick tropoelastin isoforms. From the gene to the extracellular matrix. J. Biol. Chem. 265: 3697–3702.
[6] Fazio, M.J. et al. (1991) Human elastin gene: New evidence for localization to the long arm of chromosome 7. Am. J. Human Genet. 48: 696–703.
[7] Soskel, N.T. et al. (1987) Isolation and characterisation of insoluble and soluble elastins. Methods Enzymol. 144: 196–214.

Fibrillin-1

Fibrillin-1 (FBN1) is a glycoprotein which occurs as a major component of a subset of connective tissue microfibrils. These microfibrils have a beaded appearance and a cross-sectional diameter of 10–12 nm. In elastic tissues, these structures are thought to provide the scaffold on to which elastin is assembled to form elastic fibres, although their functional role in non-elastic tissues is unclear. Defects in fibrillin on human chromosome 15 have been shown to result in the Marfan syndrome, and a tandem duplication in FBN1 is associated with the mouse tight skin mutation.

Molecular structure

The fibrillin transcribed from chromosome 15 has a molecular weight of approximately 312 000 and contains 2871 amino acids. Its high cysteine content (14%) suggests that intra-chain disulphide bonds are important in stabilizing the molecule. Five structurally distinct regions have been delineated, the largest of which contains a series of cysteine-rich repeats. Internal sequence homology reveals two distinct motifs that make up the majority of the protein; there are 41 six-cysteine epidermal growth factor (EGF) repeats and 8 eight-cysteine repeats related to motifs found in the transforming growth factor β1 (TGFβ1)-binding protein. There is a second cysteine-rich region towards the NH_2-terminus which, in addition to two EGF repeats, contains a number of unusual cysteine-rich modules that may be unique to this molecule. Separating these regions is a polypeptide domain rich in proline (42%). In contrast, the COOH-terminal 184 amino acids, which are rich in lysine and arginine, have no homology to the rest of the protein, and the extreme NH_2-terminal 29 amino acids are highly basic [1–9]. The length of the molecule has been reported to be 148 nm, but it is unclear how this soluble precursor molecule assembles into microfibrils with a periodicity of 50 nm, although a number of models have been proposed [10]. Sedimentation experiments show that the molecule is extremely stable even in 8 M urea. Fibrillin-1 is processed at the carboxyl-terminus prior to assembly. A putative RGD cell adhesion site is located in one of the eight-cysteine repeats [1–7].

Isolation

Fibrillin has been purified from culture medium by precipitation with 40% ammonium sulphate, DEAE ion-exchange chromatography, and FPLC on Mono-Q and Superose 6 columns [7]. Fibrillin-containing microfibrils have been isolated by bacterial collagenase digestion and Sepharose CL-2B chromatography [8].

Accession number

S17064

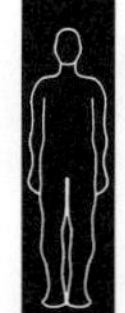

Primary structure

Ala	A	89	Cys	C	362	Asp	D	172	Glu	E	201
Phe	F	84	Gly	G	307	His	H	48	Ile	I	148
Lys	K	111	Leu	L	141	Met	M	52	Asn	N	188
Pro	P	175	Gln	Q	101	Arg	R	131	Ser	S	173
Thr	T	166	Val	V	108	Trp	W	13	Tyr	Y	94

Mol. wt (calc.) = 311 949 Residues = 2871

```
1     MRRGRLLEIA LGFTVLLASY TSHGADANLE AGNVKETRAS RAKRRGGGGH
51    DALKGPNVCG SRYNAYCCPG WKTLPGGNQC IVPICRHSCG DGFCSRPNMC
101   TCPSGQIAPS CGSRSIQHCN IRCMNGGSCS DDHCLCQKGY IGTHCGQPVC
151   ESGCLNGGRC VAPNRCACTY GFTGPQCERD YRTGPCFTVI SNQMCQGQLS
201   GIVCTKQLCC ATVGRAWGHP CEMCPAQPHP CRRGFIPNIR TGACQDVDEC
251   QAIPGLCQGG NCINTVGSFE CKCPAGHKLN EVSQKCEDID ECSTIPGICE
301   GGECTNTVSS YFCKCPPGFY TSPDGTRCID VRPGYCYTAL TNGRCSNQLP
351   QSITKMQCCC DAGRCWSPGV TVAPEMCPIR ATEDFNKLCS VPMVIPGRPE
401   YPPPPLGPIP PVLPVPPGFP PGPQIPVPRP PVEYLYPSRE PPRVLPVNVT
451   DYCQLVRYLC QNGRCIPTPG SYRCECNKGF QLDLRGECID VDECEKNPCA
501   GGECINNQGS YTCQCRAGYQ STLTRTECRD IDECLQNGRI CNNGRCINTD
551   GSFHCVCNAG FHVTRDGKNC EDMDECSIRN MCLNGMCINE DGSFKCICKP
601   GFQLASDGRY CKDINECETP GICMNGRCVN TDGSYRCECF PGLAVGLDGR
651   VCVDTHMRST CYGGYKRGQC IKPLFGAVTK SECCCASTEY AFGEPCQPCP
701   AQNSAEYQAL CSSGPGMTSA GSDINECALD PDICPNGICE NLRGTYKCIC
751   NSGYEVDSTG KNCVDINECV LNSLLCDNGQ CRNTPGSFVC TCPKGFIYKP
801   DLKTCEDIDE CESSPCINGV CKNSPGSFIC ECSSESTLDP TKTICIETIK
851   GTCWQTVICG RCEININGAT LKSQCCSSLG AAWGSPCTLC QVDPICGKGY
901   SRIKGTQCED IDECEVFPGV CKNGLCVNTR GSFKCQCPSG MTLDATGRIC
951   LDIRLETCFL RYEDEECTLP IAGRHRMDAC CCSVGAAWGT EECEECPMRN
1001  TPEYEELCPR GPGFATKEIT NGKPFFKDIN ECKMIPSLCT HGKCRNTIGS
1051  FKCRCDSGFA LDSEERNCTD IDECRISPDL CGRGQCVNTP GDFECKCDEG
1101  YESGFMMMKN CMDIDECQRD PLLCRGGVCH NTEGSYRCEC PPGHQLSPNI
1151  SACIDINECE LSAHLCPNGR CVNLIGKYQC ACNPGYHSTP DRLFCVDIDE
1201  CSIMNGGCET FCTNSEGSYE CSCQPGFALM PDQRSCTDID ECEDNPNICD
1251  GGQCTNIPGE YRCLCYDGFM ASEDMKTCVD VNECDLNPNI CLSGTCENTK
1301  GSFICHCDMG YSGKKGKTGC TDINECEIGA HNCGKHAVCT NTAGSFKCSC
1351  SPGWIGDGIK CTDLDECSNG THMCSQHADC KNTMGSYRCL CKEGYTGDGF
1401  TCTDLDECSE NLNLCGNGQC LNAPGGYRCE CDMGFVPSAD GKACEDIDEC
1451  SLPNICVFGT CHNLPGLFRC ECEIGYELDR SGGNCTDVNE CLDPTTCISG
1501  NCVNTPGSYI CDCPPDFELN PTRVGCVDTR SGNCYLDIRP RGDNGDTACS
1551  NEIGVGVSKA SCCCSLGKAW GTPCEMCPAV NTSEYKILCP GGEGFRPNPI
1601  TVILEDIDEC QELPGLCQGG KCINTFGSFQ CRCPTGYYLN EDTRVCDDVN
1651  ECETPGICGP GTCYNTVGNY TCICPPDYMQ VNGGNNCMDM RRSLCYRNYY
1701  ADNQTCDGEL LFNMTKKMCC CSYNIGRAWN KPCEQCPIPS TDEFATLCGS
1751  QRPGFVIDIY TGLPVDIDEC REIPGVCENG VCINMVGSFR CECPVGFFYN
1801  DKLLVCEDID ECQNGPVCQR NAECINTAGS YRCDCKPGYR FTSTGQCNDR
1851  NECQEIPNIC SHGQCIDTVG SFYCLCHTGF KTNDDQTMCL DINECERDAC
1901  GNGTCRNTIG SFNCRCNHGF ILSHNNDCID VDECASGNGN LCRNGQCINT
1951  VGSFQCQCNE GYEVAPDGRT CVDINECLLE PRKCAPGTCQ NLDGSYRCIC
2001  PPGYSLQNEK CEDIDECVEE PEICALGTCS NTEGSFKCLC PEGFSLSSSG
2051  RRCQDLRMSY CYAKFEGGKC SSPKSRNHSK QECCCALKGE GWGDPCELCP
2101  TEPDEAFRQI CPYGSGIIVG PDDSAVDMDE CKEPDVCKHG QCINTDGSYR
2151  CECPFGYTLA GNECVDTDEC SVGNPCGNGT CKNVIGGFEC TCEEGFEPGP
```

```
2201 MMTCEDINEC AQNPLLCAFR CVNTYGSYEC KCPVGYVLRE DRRMCKDEDE
2251 CEEGKHDCTE KQMECKNLIG TYMCICGPGY QRRPDGEGCV DENECQTKPG
2301 ICENGRCLNT RGSYTCECND GFTASPNQDE CLDNREGYCF TEVLQNMCQI
2351 GSSNRNPVTK SECCCDGGRG WGPHCEICPF QGTVAFKKLC PHGRGFMTNG
2401 ADIDECKVIH DVCRNGECVN DRGSYHCICK TGYTPDITGT SCVDLNECNQ
2451 APKPCNFICK NTEGSYQCSC PKGYILQEDG RSCKDLDECA TKQHNCQFLC
2501 VNTIGGFTCK CPPGFTQHHT SCIDNNECTS DINLCGSKGI CQNTPGSFTC
2551 ECQRGFSLDQ TGSSCEDVDE CEGNHRCQHG CQNIIGGYRC SCPQGYLQHY
2601 QWNQCVDENE CLSAHICGGA SCHNTLGSYK CMCPAGFQYE QFSGGCQDIN
2651 ECGSAQAPCS YGCSNTEGGY LCGCPPGYFR IGQGHCVSGM GMGRGNPEPP
2701 VSGEMDDNSL SPEACYECKI NGYPKRGRKR RSTNETDASN IEDQSETEAN
2751 VSLASWDVEK TAIFAFNISH VSNKVRILEL LPALTTLTNH NRYLIESGNE
2801 DGFFKINQKE GISYLHFTKK KPVAGTYSLQ ISSTPLYKKK ELNQLEDKYD
2851 KDYLSGELGD NLKMKIQVLL H
```

Structural and functional sites

Signal peptide: 1–27

EGF (6C) repeats: 245–287, 288–329, 489–528, 529–571, 572–612, 613–653, 723–764, 765–806, 807–846, 910–951, 1028–1069, 1070–1112, 1113–1154, 1155–1196, 1197–1237, 1238–1279, 1280–1321, 1322–1362, 1363–1403, 1404–1445, 1446–1486, 1487–1527, 1606–1647, 1648–1688, 1766–1807, 1808–1848, 1849–1890, 1891–1929, 1930–1972, 1973–2012, 2013–2054, 2127–2165, 2166–2205, 2206–2246, 2247–2290, 2291–2362, 2402–2443, 2444–2484, 2485–2523, 2524–2566, 2567–2607

TGFβ1 receptor repeats: 654–722, 952–1027, 1528–1606, 1689–1765, 2055–2126, 2363–2401

Potential N-linked glycosylation sites: 448, 1067, 1149, 1369, 1484, 1581, 1669, 1703, 1902, 2077, 2178, 2734, 2750, 2767

RGD cell adhesion site: 1541–1543

Gene structure

Gene for fibrillin-1 is located on human chromosomes 15q21 (FBN1). FBN1 gene is relatively large (approx. 110 kb) and contains 65 exons. The data predict a full-length pre-fibrillin mRNA of 9663 nucleotides, with untranslated sequences 134 (5′) and 916 (3′) and an open reading frame of 8613 nucleotides[2–6].

References

[1] Sakai, L.Y. et al. (1986) Fibrillin, a new 350-kD glycoprotein, is a component of extracellular microfibrils. J. Cell Biol. 103: 2499–2509.

[2] Kainulainen, K. et al. (1994) Mutations in the fibrillin gene responsible for dominant ectopia lentis and neonatal Marfan syndrome. Nature Genet. 6: 64–69.

[3] Dietz, H.C. and Pyeritz, R.E. (1995) Mutations in the human gene for fibrillin-1 (FBN1) in the Marfan syndrome and related disorders. Hum. Mol. Genet. 4: 1799–1809.

[4] Lee, B. et al. (1991) Linkage of Marfan syndrome and a phenotypically related disorder to two different fibrillin genes. Nature 352: 330–334.

[5] Maslen, C.L. et al. (1991) Partial sequence of a candidate gene for the Marfan syndrome. Nature 352: 334–337.
[6] Pereira, L. et al. (1993) Genomic organization of the sequence coding for fibrillin, the defective gene product in Marfan syndrome. Hum. Mol. Genet. 2: 961–967.
[7] Sakai, L.Y. et al. (1991) Purification and partial characterisation of fibrillin, a cysteine-rich structural component of connective tissue microfibrils. J. Biol. Chem. 266: 14763–14770.
[8] Kielty, C.M. et al. (1991) Isolation and ultrastructural analysis of microfibrillar structures from foetal bovine elastic tissues. J. Cell Sci. 99: 797–807.
[9] Siracusa, L.D. et al. (1996) A tandem duplication in the fibrillin 1 gene is associated with the mouse tight skin mutation. Genome Res. 6: 300–313.
[10] Reinhardt, D.P. et al. (1996) Fibrillin-1: Organisation in microfibrils and structural properties. J. Mol. Biol. 258: 104–116.

Fibrillin-2

Fibrillin-2 (FBN2) is highly homologous to fibrillin-1 and both fibrillin proteins localize to elastin-associated microfibrils. Immunolocalization studies show codistribution of fibrillin-1 and -2 in both elastic and non-elastic tissues, but fibrillin-2 appeared to be preferentially accumulated in elastic-rich tissues. Defects in FBN2 on human chromosome 5 have been shown to be linked to congenital contractural arachnodactyly.

Molecular structure

Determination of the primary structure of FBN2 showed that the 2889 amino acid polypeptide could be divided into five regions and was highly homologous to fibrillin-1, the only significant sequence divergence is the replacement of the short proline-rich region in fibrillin-1 with a glycine-rich region. Fibrillin-2 contains two RGD sequences and contains 11 putative N-linked glycosylation sites. Its high cysteine content (14%) suggests that intra-chain disulphide bonds are important in stabilizing the molecule [1–8].

Isolation

Fibrillin-2 has not been isolated although it may have been present when fibrillin-1 monomers were isolated [6] and may be a component of fibrillin-containing microfibrils isolated by bacterial collagenase digestion and Sepharose CL-2B chromatography [5].

Accession number

P35556

Primary structure

Ala	A	89	Cys	C	362	Asp	D	172	Glu	E	201
Phe	F	84	Gly	G	307	His	H	48	Ile	I	148
Lys	K	111	Leu	L	141	Met	M	52	Asn	N	188
Pro	P	175	Gln	Q	101	Arg	R	131	Ser	S	173
Thr	T	166	Val	V	108	Trp	W	13	Tyr	Y	94

Mol. wt (calc.) = 314 346   Residues = 2911

```
1    MGRRRRLCLQ LYFLWLGCVV LWAQGTAGQP QPPPPKPPRP QPPPQQVRSA
51   TAGSEGGFLA PEYREEGAAV ASRVRRRGQQ DVLRGPNVCG SRFHSYCCPG
101  WKTLPGGNQC IVPICRNSCG DGFCSRPNMC TCSSGQISST CGSKSIQQCS
151  VRCMNGGTCA DDHCQCQKGY IGTYCGQPVC ENGCQNGGRC IAQPCACVYG
201  FTGPQCERDY RTGPCFTQVN NQMCQGQLTG IVCTKTLCCA TTGRAWGHPC
251  EMCPAQPQPC RRGFIPNIRT GACQDVDECQ AIPGICQGGN CINTVGSFEC
301  RCPAGHKQSE TTQKCEDIDE CSIIPGICET GECSNTVGSY FCVCPRGYVT
351  STDGSRCIDQ RTGMCFSGLV NGRCAQELPG RMTKMQCCCE PGRCWGIGTI
401  PEACPVRGSE EYRRLCMDGL PMGGIPGSAG SRPGGTGGNG FAPSGNGNGY
451  GPGGTGFIPI PGGNGFSPGV GGAGVGAGGQ GPIITGLTIL NQTIDICKHH
```

```
501   ANLCLNGRCI  PTVSSYRCEC  NMGYKQDANG  DCIDVDECTS  NPCTNGDCVN
551   TPGSYYCKCH  AGFQRTPTKQ  ACIDIDECIQ  NGVLCKNGRC  VNSDGSFQCI
601   CNAGFELTTD  GKNCVDHDEC  TTTNMCLNGM  CINEDGSFKC  ICKPGFVLAP
651   NGRYCTDVDE  CQTPGICMNG  HCINSEGSFR  CDCPPGLAVG  MDGRVCVDTH
701   MRSTCYGGIK  KGVCVRPFPG  AVTKSECCCA  NPDYGFGEPC  QPCPAKNSAE
751   FHGLCSSGVG  ITVDGRDINE  CALDPDICAN  GICENLRGSY  RCNCNSGYEP
801   DASGRNCIDI  DECLVNRLLC  DNGLCRNTPG  SYSCTCPPGY  VFRTETETCE
851   DINECESNPC  VNGACRNNLG  SFNCECSPGS  KLSSTGLICI  DSLKGTCWLN
901   IQDSRCEVNI  NGATLKSECC  ATLGAAWGSP  CERCELDTAC  PRGLARIKGV
951   TCEDVNECEV  FPGVCPNGRC  VNSKGSFHCE  CPEGLTLDGT  GRVCLDIRME
1001  QCYLKWDEDE  CIHPVPGKFR  MDACCCAVGA  AWGTECEECP  KPGTKEYETL
1051  CPRGAGFANR  GDVLTGRPFY  KDINECKAFP  GMCTYGKCRN  TIGSFKCRCN
1101  SGFALDMEER  NCTDIDECRI  SPDLCGSGIC  VNTPGSFECE  CFEGYESGFM
1151  MMKNCMDIDG  CERNPLLCRG  GTCVNTEGSF  QCDCPLGHEL  SPSREDCVDI
1201  NECSLSDNLC  RNGKCVNMIG  TYQCSCNPGY  QATPDRQGCT  DIDECMIMNG
1251  GCDTQCTNSE  GSYECSCSEG  YALMPDGRSC  ADIDECENNP  DICDGGQCTN
1301  IPGEYRCLCY  DGFMASMDMK  TCIDVNECDL  NSNICMFGEC  ENTKGSFICH
1351  CQLGYSVKKG  TTGCTDVDEC  EIGAHNCDMH  ASCLNIPGSF  KCSCREGWIG
1401  NGIKCIDLDE  CSNGTHQCSI  NAQCVNTPGS  YRCACSEGFT  GDGFTCSDVD
1451  ECAENINLCE  NGQCLNVPGA  YRCECEMGFT  PASDSRSCQD  IDECSFQNIC
1501  VSGTCNNLPG  MFHCICDDGY  ELDRTGGNCT  DIDECADPIN  CVNGLCVNTP
1551  GRYECNCPPD  FQLNPTGVGC  VDNRVGNCYL  KFGPRGDGSL  SCNTEIGVGV
1601  SRSSCCCSLG  KAWGNPCETC  PPVNSTEYYT  LCPGGEGFRP  NPITIILEDI
1651  DECQELPGLC  QGGNCINTFG  SFQCECPQGY  YLSEDTRICE  DIDECFAHPG
1701  VCGPGTCYNT  LGNYTCICPP  EYMQVNGGHN  CMDMRKSFCY  RSYNGTTCEN
1751  ELPFNVTKRM  CCCTYNVGKA  GNKPCEPCPT  PGTADFKTIC  GNIPGFTFDI
1801  HTGKAVDIDE  CKEIPGICAN  GVCINQIGSF  RCECPTGFSY  NDLLLVCEDI
1851  DECSNGDNLC  QRNADCINSP  GSYRCECAAG  FKLSPNGACV  DRNECLEIPN
1901  VCSHGLCVDL  QGSYQCICHN  GFKASQDQTM  CMDVDECERH  PCGNGTCKNT
1951  VGSYNCLCYP  GFELTHNNDC  LDIDECSSFF  GQVCRNGRCF  NEIGSFKCLC
2001  NEGYELTPDG  KNCIDTNECV  ALPGSCSPGT  CQNLEGSFRC  ICPPGYEVKS
2051  ENCIDINECD  EDPNICLFGS  CTNTPGGFQC  LCPPGFVLSD  NGRRCFDTRQ
2101  SFCFTNFENG  KCSVPKAFNT  TKAKCCCSKM  PGEGWGDPCE  LCPKDDEVAF
2151  QDLCPYGHGT  VPSLHDTRED  VNECLESPGI  CSNGQCINTD  GSFRCECPMG
2201  YNLDYTGVRC  VDTDECSIGN  PCGNGTCTNV  IGSFECNCNE  GFEPGPMMNC
2251  EDINECAQNP  LLCALRCMNT  FGSYECTCPI  GYALREDQKM  CKDLDECAEG
2301  LHDCESRGMM  CKNLIGTFMC  ICPPGMARRP  DGEGCVDENE  CRTKPGICEN
2351  GRCVNIIGSY  RCECNEGFQS  SSSGTECLDN  RQGLCFAEVL  QTICQMASSS
2401  RNLVTKSECC  CDGGRGWGHQ  CELCPLPGTA  QYKKICPHGP  GYTTDGRDID
2451  ECKVMPNLCT  NGQCINTMGS  FRCFCKVGYT  TDISGTSCID  LDECSQSPKP
2501  CNYICKNTEG  SYQCSCPRGY  VLQEDGKTCK  DLDECQTKQH  NCQFLCVNTL
2551  GGFTCKCPPG  FTQHHTACID  NNECGSQPLL  CGGKGICQNT  PGSFSCECQR
2601  GFSLDATGLN  CEDVDECDGN  HRCQHGCQNI  LGGYRCGCPQ  GYIQHYQWNQ
2651  CVDENECSNP  NACGSASCYN  TLGSYKCACP  SGFSFDQFSS  ACHDVNECSS
2701  SKNPCNYGCS  NTEGGYLCGC  PPGYYRVGQG  HCVSGMGFNK  GQYLSLDTEV
2751  DEENALSPEA  CYECKINGYP  KKDSRQKRSI  HEPDPTAVEQ  ISLESVDMDS
2801  PVNMKFNLSH  LGSKEHILEL  RPAIQPLNNH  IRYVISQGND  DSVFRIHQRN
2851  GLSYLHTAKK  KLMPGTYTLE  ITSIPLYKKK  ELKKLEESNE  DDYLLGELGE
2900  ALRMRLQIQL  Y
```

Structural and functional sites

Signal peptide: 1–28

EGF (6C) repeats: 111–142, 145–176, 177–207, 225–316, 317–358, 493–533, 534–573, 574–615, 616–656, 657–697, 767–808, 809–850, 851–890, 954–995, 1072–1113, 1114–1156, 1157–1198, 1199–1240, 1241–1281, 1282–1323, 1324–1365, 1366–1406, 1407–1447, 1448–1489, 1490–1530, 1531–1571, 1649–1690, 1691–1732, 1807–1848, 1849–1890, 1891–1932, 1933–1971, 1972–2014, 2015–2054, 2055–2096, 2170–2211, 2212–2251, 2252–2292, 2293–2336, 2337–2378, 2448–2489, 2490–2530, 2531–2569, 2570–2612, 2613–2652, 2653–2693, 2694–2733
TGFβ1 receptor repeats: 359–425, 698–766, 996–1071, 1572–1648, 1733–1806, 2097–2169, 2379–2447
Potential N-linked glycosylation sites: 491, 1111, 1413, 1528, 1624, 1713, 1744, 1755, 1944, 2119, 2224, 2807
RGD cell adhesion site: 1600–1602, 1985–1987

Gene structure

Genes for fibrillin-2 are located on human chromosomes 5q23–q31 (FBN2)[3,4,8].

References

[1] Lee, B. et al. (1991) Linkage of Marfan syndrome and a phenotypically related disorder to two different fibrillin genes. Nature 352: 330–334.

[2] Tsipouras, P. et al. (1992) Genetic linkage of the Marfan syndrome, ectopia lentis, and congenital arachnodactyly to the fibrillin genes on chromosomes 15 and 5. N. Eng. J. Med. 326: 905–909.

[3] Zhang, H. et al. (1994) Structure and expression of fibrillin-2, a novel microfibrillar component located in elastic matrices. J. Cell Biol. 124: 855–863.

[4] Zhang, H. et al. (1995) Developmental expression of fibrillin genes suggests heterogeneity of extracellular microfibrils. J. Cell Biol. 129: 1165–1176.

[5] Kielty, C.M. et al. (1991) Isolation and ultrastructural analysis of microfibrillar structures from foetal bovine elastic tissues. J. Cell Sci. 99: 797–807.

[6] Sakai, L.Y. et al. (1991) Purification and partial characterization of fibrillin, a cysteine-rich structural component of connective tissue microfibrils. J. Biol. Chem. 266: 14763–14770.

[7] Mariencheck, M.C. et al. (1995) Fibrillin-1 and fibrillin-2 show temporal and tissue-specific regulation of expression in developing elastic tissues. Conn. Tiss. Res. 31: 87–97.

[8] Putnam, E.A. et al. (1995) Fibrillin-2 (FBN2) mutations result in the Marfan-like disorder, congenital contractural arachnodactyly. Nature Genet. 11: 456–458.

Fibrinogen

Fibrinogen is a soluble plasma protein that is cleaved by thrombin to produce an insoluble fibrin clot. Polymerization to form fibrin acts also as a cofactor in platelet aggregation. Thrombin action cleaves fibrinopeptides A and B from the α and β chains, respectively, and exposes the polymerization site. The initial clot formed by polymerization is converted into a reinforced clot by factor XIIIa transglutaminase which catalyses the ε-(γ-glutamyl) lysine cross-linking between the γ chains and α chains of different monomers.

Molecular structure

Fibrinogen is a hexamer that contains two sets of non-identical chains (α, β and γ) linked to each other in an anti-parallel arrangement by disulphide bonds. All NH_2-termini of the chains are contained in a central nodule, two three-chain coiled-coils extend from this central nodule to two terminal domains and are gathered together at two nodes by disulphide rings. A further set of disulphide rings terminates the coiled-coil domain. The two terminal domains are composed of the COOH two-thirds of the β and γ chains and are highly folded. A short section of the α chain is associated with these terminal domains before emerging into a highly flexible appendage that is readily cleaved by proteases. The α chain protuberance and the coiled-coils together account for half the mass of the fibrinogen molecule. Altogether, there are four carbohydrate clusters, one each on the two β and two γ chains. The α chain can be cross-linked in several places to fibronectin. Although the molecule contains two RGD sequences in its α chain, a pentapeptide QAGDV at the extreme COOH-terminus of the γ chain represents the main binding site for the platelet integrin receptor IIbIIIa (αIIbβ3). Interestingly, the γ chain can be alternatively spliced to produce a variant (γB) which lacks the COOH-terminal four residues of γA but adds 20 different residues unrelated to the integrin-binding motif. The γB chain is present in about 10% of fibrinogen molecules in plasma, but is absent from platelet fibrinogen. Variations in sequence at the fibrinopeptide A cleavage site (Arg-35) lead to α-dysfibrinogenemias [1–6].

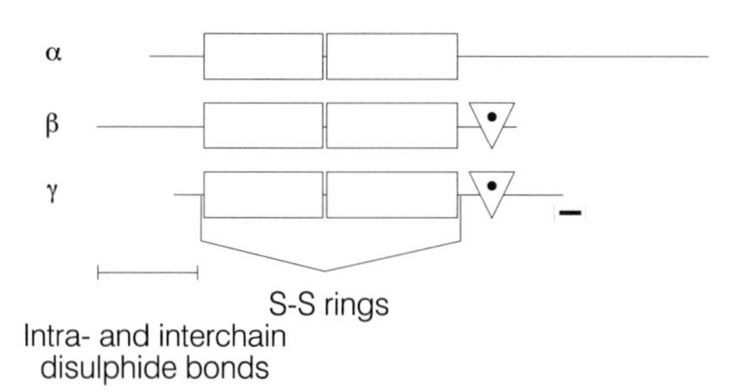

Isolation

Fibrinogen can be purified by a modified cold ethanol fractionation [7], and the individual chains separated by carboxymethylation in 6 M guanidine–HCl and CM-cellulose chromatography in 8 M urea [8].

Accession number

P02671

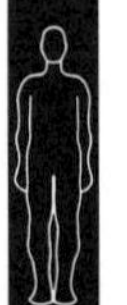

Primary structure: α chain

Ala	A	23	Cys	C	9	Asp	D	35	Glu	E	44
Phe	F	20	Gly	G	72	His	H	16	Ile	I	19
Lys	K	40	Leu	L	33	Met	M	12	Asn	N	29
Pro	P	38	Gln	Q	18	Arg	R	42	Ser	S	91
Thr	T	51	Val	V	32	Trp	W	11	Tyr	Y	9

Mol. wt (calc.) = 69 680 Residues = 644

```
1     MFSMRIVCLV  LSVVGTAWTA  DSGEGDFLAE  GGGVRGPRVV  ERHQSACKDS
51    DWPFCSDEDW  NYKCPSGCRM  KGLIDEVNQD  FTNRINKLKN  SLFEYQKNNK
101   DSHSLTTNIM  EILRGDFSSA  NNRDNTYNRV  SEDLRSRIEV  LKRKVIEKVQ
151   HIQLLQKNVR  AQLVDMKRLE  VDIDIKIRSC  RGSCSRALAR  EVDLKDYEDQ
201   QKQLEQVIAK  DLLPSRDRQH  LPLIKMKPVP  DLVPGNFKSQ  LQKVPPEWKA
251   LTDMPQMRME  LERPGGNEIT  RGGSTSYGTG  SETESPRNPS  SAGSWNSGSS
301   GPGSTGNRNP  GSSGTGGTAT  WKPGSSGPGS  TGSWNSGSSG  TGSTGNQNPG
351   SPRPGSTGTW  NPGSSERGSA  GHWTSESSVS  GSTGQWHSES  GSFRPDSPGS
401   GNARPNNPDW  GTFEEVSGNV  SPGTRREYHT  EKLVTSKGDK  ELRTGKEKVT
451   SGSTTTTRRS  CSKTVTKTVI  GPDGHKEVTK  EVVTSEDGSD  CPEAMDLGTL
501   SGIGTLDGFR  HRHPDEAAFF  DTASTGKTFP  GFFSPMLGEF  VSETESRGSE
551   SGIFTNTKES  SSHHPGIAEF  PSRGKSSSYS  KQFTSSTSYN  RGDSTFESKS
601   YKMADEAGSE  ADHEGTHSTK  RGHAKSRPVR  GIHTSPLGKP  SLSP
```

Structural and functional sites

Signal peptide: 1–19
Fibrinopeptide: 20–35
Phosphorylation site: 22
Thrombin cleavage site: 35–36
Polymerization site: 36–38
RGD cell adhesion sites: 114–116, 591–593
α2-Plasmin inhibitor binding site: 322
Acceptor cross-linking sites: 347, 385

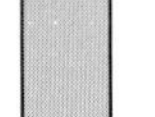

Accession number

P02675

Primary structure: β chain

Ala	A	24	Cys	C	12	Asp	D	28	Glu	E	30
Phe	F	12	Gly	G	42	His	H	9	Ile	I	16
Lys	K	36	Leu	L	36	Met	M	18	Asn	N	32
Pro	P	22	Gln	Q	26	Arg	R	28	Ser	S	34
Thr	T	23	Val	V	28	Trp	W	14	Tyr	Y	21

Mol. wt (calc.) = 55 839 Residues = 491

```
1     MKRMVSWSFH KLKTMKHLLL LLLCVFLVKS QGVNDNEEGF FSARGHRPLD
51    KKREEAPSLR PAPPPISGGG YRARPAKAAA TQKKVERKAP DAGGCLHADP
101   DLGVLCPTGC QLQEALLQQE RPIRNSVDEL NNNVEAVSQT SSSSFQYMYL
151   LKDLWQKRQK QVKDNENVVN EYSSELEKHQ LYIDETVNSN IATNLRVLRS
201   ILENLRSKIQ KLESDVSAQM EYCRTPCTVS CNIPVVSGKE CEEIIRKGGE
251   TSEMYLIQPD SSVKPYRVYC DMNTENGGWT VIQNRQDGSV DFGRKWDPYK
301   QGFGNVATNT DGKNYCGLPG EYWLGNDKIS QLTRMGPTEL LIEMEDWKGD
351   KVKAHYGGFT VQNEANKYQI SVNKYRGTAG NALMDGASQL MGENRTMTIH
401   NGMFFSTYDR DNDGWLTSDP RKQCSKEDGG GWWYNRCHAA NPNGRYYWGG
451   QYTWDMAKHG TDDGVVWMNW KGSWYSMRKM SMKIRPFFPQ Q
```

Structural and functional sites

Signal peptide: 1–30
Fibrinopeptide: 31–44
Pyrrolidone carboxylic acid: 31
Potential N-linked glycosylation site: 394
Thrombin cleavage site: 44–45
Possible polymerization site: 45–47

Accession number

P02679, P04469, P04470

Primary structure: γ chain

γA variant:

Ala	A	28	Cys	C	11	Asp	D	32	Glu	E	22
Phe	F	20	Gly	G	35	His	H	11	Ile	I	27
Lys	K	33	Leu	L	32	Met	M	9	Asn	N	24
Pro	P	12	Gln	Q	24	Arg	R	11	Ser	S	29
Thr	T	29	Val	V	15	Trp	W	11	Tyr	Y	22

Mol. wt (calc.) = 49 426 Residues = 437

γB variant:

Ala	A	28	Cys	C	11	Asp	D	34	Glu	E	26
Phe	F	20	Gly	G	34	His	H	12	Ile	I	27
Lys	K	33	Leu	L	34	Met	M	9	Asn	N	24
Pro	P	15	Gln	Q	24	Arg	R	12	Ser	S	30
Thr	T	30	Val	V	15	Trp	W	11	Tyr	Y	24

Mol. wt (calc.) = 51 439 Residues = 453

```
1     MSWSLHPRNL ILYFYALLFL SSTCVAYVAT RDNCCILDER FGSYCPTTCG
51    IADFLSTYQT KVDKDLQSLE DILHQVENKT SEVKQLIKAI QLTYNPDESS
101   KPNMIDAATL KSRIMLEEIM KYEASILTHD SSIRYLQEIY NSNNQKIVNL
151   KEKVAQLEAQ CQEPCKDTVQ IHDITGKDCQ DIANKGAKQS GLYFIKPLKA
201   NQQFLVYCEI DGSGNGWTVF QKRLDGSVDF KKNWIQYKEG FGHLSPTGTT
251   EFWLGNEKIH LISTQSAIPY ALRVELEDWN GRTSTADYAM FKVGPEADKY
301   RLTYAYFAGG DAGDAFDGFD FGDDPSDKFF TSHNGMQFST WDNDNDKFEG
351   NCAEQDGSGW WMNKCHAGHL NGVYYQGGTY SKASTPNGYD NGIIWATWKT
401   RWYSMKKTTM KIIPFNRLTI GEGQQHHLGG AKQ
```

γA variant continues:
```
434                                AGDV
```

γB variant continues:
```
434                                VRPEHPA   ETEYDSLYPE
451   DDL
```

Structural and functional sites

Signal peptide: 1–26
Potential N-linked glycosylation site: 78
Calcium binding sites: 341, 355
Possible polymerization sites: 400, 422
QAGDV cell adhesion site: 433–437 (in γA variant only)
Cross-linking sites: 424, 432

Gene structure

Single copies of the genes for all three chains are closely linked on human chromosome 4. Two types of γ chain mRNAs can be produced in the human by alternative splicing[9].

References

1. Crabtree, G.R. and Kant, J.A. (1981) Molecular cloning of cDNA for the α, β and γ chains of rat fibrinogen. J. Biol. Chem. 257: 7277–7279.
2. Kant, J.A. et al. (1983) Partial mRNA sequences for human Aα, Bβ and γ-fibrinogen chains: Evolutionary and functional implications. Proc. Natl Acad. Sci. USA 80: 3953–3957.
3. Doolittle, R.F. (1984) Fibrinogen and fibrin. Annu. Rev. Biochem. 53: 195–229.
4. Kloczewiak, M. et al. (1984) Platelet receptor recognition site on human fibrinogen. Synthesis and structure–function relationship of peptides corresponding to the carboxy-terminal segment of the gamma chain. Biochemistry 23: 1767–1774.
5. Kimura, S. and Aoki, N. (1986) Cross-linking site in fibrinogen for α2-plasmin inhibitor. J. Biol. Chem. 261: 15591–15595.
6. Doolittle, R.F. (1987) Fibrinogen and fibrin. In: Haemostasis and Thrombosis, 2nd edition, ed. Bloom, A.L. and Thomas, D.P., Churchill Livingstone, Edinburgh, pp. 192–215.
7. Doolittle, R.F. et al. (1967) Amino acid sequence studies on artiodactyl fibrinopeptides. Arch. Biochem. Biophys. 118: 456–467.
8. Doolittle, R.F. et al. (1977) Amino acid sequence studies on the α chain of human fibrinogen. Characterization of 11 cyanogen bromide fragments. Biochemistry 16: 1703–1709.
9. Oliasen, B. et al. (1982) Fibrinogen gamma chain locus is on chromosome 4 in man. Hum. Genet. 61: 24–26

Fibromodulin

Fibromodulin is a member of the small chondroitin sulphate/dermatan sulphate proteoglycans with leucine-rich repeat core proteins. It was originally described as a 59 kDa connective tissue protein in cartilage and subsequently renamed to describe its ability to modulate collagen fibre formation. It is present in most tissues, including tendon, skin, sclera and cornea, but is somewhat more abundant in cartilage where it represents 0.1–0.3% of tissue wet weight. Like decorin, fibromodulin binds types I and II collagen fibrils *in vitro* and inhibits collagen fibril assembly. Fibromodulin is substituted with keratan sulphate glycosaminoglycan chains and has an NH_2-terminal tail of sulphated tyrosine residues.

Molecular structure

Fibromodulin has a characteristic amino acid composition, with 14% of its residues being made up of leucine. Most of the protein consists of ten repeats of 23 residues. The leucine-rich sequence shares homology with sequences in other proteins including decorin, biglycan, the serum protein LRG, platelet surface protein GPIb, ribonuclease/angiotensin inhibitor, chaoptin, toll protein and adenylate cyclase. Decorin, biglycan and fibromodulin are similar in size and the cysteine residues are located in conserved positions. The core protein is substituted with keratan sulphate side-chains and contains sulphated tyrosine [1–3].

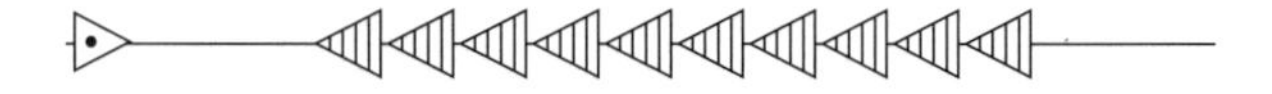

Isolation

Fibromodulin can be isolated from cartilage by 4 M guanidine–HCl extraction and purified by ion exchange chromatography and gel filtration [1].

Accession number

P13605

Primary structure

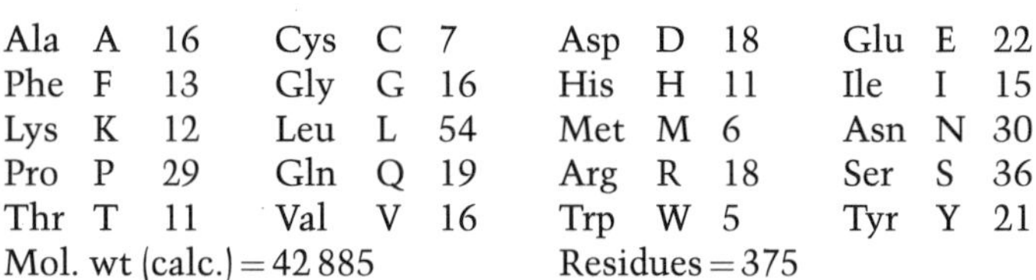

Ala	A	16	Cys	C	7	Asp	D	18	Glu	E	22
Phe	F	13	Gly	G	16	His	H	11	Ile	I	15
Lys	K	12	Leu	L	54	Met	M	6	Asn	N	30
Pro	P	29	Gln	Q	19	Arg	R	18	Ser	S	36
Thr	T	11	Val	V	16	Trp	W	5	Tyr	Y	21

Mol. wt (calc.) = 42 885 Residues = 375

```
1     MQWASILLLA  GLCSLSWAQY  EEDSHWWFQF  LRNQQSTYDD  PYDPYPYEPY
51    EPYPTGEEGP  AYAYGSPPQP  EPRDCPQECD  CPPNFPTAMY  CDNRNLKYLP
101   FVPSRMKYVY  FQNNQISSIQ  EGVFDNATGL  LWIALHGNQI  TSDKVGKKVF
151   SKLRHLERLY  LDHNNLTRIP  SPLPRSLREL  HLDHNQISRV  PNNALEGLEN
201   LTALYLHHNE  IQEVGSSMKG  LRSLILLDLS  YNHLRKVPDG  LPSALEQLYL
```

```
251   EHNNVFSVPD  SYFRGSPKLL  YVRLSHNSLT  NNGLASNTFN  SSSLLELDLS
301   YNQLQKIPPV  STNLENLYLQ  GNRINEFSIS  SFCTVVDVMN  FSKLQVQRLD
351   GNEIKRSAMP  ADAPLCLRLA  SLIEI
```

Structural and functional sites

Signal peptide: 1–18
Leucine-rich repeats: 114–137, 138–163, 164–184, 185–208, 209–231, 232–252, 253–276, 277–301, 302–321, 322–344
Potential N-linked glycosylation sites: 126, 165, 200, 290, 340
Potential sites for tyrosine sulphation: 20, 28, 32, 35, 37, 40, 43

Gene structure

The human fibromodulin gene spans approximately 8.5 kb.

References

[1] Heinegård, D. et al. (1986) Two novel matrix proteins isolated from articular cartilage show wide distributions among connective tissues. J. Biol. Chem. 261: 3866–3872.

[2] Hedbom, E. and Heinegård, D. (1989) Interaction of a 59-kDa connective tissue matrix protein with collagen I and collagen II. J. Biol. Chem. 264: 6898–6905.

[3] Oldberg, Å. et al. (1989) A collagen-binding 59-kd protein (fibromodulin) is structurally related to the small interstitial proteoglycans PG-S1 and PG-S2 (decorin). EMBO J. 8: 2601–2604.

Fibronectin

Fibronectin is a widely distributed glycoprotein present at high concentrations in most extracellular matrices, in plasma (300 μg/ml), and in other body fluids. Fibronectin is a prominent adhesive protein and mediates various aspects of cellular interactions with extracellular matrices including migration. Its principal functions appear to be in cellular migration during development and wound healing, regulation of cell growth and differentiation, and haemostasis/thrombosis.

Molecular structure

Fibronectin is a dimer of two non-identical subunits covalently linked near their COOH-termini by a pair of disulphide bonds. The difference between the subunits is determined by alternative splicing of the IIICS (or V) region. In the insoluble, matrix form of fibronectin, the dimer associates into disulphide-bonded oligomers and fibrils, while soluble, body fluid fibronectin is predominantly dimeric. Three regions of fibronectin are subject to alternative splicing and in general the matrix form of the molecule has a higher content of these segments than the soluble form. The human IIICS region has five potential variations, while the rat, bovine and chicken sequences have three, three and two, respectively. Each subunit is composed of a series of structurally independent domains linked by flexible polypeptide segments. At the primary sequence level, the origin of the majority of the fibronectin molecule can be accounted for by endo-duplication of three types of polypeptide repeat. Different fibronectin domains are specialized for binding extracellular matrix macromolecules or bacterial or eukaryotic membrane receptors. The central cell-binding domain is recognized by most adherent cells via the integrin receptors $\alpha3\beta1$, $\alpha5\beta1$, $\alpha V\beta1$, $\alpha IIb\beta3$, $\alpha V\beta3$, $\alpha V\beta5$ and $\alpha V\beta6$. The IIICS/HepII cell-binding domain is recognized by lymphoid cells, neural crest derivatives and myoblasts via the integrins $\alpha4\beta1$ and $\alpha4\beta7$. Several peptide active sites have been identified in these domains[1–9].

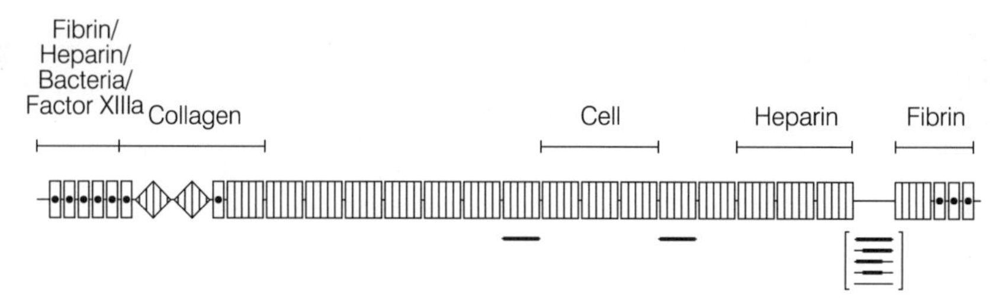

Isolation

Plasma fibronectin can be purified by a combination of gelatin and heparin affinity chromatography[10,11]. Cell-associated fibronectin can be extracted from culture monolayers with 1 M urea[11].

Accession number

P02751

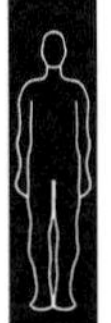

Primary structure

Ala	A	100	Cys	C	63	Asp	D	126	Glu	E	145
Phe	F	54	Gly	G	208	His	H	51	Ile	I	121
Lys	K	78	Leu	L	136	Met	M	27	Asn	N	101
Pro	P	195	Gln	Q	133	Arg	R	126	Ser	S	200
Thr	T	268	Val	V	200	Trp	W	40	Tyr	Y	105

Mol. wt (calc.) = 273 715   Residues = 2477

1	MLRGPGPGLL	LLAVQCLGTA	VPSTGASKSK	RQAQQMVQPQ	SPVAVSQSKP
51	GCYDNGKHYQ	INQQWERTYL	GNVLVCTCYG	GSRGFNCESK	PEAEETCFDK
101	YTGNTYRVGD	TYERPKDSMI	WDCTCIGAGR	GRISCTIANR	CHEGGQSYKI
151	GDTWRRPHET	GGYMLECVCL	GNGKGEWTCK	PIAEKCFDHA	AGTSYVVGET
201	WEKPYQGWMM	VDCTCLGEGS	GRITCTSRNR	CNDQDTRTSY	RIGDTWSKKD
251	NRGNLLQCIC	TGNGRGEWKC	ERHTSVQTTS	SGSGPFTDVR	AAVYQPQPHP
301	QPPPYGHCVT	DSGVVYSVGM	QWLKTQGNKQ	MLCTCLGNGV	SCQETAVTQT
351	YGGNSNGEPC	VLPFTYNGRT	FYSCTTEGRQ	DGHLWCSTTS	NYEQDQKYSF
401	CTDHTVLVQT	QGGNSNGALC	HFPFLYNNHN	YTDCTSEGRR	DNMKWCGTTQ
451	NYDADQKFGF	CPMAAHEEIC	TTNEGVMYRI	GDQWDKQHDM	GHMMRCTCVG
501	NGRGEWTCYA	YSQLRDQCIV	DDITYNVNDT	FHKRHEEGHM	LNCTCFGQGR
551	GRWKCDPVDQ	CQDSETGTFY	QIGDSWEKYV	HGVRYQCYCY	GRGIGEWHCQ
601	PLQTYPSSSG	PVEVFITETP	SQPNSHPIQW	NAPQPSHISK	YILRWRPKNS
651	VGRWKEATIP	GHLNSYTIKG	LKPGVVYEGQ	LISIQQYGHQ	EVTRFDFTTT
701	STSTPVTSNT	VTGETTPFSP	LVATSESVTE	ITASSFVVSW	VSASDTVSGF
751	RVEYELSEEG	DEPQYLDLPS	TATSVNIPDL	LPGRKYIVNV	YQISEDGEQS
801	LILSTSQTTA	PDAPPDPTVD	QVDDTSIVVR	WSRPQAPITG	YRIVYSPSVE
851	GSSTELNLPE	TANSVTLSDL	QPGVQYNITI	YAVEENQEST	PVVIQQETTG
901	TPRSDTVPSP	RDLQFVEVTD	VKVTIMWTPP	ESAVTGYRVD	VIPVNLPGEH
951	GQRLPISRNT	FAEVTGLSPG	VTYYFKVFAV	SHGRESKPLT	AQQTTKLDAP
1001	TNLQFVNETD	STVLVRWTPP	RAQITGYRLT	VGLTRRGQPR	QYNVGPSVSK
1051	YPLRNLQPAS	EYTVSLVAIK	GNQESPKATG	VFTTLQPGSS	IPPYNTEVTE
1101	TTIVITWTPA	PRIGFKLGVR	PSQGGEAPRE	VTSDSGSIVV	SGLTPGVEYV
1151	YTIQVLRDGQ	ERDAPIVNKV	VTPLSPPTNL	HLEANPDTGV	LTVSWERSTT
1201	PDITGYRITT	TPTNGQQGNS	LEEVVHADQS	SCTFDNLSPG	LEYNVSVYTV
1251	KDDKESVPIS	DTIIPEVPQL	TDLSFVDITD	SSIGLRWTPL	NSSTIIGYRI
1301	TVVAAGEGIP	IFEDFVDSSV	GYYTVTGLEP	GIDYDISVIT	LINGGESAPT
1351	TLTQQTAVPP	PTDLRFTNIG	PDTMRVTWAP	PPSIDLTNFL	VRYSPVKNEE
1401	DVAELSISPS	DNAVVLTNLL	PGTEYVVSVS	SVYEQHESTP	LRGRQKTGLD
1451	SPTGIDFSDI	TANSFTVHWI	APRATITGYR	IRHHPEHFSG	RPREDRVPHS
1501	RNSITLTNLT	PGTEYVVSIV	ALNGREESPL	LIGQQSTVSD	VPRDLEVVAA
1551	TPTSLLISWD	APAVTVRYYR	ITYGETGGNS	PVQEFTVPGS	KSTATISGLK
1601	PGVDYTITVY	AVTGRGDSPA	SSKPISINYR	TEIDKPSQMQ	VTDVQDNSIS
1651	VKWLPSSSPV	TGYRVTTTPK	NGPGPTKTKT	AGPDQTEMTI	EGLQPTVEYV
1701	VSVYAQNPSG	ESQPLVQTAV	TNIDRPKGLA	FTDVDVDSIK	IAWESPQGQV
1751	SRYRVTYSSP	EDGIHELFPA	PDGEEDTAEL	QGLRPGSEYT	VSVVALHDDM
1801	ESQPLIGTQS	TAIPAPTDLK	FTQVTPTSLS	AQWTPPNVQL	TGYRVRVTPK
1851	EKTGPMKEIN	LAPDSSSVVV	SGLMVATKYE	VSVYALKDTL	TSRPAQGVVT
1901	TLENVSPPRR	ARVTDATETT	ITISWRTKTE	TITGFQVDAV	PANGQTPIQR
1951	TIKPDVRSYT	ITGLQPGTDY	KIYLYTLNDN	ARSSPVVIDA	STAIDAPSNL
2001	RFLATTPNSL	LVSWQPPRAR	ITGYIIKYEK	PGSPPREVVP	RPRPGVTEAT
2051	ITGLEPGTEY	TIYVIALKNN	QKSEPLIGRK	KTDELPQLVT	LPHPNLHGPE
2101	ILDVPSTVQK	TPFVTHPGYD	TGNGIQLPGT	SGQQPSVGQQ	MIFEEHGFRR
2151	TTPPTTATPI	RHRPRPYPPN	VGEEIQIGHI	PREDVDYHLY	PHGPGLNPNA

```
2201  STGQEALSQT  TISWAPFQDT  SEYIISCHPV  GTDEEPLQFR  VPGTSTSATL
2251  TGLTRGATYN  IIVEALKDQQ  RHKVREEVVT  VGNSVNEGLN  QPTDDSCFDP
2301  YTVSHYAVGD  EWERMSESGF  KLLCQCLGFG  SGHFRCDSSR  WCHDNGVNYK
2351  IGEKWDRQGE  NGQMMSCTCL  GNGKGEFKCD  PHEATCYDDG  KTYHVGEQWQ
2401  KEYLGAICSC  TCFGGQRGWR  CDNCRRPGGE  PSPEGTTGQS  YNQYSQRYHQ
2451  RTNTNVNCPI  ECFMPLDVQA  DREDSRE
```

Structural and functional sites

Signal peptide: 1–20
Propeptide: 21–31
Type I repeats: 52–96, 97–140, 141–185, 186–230, 231–272, 308–344, 470–517, 518–560, 561–608, 2297–2341, 2342–2385, 2386–2428
Type II repeats: 345–404, 405–469
Type III repeats: 609–700, 719–809, 810–905, 906–995, 996–1085, 1086–1172, 1173–1265, 1357–1447, 1448–1537, 1538–1631, 1632–1721, 1812–1903, 1904–1992, 1993–2082, 2203–2273
Alternatively spliced domains: 1722–1811 (ED-A), 1266–1356 (ED-B), 2083–2202 (IIICS)
Potential N-linked glycosylation sites: 430, 528, 542, 877, 1007, 1244, 1291, 1904, 2199
O-linked glycosylation site: 2155
Inter-chain disulphide bond residues: 2458, 2462
RGD cell adhesion site: 1615–1618
IDAPS cell adhesion site: 1994–1998
LDV cell adhesion site: 2102–2104
REDV cell adhesion site: 2182–2185
Heparin-binding sites: 2028–2046 (FN-C/H I), 2068–2082 (FN-C/H II)
Factor XIIIa transglutaminase cross-linking site: 34

Gene structure

Fibronectin is encoded by a single gene found on human chromosome 2 (at locus p14–16 or q34–36). The human gene is not fully characterized but contains approximately 50 exons. The chicken and rat genes span approximately 50 kb and 70 kb, respectively. Type I and II repeats correspond to single exons, type III repeats to two exons (except for III9). ED-A and ED-B are single exons. The IIICS is combined with the first half of III15 in one exon.

References

[1] Pierschbacher, M.D. and Ruoslahti, E. (1984) The cell attachment activity of fibronectin can be duplicated by small synthetic fragments of the molecule. Nature 309: 30–33.

[2] Yamada, K.M. and Kennedy, D.W. (1984) Dualistic nature of adhesive protein function: Fibronectin and its biologically active peptide fragments can autoinhibit fibronectin function. J. Cell Biol. 99: 29–36.

[3] Kornblihtt, A.R. et al. (1985) Primary structure of human fibronectin: Differential splicing may generate at least 10 polypeptides from a single gene. EMBO J. 4: 1755–1759.

[4] Humphries, M.J. et al. (1986) Identification of an alternatively spliced site in human plasma fibronectin that mediates cell type-specific adhesion. J. Cell Biol. 103: 2637–2647.
[5] Humphries, M.J. et al. (1987) Identification of two distinct regions of the type III connecting segment of human plasma fibronectin that promote cell type-specific adhesion. J. Biol. Chem. 262: 6886–6892.
[6] McCarthy, J.B. et al. (1988) Localization and chemical synthesis of fibronectin peptides with melanoma adhesion and heparin binding activities. Biochemistry 27: 1380–1388.
[7] Mosher, D.F. (1989) Fibronectin, Academic Press, New York.
[8] Hynes, R.O. (1990) Fibronectins, Springer-Verlag, New York.
[9] Mould, A.P. and Humphries, M.J. (1991) Identification of a novel recognition sequence for the integrin $\alpha 4\beta 1$ in the COOH-terminal heparin-binding domain of fibronectin. EMBO J. 10: 4089–4095.
[10] Engvall, E. and Ruoslahti, E. (1977) Binding of soluble form of fibroblast surface protein, fibronectin, to collagen. Int. J. Cancer 20: 1–5.
[11] Yamada, K.M. (1983) Isolation of fibronectin from plasma and cells. In: Immunochemistry of the Extracellular Matrix, ed. Furthmayr, H., CRC Press, Boca Raton, FL, pp. 111–123.

Fibulin-1

fibulin, BM-90

Fibulin-1 is a glycoprotein found in the extracellular matrix and at moderate concentration in plasma (33 μg/ml). Initially, fibulin was suspected to be an intracellular protein interacting with the cytoplasmic domain of β1 integrins. Subsequent work showed it to be secreted by fibroblasts and incorporated into the extracellular matrix in a similar fashion to fibronectin. Apart from its calcium- and fibronectin-binding capacity, little is known about its role in the extracellular matrix or plasma.

Molecular structure

Fibulin-1 is a single-chain polypeptide (calculated molecular weight of approximately 71 500), rich in cysteine (11%), which probably contains both N- and O-linked oligosaccharides. cDNA cloning indicates that three forms of fibulin exist, encoded by three transcripts that are likely to be derived from a common precursor mRNA. Fibulin can self-associate and bind to fibronectin. The NH_2-terminal portion of the molecule contains three repeated motifs that have potential disulphide loop structure and resemble the complement component anaphylatoxins C3a, C4a and C5a, as well as members of the albumin gene family. The bulk of the remainder of the molecule consists of nine cysteine-containing EGF repeats. Separating these two repeat domains is a 33-residue segment containing 12 acidic amino acids [1–7].

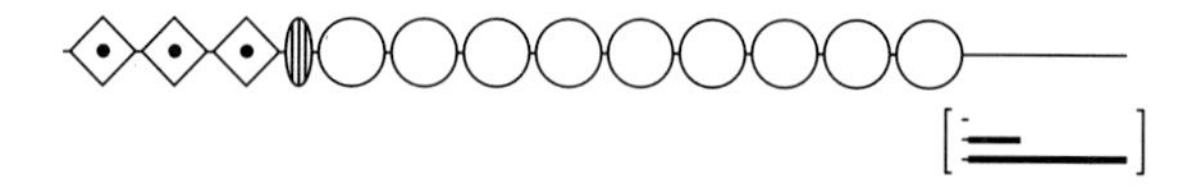

Isolation

Fibulin has been purified from placental extracts by affinity chromatography on a synthetic integrin β1 subunit cytoplasmic domain peptide–Sepharose column [1] or by monoclonal antibody affinity chromatography [4].

Accession number

P23142; P23143; P23144

Primary structure

A variant:

Ala	A	30	Cys	C	69	Asp	D	34	Glu	E	43
Phe	F	14	Gly	G	53	His	H	14	Ile	I	24
Lys	K	13	Leu	L	39	Met	M	5	Asn	N	26
Pro	P	23	Gln	Q	28	Arg	R	34	Ser	S	46
Thr	T	24	Val	V	28	Trp	W	0	Tyr	Y	19

Mol. wt (calc.) = 61 525 Residues = 566

B variant:

Ala	A	32	Cys	C	70	Asp	D	37	Glu	E	45
Phe	F	14	Gly	G	56	His	H	15	Ile	I	24
Lys	K	18	Leu	L	42	Met	M	5	Asn	N	27
Pro	P	25	Gln	Q	31	Arg	R	36	Ser	S	49
Thr	T	26	Val	V	29	Trp	W	1	Tyr	Y	19

Mol. wt (calc.) = 65 412 Residues = 601

C variant:

Ala	A	37	Cys	C	72	Asp	D	37	Glu	E	50
Phe	F	21	Gly	G	60	His	H	18	Ile	I	28
Lys	K	18	Leu	L	51	Met	M	8	Asn	N	29
Pro	P	33	Gln	Q	30	Arg	R	42	Ser	S	57
Thr	T	32	Val	V	39	Trp	W	0	Tyr	Y	21

Mol. wt (calc.) = 74 392 Residues = 683

A variant:

```
1     MERAAPSRRV PLPLLLLGGL ALLAAGVDAD VLLEACCADG HRMATHQKDC
51    SLPYATESKE CRMVQEQCCH SQLEELHCAT GISLANEQDR CATPHGDNAS
101   LEATFVKRCC HCCLLGRAAQ AQGQSCEYSL MVGYQCGQVF RACCVKSQET
151   GDLDVGGLQE TDKIIEVEEE QEDPYLNDRC RGGGPCKQQC RDTGDEVVCS
201   CFVGYQLLSD GVSCEDVNEC ITGSHSCRLG ESCINTVGSF RCQRDSSCGT
251   GYELTEDNSC KDIDECESGI HNCLPDFICQ NTLGSFRCRP KLQCKSGFIQ
301   DALGNCIDIN ECLSISAPCP IGHTCINTEG SYTCQKNVPN CGRGYHLNEE
351   GTRCVDVDEC APPAEPCGKG HRCVNSPGSF RCECKTGYYF DGISRMCVDV
401   NECQRYPGRL CGHKCENTLG SYLCSCSVGF RLSVDGRSCE DINECSSSPC
451   SQECANVYGS YQCYCRRGYQ LSDVDGVTCE DIDECALPTG GHICSYRCIN
501   IPGSFQCSCP SSGYRLAPNG RNCQDIDECV TGIHNCSINE TCFNIQGAFR
551   CLAFECPENY RRSAAT
```

B variant continues after A:

```
567                          QKSK KGRQNTPAGS SKEDCRVLPW KQGLEDTHLD
601   A
```

C variant continues after A:

```
567                          RCER LPCHENRECS KLPLRITYYH LSFPTNIQAP
601   AVVFRMGPSS AVPGDSMQLA ITGGNEEGFF TTRKVSPHSG VVALTKPVPE
651   PRDLLLTVKM DLSRHGTVSS FVAKLFIFVS AEL
```

Structural and functional sites

Signal peptide: 1–29
Type I repeats: 36–76, 77–111, 112–144
EGF (6C) repeats: 179–219, 220–265, 266–311, 312–359, 360–402, 403–444, 445–485, 486–528, 529–566
Potential N-linked glycosylation sites: 98, 535, 539
Alternative splicing sites: 566

Gene structure

cDNA cloning indicates that three forms of fibulin exist encoded by three transcripts that are likely to be derived from a common mRNA. Northern hybridization of several tissues revealed two mRNAs of 2.3 and 2.7 kb. The human gene is located on chromosome 22 [5].

References

[1] Argraves, W.S. et al. (1989) Fibulin, a novel protein that interacts with the fibronectin receptor β subunit cytoplasmic domain. Cell 58: 623–629.
[2] Argraves, W.S. et al. (1990) Fibulin is an extracellular matrix and plasma glycoprotein with repeated domain structure. J. Cell Biol. 111: 3155–3164.
[3] Kluge, M. et al. (1990) Characterization of a novel calcium-binding, 90-kDa glycoprotein (BM-90) shared by basement membranes and serum. Eur. J. Biochem. 193: 651–659.
[4] Balbona, K. et al. (1992) Fibulin binds to itself and to the carboxyl-terminal heparin-binding region of fibronectin. J. Biol. Chem. 267: 20120–20125.
[5] Mattei, M.G. et al. (1994) The fibulin-1 gene (FBLN1) is located on human chromosome 22 and mouse chromosome 15. Genomics 22: 437–438.
[6] Saski, T. et al. (1995) Structural characterization of two variants of fibulin-1 that differ in nidogen affinity. J. Mol. Biol. 245: 241–250.
[7] Spence, S.G. et al. (1992) Fibulin is localized at sites of epithelial-mesenchymal transition in the early avian embryo. Dev. Biol. 151: 473–484.

Fibulin-2

Fibulin-2 is another member of the fibulin family of extracellular, multi-domain proteins. Originally discovered by cDNA cloning, the protein has so far only been obtained in recombinant form. Preliminary studies indicate similar but different binding characteristics to those shown by fibulin-1, and suggest it also plays a role in the elaboration of matrices. High expression of fibulin-2 is shown in heart tissue, placenta and ovaries.

Molecular structure

Fibulin-2 is a single polypeptide chain of 1157 amino acids and is larger than fibulin-1 by virtue of possessing a 408 amino acid amino-terminal domain structure not found in fibulin-1. This region contains a cysteine-rich segment of 150 residues, and a cysteine-free segment containing a stretch of acidic amino acids. The remaining C-terminal region of fibulin-2 shows 45% sequence identity to fibulin-1 and comprises three anaphylatoxin-like motifs, 10 EGF-like motifs and a C-terminal region similar to the C-variant of fibulin-1. Rotary shadowing showed a trimeric rod structure, each rod being 40–45 nm long [1–5].

Isolation

Recombinant protein has been purified from culture medium by ammonium sulphate precipitation, Sepharose CL-6B and MonoQ chromatography, and finally purified on a Superose 6 column [1].

Accession number

P98095

Primary structure

Ala	A	114	Cys	C	101	Asp	D	55	Glu	E	103
Phe	F	30	Gly	G	113	His	H	33	Ile	I	35
Lys	K	24	Leu	L	84	Met	M	12	Asn	N	40
Pro	P	94	Gln	Q	59	Arg	R	56	Ser	S	67
Thr	T	62	Val	V	63	Trp	W	7	Tyr	Y	32

Mol. wt (calc.) = 126 543　　Residues = 1184

```
1     MVLLWEPAGA  WLALGLALAL  GPSVAAAAPR  QDCTGVECPP  LENCIEEALE
51    PGACCATCVQ  QGCACEGYQY  YDCLQGGFVR  GRVPAGQSYF  VDFGSTECSC
101   PPGGGKISCQ  FMLCPELPPN  CIEAVVVADS  CPQCGQVGCV  HAGHEYAAGH
151   TVHLPPCRAC  HCPDAGGELI  CYQLPGCHGN  FSDAEEGDPE  RHYEDPYSYD
201   QEVAEVEAAT  ALGGEVQAGA  VQAGAGGPPA  ALGGGSQPLS  TIQAPPWPAV
251   LPRPTAAAAL  GPPAPVQAKA  RRVTEDSEEE  EEEEEEREEM  AVTEQLAAGG
301   HRGLDGLPTT  APAGPSLPIQ  EERAEAGARA  EAGARPEENL  ILDAQATSRS
351   TGPEGVTHAP  SLGKAALVPT  QAVPGSPRDP  VKPSPHNILS  TSLPDAAWIP
```

```
401   PTREVPRKPQ VLPHSHVEED TDPNSVHSIP RSSPEGSTKD LIETCCAAGQ
451   QWAIDNDECL EIPESGTEDN VCRTAQRHCC VSYLQEKSCM AGVLGAKEGE
501   TCGAEDNDSC GISLYKQCCD CCGLGLRVRA EGQSCESNPN LGYPCNHVML
551   SCCEGEEPLI VPEVRRPPEP AAAPRRVSEA EMAGREALSL GTEAELPNSL
601   PGDDQDECLL LPGELCQHLC INTVGSYHCA CFPGFSLQDD GRTCRPEGHP
651   PQPEAPQEPA LKSEFSQVAS NTIPLPLPQP NTCKDNGPCK QVCSTVGGSA
701   ICSCFPGYAI MADGVSCEDI NECVTDLHTC SRGEHCVNTL GSFHCYKALT
751   CEPGYALKDG ECEDVDECAM GTHTCQPGFL CQNTKGSFYC QARQRCMDGF
801   LQDPEGNCVD INECTSLSEP CRPGFSCINT VGSYTCQRNP LICARGYHAS
851   DDGAKCVDVN ECETGVHRCG EGQVCHNLPG SYRCDCKAGF QRDAFGRGCI
901   DVNECWASPG RLCQHTCENT LGSYRCSCAS GFLLAADGKR CEDVNECEAQ
951   RCSQECANIY GSYQCYCRQG YQLAEDGHTC TDIDECAQGA GILCTFRCLN
1001  VPGSYQCACP EQGYTMTANG RSCKDVDECA LGTHNCSEAE TCHNIQGSFR
1051  CLRFECPPNY VQVSKTKCER TTCHDFLECQ NSPARITHYQ LNFQTGLLVP
1101  AHIFRIGPAP AFTGDTIALN IIKGNEEGYF GTRRLNAYTG VVYLQRAVLE
1151  PRDFALDVEM KLWRQGSVTT' FLAKMHIFFT TFAL
```

Structural and functional sites

Signal peptide: 1–27
Type I repeats: 445–480, 488–519, 521–553
EGF (6C) repeats: 604–645, 679–718, 719–763, 764–809, 810–857, 858–900, 901–942, 943–981, 982–1024, 1025–1069
Potential N-linked glycosylation sites: 153, 480, 1008

Gene structure

Gene is on the p24–p25 region of human chromosome 3. Northern blotting revealed a 4.5 kb transcript in a variety of human tissues [2].

References

[1] Pan, T.-C. et al. (1993) Structure and expression of fibulin-2, a novel extracellular matrix protein with multiple EGF-like repeats and consensus motifs for calcium binding. J. Cell Biol. 123: 1269–1277.

[2] Zhang, R.-Z. et al. (1994) Fibulin-2 (FBLN2): Human cDNA sequence, mRNA expression, and mapping of the gene on human and mouse chromosomes. Genomics 22: 425–430.

[3] Zhang, H.-Y. et al. (1995) Extracellular matrix protein fibulin-2 is expressed in the embryonic endocardial cushion tissue and is a prominent component of valves in adult heart. Develop. Biol. 167: 18–26.

[4] Sasaki, T. et al. (1995) Structural characterisation of two variants of fibulin-1 that differ in nidogen affinity. J. Mol. Biol. 245: 241–250.

[5] Sasaki, T. et al. (1995) Binding of mouse and human fibulin-2 to extracellular matrix ligands. J. Mol. Biol. 254: 892–899.

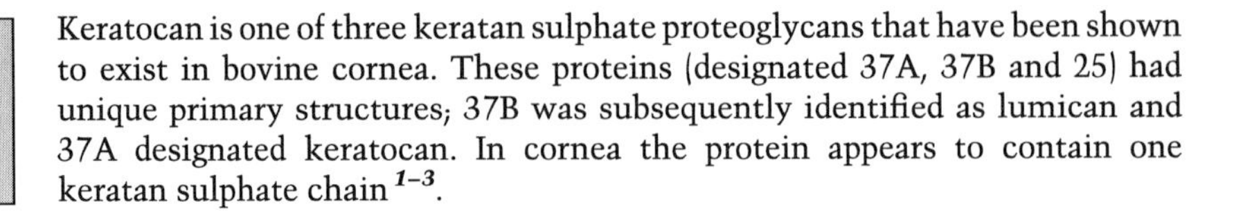

Keratocan

corneal keratan sulphate proteoglycan, 37A

Keratocan is one of three keratan sulphate proteoglycans that have been shown to exist in bovine cornea. These proteins (designated 37A, 37B and 25) had unique primary structures; 37B was subsequently identified as lumican and 37A designated keratocan. In cornea the protein appears to contain one keratan sulphate chain [1–3].

Molecular structure

Keratocan is a member of the leucine-rich repeat family of proteins. It contains 11 repeats of the sequence LXXLXLXXNXL/I where X is any amino acid, and conserved cysteines at the N- and C-terminal ends. The mature protein contains 332 amino acids with a molecular mass of 38 047 Da. There are five potential *N*-glycosylation sites, a consensus sequence for tyrosine sulphation at the N-terminus [3].

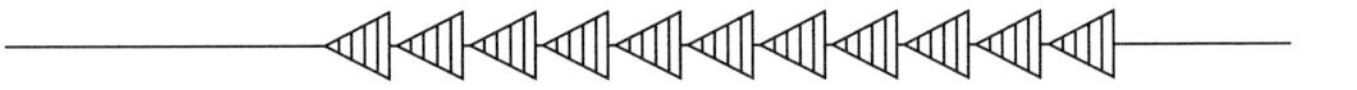

Isolation

Keratocan has been isolated from cornea by 4 M GuCl extraction and fractionation on DEAE–cellulose, after endo-β-glycosidase cleavage the protein could be separated on DEAE–Sephacel [1].

Accession number

U48360

Primary structure (cow)

Ala	A	11	Cys	C	7	Asp	D	19	Glu	E	13
Phe	F	15	Gly	G	17	His	H	8	Ile	I	22
Lys	K	22	Leu	L	55	Met	M	5	Asn	N	34
Pro	P	20	Gln	Q	12	Arg	R	6	Ser	S	24
Thr	T	20	Val	V	16	Trp	W	1	Tyr	Y	16

Mol. wt (calc.) = 40 405 Residues = 352

```
1     MASTICFILW VVFVTDTVWT RSVRQVYEAS DPEDWTMHDF DCPRECFCPP
51    SFPTALYCEN RGLKEIPAIP SRIWYLYLEN NLIETIPEKP FENATQLRWI
101   NLNKNKITNY GIEKGALSQL KKLLFLFLED NELEEVPSPL PRSLEQLQLA
151   RNKVSRIPQG TFSNLENLTL LDLQHNKLLD NAFQRDTFKG LKNLMQLNMA
201   KNALRNMPPR LPANTMQVFL DNNSIEGIPE NYFNVIPKVA FLRLNHNKLS
251   DAGLPSSGFN VSSILDLQLS HNQLTKVPKI SAHLQHLHLD HNKIRNVNVS
301   VICPSTPTTL PVEDSFSYGP HLRYLRLDGN EIKPPIPMDL MTCFRLLQAV
351   II
```

Structural and functional sites

Signal peptide: 1–20

Leucine-rich repeats: 73–83, 97–107, 123–133, 144–154, 168–178, 193–203, 215–225, 239–249, 265–275, 284–294, 322–332
Potential N-linked glycosylation sites: 93, 167, 222, 260, 298

Gene structure

Northern blot analysis shows a 2.5 kb transcript.

References

1 Funderburgh, J.L. et al. (1991) Unique glycosylation of 3 keratan sulfate proteoglycan isoforms. J. Biol. Chem. 266: 14226–14231.

2 Bengtsson, E. et al. (1995) The primary structure of a basic leucine-rich repeat protein, PRELP, found in connective tissues. J. Biol. Chem. 270: 25639–25644.

3 Corpuz, L.M. et al. (1996) Molecular cloning and tissue distribution of keratocan. J. Biol. Chem. 271: 9759–9763.

Laminins

Laminins are a family of large glycoproteins that are distributed ubiquitously in basement membranes. This family of molecules are multi-functional, performing key roles in development, differentiation and migration through their ability to interact with cells via cell-surface receptors, including the integrins $\alpha1\beta1$, $\alpha2\beta1$, $\alpha3\beta1$, $\alpha6\beta1$, $\alpha7\beta1$ and $\alpha v\beta3$, and with other basement membrane components such as type IV collagen, entactin/nidogen and heparan sulphate proteoglycan [1]. Studies on proteolytically derived fragments of laminin-1 [derived from the Engelbreth–Holm–Swarm (EHS) tumour] have localized binding sites for collagen, heparin and entactin/nidogen, and defined domains of the molecule responsible for distinct biological activities such as cell attachment and neurite outgrowth promotion.

Molecular structure

A new and systematic nomenclature for the individual laminin chains and for the trimeric laminin molecules has been devised [2] and will be used throughout this book. Laminins are composed of three genetically distinct chains (α, β and γ chains) that assemble into a cruciform-shaped molecule with one long arm and three short arms (molecular weights 500 000–800 000). The long arm is formed by the association of the α, β and γ chains into a triple coiled-coil that is stabilized by inter-chain disulphide bridges. The three short arms represent the N-terminal regions of each chain. Five α chains (α1–5), three β chains (β1–3) and two γ chains (γ1,2) have been cloned and sequenced to date. These assemble into at least seven isomeric forms of laminin, namely: laminin-1 ($\alpha1,\beta1,\gamma1$); laminin-2 ($\alpha2,\beta1,\gamma1$); laminin-3 ($\alpha1,\beta2,\gamma1$); laminin-4 ($\alpha2,\beta2,\gamma1$); laminin-5 ($\alpha3,\beta3,\gamma2$); laminin-6 ($\alpha1,\beta3,\gamma1$); laminin-7 ($\alpha3,\beta2,\gamma1$). The chain compositions of the molecules containing the $\alpha4$ and $\alpha5$ chains have yet to be determined. Isomeric forms of laminin exhibit developmental stage- and tissue-specific patterns of expression based on mRNA and immunolocalization studies.

Laminin chains share a similar domain structure. Domains I and II are located at the COOH-terminal ends of the β and γ chains, and in an equivalent position in the α chains. They consist of a series of heptad repeats with a predicted α-helical conformation which form the long arm of the assembled laminin molecule. Domains I and II are interrupted in the β chains by a short cysteine-rich domain (α). Domains III and V consist of homologous repeats of approximately 50 amino acid residues that are rich in glycine and contain eight cysteine residues arranged in a fashion that resembles the six-cysteine residue motif present in epidermal growth factor (EGF) and transforming growth factor α (TGFα). Domains IV and VI are thought to form the globular domain(s) in the short arms of the laminin molecule. Laminin $\alpha1$, $\alpha2$ and $\alpha5$ chains contain an additional domain III and IV (the IIIa and IVa domains). Truncated forms of laminin chains (α3a, $\alpha4$, $\beta3$ and $\gamma2$) are missing combinations of the short arm domains (III, IV, V and VI). The IIIa domain of the α3a chain appears to be removed in an extracellular processing event. Laminin α chains contain an extra domain at their COOH-terminus (domain G) that forms the large globular domain seen by electron microscopy at the end of the long arm. Part of the α3a chain G domain appears to be removed in an extracellular processing event.

Laminins contain a variety of potential cell-binding sequences, although most of these are incompletely characterized. A non-conserved RGD site in the α1 chain interacts with the integrin $\alpha V\beta 3$, the GD-6 peptide, again in the α1 chain, interacts with $\alpha 3\beta 1$, and two β1 chain peptides, YIGSR and LGTIPG, probably interact with a 67-kDa binding protein[1–31].

α1
α2
α3
α4
β1
β2
β3
γ1
γ2

Intermolecular disulphide bonding

Chain trimerisation

Isolation

The most studied isoform of laminin (laminin-1) is obtained in large amounts by 0.5 M NaCl extraction of the mouse EHS tumour[3] and is the only form of laminin available commercially. Alternatively, laminin associated with entactin/nidogen can be extracted from the EHS tumour and other tissues in a buffer containing 10 mM EDTA, 150 mM NaCl, 50 mM Tris–HCl, pH 7.4 in the presence of protease inhibitors[11].

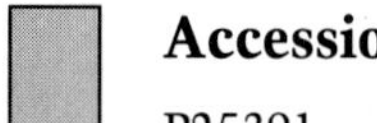

Accession number

P25391

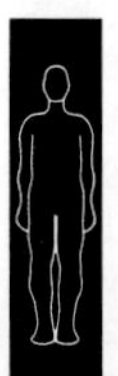

Primary structure: α1 chain

Ala	A	193	Cys	C	166	Asp	D	171	Glu	E	187
Phe	F	86	Gly	G	256	His	H	91	Ile	I	122
Lys	K	141	Leu	L	289	Met	M	39	Asn	N	154
Pro	P	145	Gln	Q	132	Arg	R	161	Ser	S	244
Thr	T	173	Val	V	203	Trp	W	24	Tyr	Y	98

Mol. wt (calc.) = 336 777 Residues = 3075

```
1     MRGGVLLVLL  LCVAAQCRQR  GLFPAILNLA  SNAHISTNAT  CGEKGPEMFC
51    KLVEHVPGRP  VRNPQCRICD  GNSANPRERH  PISHAIDGTN  NWWQSPSIQN
101   GREYHWVTIT  LDLRQVFQVA  YVIIKAANAP  RPGNWILERS  LDGTTFSPWQ
151   YYAVSDSECL  SRYNITPRRG  PPTYRADDEV  ICTSYYSRLV  PLEHGEIHTS
201   LINGRPSADD  LSPKLLEFTS  ARYIRLRLQR  IRTLNADLMT  LSHREPKELD
251   PIVTRRYYYS  IKDISVGGMC  ICYGHASSCP  WDETTKKLQC  QCEHNTCGES
301   CNRCCPGYHQ  QPWRPGTVSS  GNTCEACNCH  NKAKDCYYDE  SVAKQKKSLN
351   TAGQFRGGGV  CINCLQNTMG  INCETCIDGY  YRPHKVSPYE  DEPCRPCNCD
401   PVGSLSSVCI  KDDLHSDLHN  GKQPGQCPCK  EGYTGEKCDR  CQLGYKDYPT
451   CVSCGCNPVG  SASDEPCTGP  CVCKENVEGK  ACDRCKPGFY  NLKEKNPRGC
501   SECFCFGVSD  VCSSLSWPVG  QVNSMSGWLV  TDLISPRKIP  SQQDALGGRH
551   QVSINNTAVM  QRLAPKYYWA  APEAYLGNKL  TAFGGFLKYT  VSYDIPVETV
601   DSNLMSHADV  IIKGNGLTLS  TQAEGLSLQP  YEEYLNVVRL  VPENFQDFHS
651   KRQIDRDQLM  TVLANVTHLL  IRATYNSAKM  ALYRLESVSL  DIASSNAIDL
701   VVAADVEHCE  CPQGYTGTSC  ESCLSGYYRV  DGILFGGICQ  PCECHGHAAE
751   CNVHGVCIAC  AHNTTGVHCE  QCLPGFYGEP  SRGTPGDCQP  CACPLTIASN
801   NFSPTCHLND  GDEVVCDWCA  PGYSGAWCER  CADGYYGNPT  VPGESCVPCD
851   CSGNVDPSEA  GHCDSVTGEC  LKCLGNTDGA  HCERCADGFY  GDAVTAKNCR
901   ACECHVKGSH  SAVCHLETGL  CDCKPNVTGQ  QCDQCLHGYY  GLDSGHGCRP
951   CNCSVAGSVS  DGCTDEGQCH  CVPGVAGKRC  DRCAHGFYAY  QDGSCTPCDC
1001  PHTQNTCDPE  TGECVCPPHT  QGGKCEECED  GHWGYDAEVG  CQACNCSLVG
1051  STHHRCDVVT  GHCQCKSKFG  GRACDQCSLG  YRDFPDCVPC  DCDLRGTSGD
1101  ACNLEQGLCG  CVEETGACPC  KENVFGPQCN  ECREGTFALR  ADNPLGCSPC
1151  FCSGLSHLCS  ELEDYVRTPV  TLGSDQPLLR  VVSQSNLRGT  TEGVYYQAPD
1201  FLLDAATVRQ  HIRAEPFYWR  LPQQFQGDQL  MAYGGKLKYS  VAFYSLDGVG
1251  TSNFEPQVLI  KGGRIRKQVI  YMDAPAPENG  VRQEQEVAMR  ENFWKYFNSV
1301  SEKPVTREDF  MSVLSDIEYI  LIKASYGQGL  QQSRISDISV  EVGRKAEKLH
1351  PEEEVASLLE  NCVCPPGTVG  FSCQDCAPGY  HRGKLPAGSD  RGPRPLVAPC
1401  VPCSCNNHSD  TCDPNTGKCL  NCGDNTAGDH  CDVCTSGYYG  KVTGSASDCA
1451  LCACPHSPPA  SFSPTCVLEG  DHDFRCDACL  LGYEGKHCER  CSSSYYGNPQ
1501  TPGGSCQKCD  CNRHGSVHGD  CDRTSGQCVC  RLGASGLRCD  ECEPRHILME
1551  TDCVSCDDEC  VGVLLNDLDE  IGDAVLSLNL  TGIIPVPYGI  LSNLENTTKY
1601  LQESLLKENM  QKDLGKIKLE  GVAEETDNLQ  KKLTRMLAST  QKVNRATERI
1651  FKESQDLAVA  IERLQMSITE  IMEKTTLNQT  LDEDFLLPNS  TLQNMQQNGT
1701  SLLEIMQIRD  FTQLHQNATL  ELKAAEDLLS  QIQENYQKPL  EELEVLKEAA
1751  SHVLSKHNNE  LKAAEALVRE  AEAKMQESNH  LLLMVNANLR  EFSDKKLHVQ
1801  EEQNLTSELI  VQGRGLIDAA  AAQTDAVQDA  LEHLEDHQDK  LLLWSAKIRH
1851  HIDDLVMHMS  QRNAVDLVYR  AEDHATEFQR  LADVLYSGLE  NIRNVSLNAT
1901  SAAYVHYNIQ  SLIEESEELA  RDAHRTVTET  SLLSESLVSN  GKAAVQRSSR
1951  FLKEGNNLSR  KLPGIALELS  ELRNKTNRFQ  ENAVEITRQT  NESLLILRAI
2001  PEGIRDKGAK  TKELATSASQ  SAVSTLRDVA  GLSQELLNTS  ASLSRVNTTL
2051  RETHQLLQDS  TMATLLAGRK  VKDVEIQANL  LFDRLKPLKM  LEENLSRNLS
2101  EIKLLISQAR  KQAASIKVAV  SADRDCIRAY  QPQISSTNYN  TLTLNVKTQE
2151  PDNLLFYLGS  STASDFLAVE  MRRGRVAFLW  DLGSGSTRLE  FPDFPIDDNR
2201  WHSIHVARFG  NIGSLSVKEM  SSNQKSPTKT  SKSPGTANVL  DVNNSTLMFV
2251  GGLGGQIKKS  PAVKVTHFKG  CLGEAFLNGK  SIGLWNYIER  EGKCRGCFGS
2301  SQNEDPSFHF  DGSGYSVVEK  SLPATVTQII  MLFNTFSPNG  LLLYLGSYGT
2351  KDFLSIELFR  GRVKVMTDLG  SGPITLLTDR  RYNNGTWYKI  AFQRNRKQGV
2401  LAVIDAYNTS  NKETKQGETP  GASSDLNRLD  KDPIYVGGLP  RSRVVRRGVT
2451  TKSFVGCIKN  LEISRSTFDL  LRNSYGVRKG  CLLEPIRSVS  FLKGGYIELP
2501  PKSLSPESEW  LVTFATTNSS  GIILAALGGD  VEKRGDREEA  HVPFFSVMLI
2551  GGNIEVHVNP  GDGTGLRKAL  LHAPTGTCSD  GQAHSISLVR  NRRIITVQLD
2601  ENNPVEMKLG  TLVESRTINV  SNLYVGGIPE  GEGTSLLTMR  RSFHGCIKNL
2651  IFNLELLDFN  SAVGHEQVDL  DTCWLSERPK  LAPDAEDSKL  LREPRAFPEQ
```

```
2701 CVVDAALEYV PGAHQFGLTQ NSHFILPFNQ SAVRKKLSVE LSIRTFASSG
2751 LIYYMAHQNQ ADYAVLQLHG GRLHFMFDLG KGRTKVSHPA LLSDGKWHTV
2801 KTDYVKRKGF ITVDGRESPM VTVVGDGTML DVEGLFYLGG LPSQYQARKI
2851 GNITHSIPAC IGDVTVNSKQ LDKDSPVSAF TVNRCYAVAQ EGTYFDGSGY
2901 AALVKEGYKV QSDVNITLEF RTSSQNGVLL GISTAKVDAI GLELVDGKVL
2951 FHVNNGAGRI TPAYEPKTAT VLCDGKWHTL QANKSKHRIT LIVDGNAVGA
3001 ESPHTQSTSV DTNNPIYVGG YPAGVKQKCL RSQTSFRGCL RKLALIKSPQ
3051 VQSFDFSRAF ELHGVFLHSC PGTES
```

Structural and functional sites

Signal peptide: 1–17
EGF (8C) repeats: 270–301, 327–394, 397–451, 454–500, 503–512 (partial), 709–739 (partial), 742–788, 791–846, 849–899, 902–948, 951–995, 998–1041, 1044–1087, 1090–1109 (partial), 1111–1147 (partial), 1150–1159 (partial), 1362–1400 (partial), 1403–1449, 1452–1506, 1509–1553
G repeats: 2140–2327, 2328–2509, 2510–2736, 2737–2913, 2914–3075
Potential N-linked glycosylation sites: 38, 164, 555, 665, 763, 801, 838, 926, 952, 1045, 1407, 1579, 1596, 1678, 1689, 1698, 1717, 1804, 1894, 1898, 1957, 1974, 1991, 2038, 2047, 2094, 2098, 2243, 2244, 2384, 2408, 2518, 2619, 2729, 2852, 2915, 2983
IKVAV cell adhesion site: 2116–2120
RGD cell adhesion site: 2534–2536 (not conserved in murine sequence)
GD-6 cell adhesion peptide: 3026–3047 (from analogous murine sequence)

Accession number

Z26653

Primary structure: $\alpha 2$ chain

Ala	A	205	Cys	C	162	Asp	D	183	Glu	E	202
Phe	F	103	Gly	G	261	His	H	71	Ile	I	166
Lys	K	184	Leu	L	246	Met	M	46	Asn	N	162
Pro	P	173	Gln	Q	119	Arg	R	159	Ser	S	194
Thr	T	193	Val	V	156	Trp	W	29	Tyr	Y	96

Mol. wt (calc.) = 342 770 Residues = 3110

```
1   MPGAAGVLLL LLLSGGLGGV QAQRPQQQRQ SQAHQQRGLF PAVLNLASNA
51  LITTNATCGE KGPEMYCKLV EHVPGQPVRN PQCRICNQNS SNPNQRHPIT
101 NAIDGKNTWW QSPSIKNGIE YHYVTITLDL QQVFQIAYVI VKAANSPRPG
151 NWILERSLDD VEYKPWQYHA VTDTECLTLY NIYPRTGPPS YAKDDEVICT
201 SFYSKIHPLE NGEIHISLIN GRPSADDPSP ELLEFTSARY IRLRFQRIRT
251 LNADLMMFAH KDPREIDPIV TRRYYYSVKD ISVGGMCICY GHARACPLDP
301 ATNKSRCECE HNTCGDSCDQ CCPGFHQKPW RAGTFLTKTE CEACNCHGKA
351 EECYYDENVA RRNLSLNIRG KYIGGGVCIN CTQNTAGINC ETCTDGFFRP
401 KGVSPNYPRP CQPCHCDPIG SLNEVCVKDE KHARRGLAPG SCHCKTGFGG
```

```
451   VSCDRCARGY  TGYPDCKACN  CSGLGSKNED  PCFGPCICKE  NVEGGDCSRC
501   KSGFFNLQED  NWKGCDECFC  SGVSNRCQSS  YWTYGKIQDM  SGWYLTDLPG
551   RIRVAPQQDD  LDSPQQISIS  NAEARQALPH  SYYWSAPAPY  LGNKLPAVGG
601   QLTFTISYDL  EEEEEDTERV  LQLMIILEGN  DLSISTAQDE  VYLHPSEEHT
651   NVLLLKEESF  TIHGTHFPVR  RKEFMTVLAN  LKRVLLQITY  SFGMDAIFRL
701   SSVNLESAVS  YPTDGSIAAA  VEVCQCPPGY  TGSSCESCWP  RHRRVNGTIF
751   GGICEPCQCF  GHAESCDDVT  GECLNCKDHT  GGPYCDKCLP  GFYGEPTKGT
801   SEDCQPCACP  LNIPSNNFSP  TCHLDRSLGL  ICDGCPVGYT  GPRCERCAEG
851   YFGQPSVPGG  SCQPCQCNDN  LDFSIPGSCD  SLSGSCLICK  PGTTGRYCEL
901   CADGYFGDAV  DAKNCQPCRC  NAGGSFSEVC  HSQTGQCECR  ANVQGQRCDK
951   CKAGTFGLQS  ARGCVPCNCN  SFGSKSFDCE  ESGQCWCQPG  VTGKKCDRCA
1001  HGYFNFQEGG  CTACECSHLG  NNCDPKTGRC  ICPPNTIGEK  CSKCAPNTWG
1051  HSITTGCKAC  NCSTVGSLDF  QCNVNTGQCN  CHPKFSGAKC  TECSRGHWNY
1101  PRCNLCDCFL  PGTDATTCDS  ETKKCSCSDQ  TGQCTCKVNV  EGIHCDRCRP
1151  GKFGLDAKNP  LGCSSCYCFG  TTTQCSEAKG  LIRTWVTLKA  EQTILPLVDE
1201  ALQHTTTKGI  VFQHPEIVAH  MDLMREDLHL  EPFYWKLPEQ  FEGKKLMAYG
1251  GKLKYAIYFE  AREETGFSTY  NPQVIIRGGT  PTHARIIVRH  MAAPLIGQLT
1301  RHEIEMTEKE  WKYYGDDPRV  HRTVTREDFL  DILYDIHYIL  IKATYGNFMR
1351  QSRISEISME  VAEQGRGTTM  TPPADLIEKC  DCPLGYSGLS  CEACLPGFYR
1401  LRSQPGGRTP  GPTLGTCVPC  QCNGHSSLCD  PETSICQNCQ  HHTAGDFCER
1451  CALGYYGIVK  GLPNDCQQCA  CPLISSSNNF  SPSCVAEGLD  DYRCTACPRG
1501  YEGQYCERCA  PGYTGSPGNP  GGSCQECECD  PYGSLPVPCD  PVTGFCTCRP
1551  GATGRKCDGC  KHWHAREGWE  CVFCGDECTG  LLLGDLARLE  QMVMSINLTG
1601  PLPAPYKMLY  GLENMTQELK  HLLSPQRAPE  RLIQLAEGNL  NTLVTEMNEL
1651  LTRATKVTAD  GEQTGQDAER  TNTRAKSLGE  FIKELARDAE  AVNEKAIKLN
1701  ETLGTRDEAF  ERNLEGLQKE  IDQMIKELRR  KNLETQKEIA  EDELVAAEAL
1751  LKKVKKLFGE  SRGENEEMEK  DLREKLADYK  NKVDDAWDLL  REATDKIREA
1801  NRLFAVNQKN  MTALEKKKEA  VESGKRQIEN  TLKEGNDILD  EANRLADEIN
1851  SIIDYVEDIQ  TKLPPMSEEL  NDKIDDLSQE  IKDRKLAEKV  SQAESHAAQL
1901  NDSSAVLDGI  LDEAKNISFN  ATAAFKAYSN  IKDYIDEAEK  VAKEAKDLAH
1951  EATKLATGPR  GLLKEDAKGC  LQKSFRILNE  AKKLANDVKE  NEDHLNGLKT
2001  RIENADARNG  DLLRTLNDTL  GKLSAIPNDT  AAKLQAVKDK  ARQANDTAKD
2051  VLAQITELHQ  NLDGLKKNYN  KLADSVAKTN  AVVKDPSKNK  IIADADATVK
2101  NLEQEADRLI  DKLKPIKELE  DNLKKNISEI  KELINQARKQ  ANSIKVSVSS
2151  GGDCIRTYKP  EIKKGSYNNI  VVNVKTAVAD  NLLFYLGSAK  FIDFLAIEMR
2201  KGKVSFLWDV  GSGVGRVEYP  DLTIDDSYWY  RIVASRTGRN  GTISVRALDG
2251  PKASIVPSTH  HSTSPPGYTI  LDVDANAMLF  VGGLTGKLKK  ADAVRVITFT
2301  GCMGETYFDN  KPIGLWNFRE  KEGDCKGCTV  SPQVEDSEGT  IQFDGEGYAL
2351  VSRPIRWYPN  ISTVMFKFRT  FSSSALLMYL  ATRDLRDFMS  VELTDGHIKV
2401  SYDLGSGMAS  VVSNQNHNDG  KWKSFTLSRI  QKQANISIVD  IDTNQEENIA
2451  TSSSGNNFGL  DLKADDKIYF  GGLPTLRNLS  MKARPEVNLK  KYSGCLKDIE
2501  ISRTPYNILS  SPDYVGVTKG  CSLENVYTVS  FPKPGFVELS  PVPIDVGTEI
2551  NLSFSTKNES  GIILLGSGGT  PAPPRRKRRQ  TGQAYYVILL  NRGRLEVHLS
2601  TGARTMRKIV  IRPEPNLFHD  GREHSVHVER  TRGIFTVQVD  ENRRYMQNLT
2651  VEQPIEVKKL  FVGGAPPEFQ  PSPLRNIPPF  EGCIWNLVIN  SVPMDFARPV
2701  SFKNADIGRC  AHQKLREDED  GAAPAEIVIQ  PEPVPTPAFP  TPTPVLTHGP
2751  CAAESEPALL  IGSKQFGLSR  NSHIAIAFDD  TKVKNRLTIE  LEVRTEAESG
2801  LLFYMAAINH  ADFATVQLRN  GLPYFSYDLG  SGDTHTMIPT  KINDGQWHKI
2851  KIMRSKQEGI  LYVDGASNRT  ISPKKADILD  VVGMLYVGGL  PINYTTRRIG
2901  PVTYSIDGCV  RNLHMAEAPA  DLEQPTSSFH  VGTCFANAQR  GTYFDGTGFA
2951  KAVGGFKVGL  DLLVEFEFAT  TTTTGVLLGI  SSQKMDGMGI  EMIDEKLMFH
3001  VDNGAGRFTA  VYDAGVPGHL  CDGQWHKVTA  NKIKHRIELT  VDGNQVEAQS
3051  PNPASTSADT  NDPVFVGGFP  DDLKQFGLTT  SIPFRGCIRS  LKLTKGTASH
3101  WRLILPRPWN
```

Structural and functional sites

Signal peptide: 1–22
EGF (8C) repeats: 287–321, 344–411, 414–466, 469–515, 518–527 (partial), 724–754 (partial), 757–804, 807–862, 865–915, 918–964, 967–1011, 1014–1057, 1060–1103, 1106–1125 (partial), 1127–1163 (partial), 1166–1175 (partial), 1380–1417 (partial), 1420–1466, 1469–1524, 1527–1571
G repeats: 2168–2358, 2359–2550, 2551–2786, 2787–2961, 2962–3110
Potential N-linked glycosylation sites: 55, 89, 303, 363, 380, 470, 746, 1061, 1597, 1614, 1700, 1810, 1901, 1916, 1920, 2017, 2028, 2045, 2126, 2240, 2360, 2435, 2478, 2551, 2558, 2648, 2868, 2893

Accession number

L34155

Primary structure: $\alpha 3$ ($\alpha 3_a$) chain

Ala	A	101	Cys	C	42	Asp	D	91	Glu	E	98
Phe	F	57	Gly	G	135	His	H	35	Ile	I	73
Lys	K	105	Leu	L	184	Met	M	26	Asn	N	105
Pro	P	77	Gln	Q	99	Arg	R	97	Ser	S	145
Thr	T	97	Val	V	94	Trp	W	12	Tyr	Y	40

Mol. wt (calc.) = 189 305 Residues = 1713

```
1     MGWLWIFGAA LGQCLGYSSQ QQRVPFLQPP GQSQLQASYV EFRPSQGCSP
51    GYYRDHKGLY TGRCVPCNCN GHSNQCQDGS GICVNCQHNT AGEHCERCQE
101   GYYGNAVHGS CRACPCPHTN SFATGCVVNG GDVRCSCKAG YTGTQCERCA
151   PGYFGNPQKF GGSCQPCSCN SNGQLGSCHP LTGDCINQEP KDSSPAEECD
201   DCDSCVMTLL NDLATMGEQL RLVKSQLQGL SASAGLLEQM RHMETQAKDL
251   RNQLLNYRSA ISNHGSKIEG LERELTDLNQ EFETLQEKAQ VNSRKAQTLN
301   NNVNRATQSA KELDVKIKNV IRNVHILLKQ ISGTDGEGNN VPSGDFSREW
351   AEAQRMMREL RNRNFGKHLR EAEADKRESQ LLLNRIRTWQ KTHQGENNGL
401   ANSIRDSLNE YEAKLSDLRA RLQEAAAQAK QANGLNQENE RALGAIQRQV
451   KEINSLQSDF TKYLTTADSS LLQTNIALQL MEKSQKEYEK LAASLNEARQ
501   ELSDKVRELS RSAGKTSLVE EAEKHARSLQ ELAKQLEEIK RNASGDELVR
551   CAVDAATAYE NILNAIKAAE DAANRAASAS ESALQTVIKE DLPRKAKTLS
601   SNSDKLLNEA KMTQKKLKQE VSPALNNLQQ TLNIVTVQKE VIDTNLTTLR
651   DGLHGIQRGD IDAMISSAKS MVRKANDITD EVLDGLNPIQ TDVERIKDTY
701   GRTQNEDFKK ALTDADNSVN KLTNKLPDLW RKIESINQQL LPLGNISDNM
751   DRIRELIQQA RDAASKVAVP MRFNGKSGVE VRLPNDLEDL KGYTSLSLFL
801   QRPNSRENGG TENMFVMYLG NKDASRDYIG MAVVDGQLTC VYNLGDREAE
851   LQVDQILTKS ETKEAVMDRV KFQRIYQFAR LNYTKGATSS KPETPGVYDM
901   DGRNSNTLLN LDPENVVFYV GGYPPDFKLP SRLSFPPYKG CIELDDLNEN
951   VLSLYNFKKT FNLNTTEVEP CRRRKEESDK NYFEGTGYAR VPTQPHAPIP
1001  TFGQTIQTTV DRGLLFFAEN GDRFISLNIE DGKLMVRYKL NSELPKERGV
1051  GDAINNGRDH SIQIKIGKLQ KRMWINVDVQ NTIIDGEVFD FSTYYLGGIP
1101  IAIRERFNIS TPAFRGCMKN LKKTSGVVRL NDTVGVTKKC SEDWKLVRSA
1151  SFSRGGQLSF TDLGLPPTDH LQASFGFQTF QPSGILLDHQ TWTRNLQVTL
1201  EDGYIELSTS DSGGPIFKSP QTYMDGLLHY VSVISDNSGL RLLIDDQLLR
```

```
1251 NSKRLKHISS SRQSLRLGGS NFEGCISNVF VQRLSLSPEV LDLTSNSLKR
1301 DVSLGGCSLN KPPFLMLLKG STRFNKTKTF RINQLLQDTP VASPRSVKVW
1351 QDACSPLPKT QANHGALQFG DIPTSHLLFK LPQELLKPRS QFAVDMQTTS
1401 SRGLVFHTGT KNSFMALYLS KGRLVFALGT DGKKLRIKSK EKCNDGKWHT
1451 VVFGHDGEKG RLVVDGLRAR EGSLPGNSTI SIRAPVYLGS PPSGKPKSLP
1501 TNSFVGCLKN FQLDSKPLYT PSSSFGVSSC LGGPLEKGIY FSEEGGHVVL
1551 AHSVLLGPEF KLVFSIRPRS LTGILIHIGS QPGKHLCVYL EAGKVTASMD
1601 SGAGGTSTSV TPKQSLCDGQ WHSVAVTIKQ HILHLELDTD SSYTAGQIPF
1651 PPASTQEPLH LGGAPANLTT LRIPVWKSFF GCLRNIHVNH IPVPVTEALE
1701 VQGPVSLNGC PDQ
```

Structural and functional sites

Signal peptide: 1–20
EGF (8C) repeats: 67–111, 114–164, 167–199 (partial)
G repeats: 794–970, 971–1139, 1140–1353, 1354–1529, 1530–1713
Potential N-linked glycosylation sites: 645, 745, 882, 924, 1108, 1131, 1325, 1477, 1667

Accession number

X85108

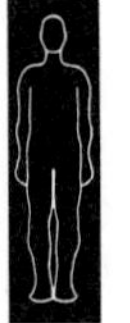

Primary structure: $\alpha3$ ($\alpha3_b$) chain

Shares the same amino acid sequence as the $\alpha3_a$ chain from residue 46 of the $\alpha3_a$ chain to the carboxyl terminus. Only partial sequence amino terminal to this residue has been published to date [14].

Accession number

S68960; S65926; I53516

Primary structure: $\alpha4$ chain

Ala	A	127	Cys	C	45	Asp	D	106	Glu	E	124
Phe	F	76	Gly	G	120	His	H	45	Ile	I	88
Lys	K	103	Leu	L	158	Met	M	31	Asn	N	103
Pro	P	77	Gln	Q	79	Arg	R	98	Ser	S	164
Thr	T	84	Val	V	120	Trp	W	11	Tyr	Y	57

Mol. wt (calc.) = 201 883 Residues = 1816

```
1   MALSSAWRSV LPLWLLWSAA CSRAASGDDN AFPFDIEGSS AVGRQDPPET
51  SEPRVALGRL PPAAEKCNAG FFHTLSGECV PCDCNGNSNE CLDGSGYCVH
101 CQRNTTGEHC EKCLDGYIGD SIRGAPQFCQ PCPCPLPHLA NFAESCYRKN
151 GAVRCICNEN YAGPNCERCA PGYYGNPLLI GSTCKKCDCS GNSDPNLIFE
201 DCDEVTGQCR NCLRNTTGFK CERCAPGYYG DARIAKNCAV CNCGGGPCDS
251 VTGECLEEGF EPPTGCDKCV WDLTDDLRLA ALSIEEGKSG VLSVSSGAAA
301 HRHVNEINAT IYLLKTKLSE RENQYALRKI QINNAENTMK SLLSDVEELV
```

```
351   EKENQASRKG QLVQKESMDT INHASQLVEQ AHDMRDKIQE INNKMLYYGE
401   EHELSPKEIS EKLVLAQKML EEIRSRQPFF TQRELVDEEA DEAYELLSQA
451   ESWQRLHNET RTLFPVVLEQ LDDYNAKLSD LQEALDQALN HVRDAEDMNR
501   ATAARQRDHE KQQERVREQM EVVNMSLSTS ADSLTTPRLT LSELDDIIKN
551   ASGIYAEIDG AKSELQVKLS NLSNLSHDLV QEAIDHAQDL QQEANELSRK
601   LHSSDMNGLV QKALDASNVY ENIVNYVSEA NETAEFALNT TDRIYDAVSG
651   IDTQIIYHKD ESENLLNQAR ELQAKAESSS DEAVADTSRR VGGALARKSA
701   LKTRLSDAVK QLQAAERGDA QQRLGQSRLI TEEANRTTME VQQATAPMAN
751   NLTNWSQNLQ HFDSSAYNTA VNSARDAVRN LTEVVPQLLD QLRTVEQKRP
801   ASNVSASIQR IREIAQTRSV ASKIQVSMMF DGQSAVEVH  SRTSMDDLKA
851   FTSLSLYMKP PVKRPELTET ADQFILYLGS KNAKKEYMGL AIKNDNLVYV
901   YNLGTKDVEI PLDSKPVSSW PAYFSIVKIE RVGKHGKVFL TVPSLSSTAE
951   EKFIKKCEFS GDDSLLDLDP EDTVFYVGGV PSNFKLPTSL NLPGFVGCLE
1001  LATLNNDVIS LYNFKHIYNM DPSTSVPCAR DKLAFTQSRA ASYFFDGSGY
1051  AVVRDITRRG KFGQVTRFDI EVRTPADNGL ILLMVNGSMF FRLEMRNGYL
1101  HVFYDFGFSS GRVHLEDTLK KAQINDAKYH EISIIYHNDK KMILVVDRRH
1151  VKSMDNEKMK IPFTDIYIGG APPEILQSRA LRAHLPLDIN FRGCMKGFQF
1201  QKKDFNLLEQ TETLGVGYGC PEDSLISRRA YFNGQSFIAS IQKISFFDGF
1251  EGGFNFRTLQ PNGLLFYYAS GSDVFSISLD NGTVIMDVKG IKVQSVDKQY
1301  NDGLSHFVIS SVSPTRYELI VDKSRVGSKN PTKGKIEQTQ ASEKKFYFGG
1351  SPISAQYANF TGCISNAYFT RVDRDVEVED FQRYTEKVHT SLYECPIESS
1401  PLFLLHKKGK NLSKPKASQN KKGGKSKDAP SWDPVALKLP ERNTPRNSHC
1451  HLSNSPRAIE HAYQYGGTAN SRQEFEHLKG DFGAKSQFSI RLRTRSSHGM
1501  IFYVSDQEEN DFMTLFLAHG RLVYMFNVGH KKLKIRSQEK YNDGLWHDVI
1551  FIRERSSGRL VIDGLRVLEE SLPPTEATWK IKGPIYLGGV APGKAVKNVQ
1601  INSIYSFSGC LSNLQLNGAS ITSASQTFSV TPCFEGPMET GTYFSTEGGY
1651  VVLDESFNIG LKFEIAFEVR PRSSSGTLVH GHSVNGEYLN VHMKNGQVIV
1701  KVNNGIRDFS TSVTPKQSLC DGRWHRITVI RDSNVVQLDV DSEVNHVVGP
1751  LNPKPIDHRE PVFVGGVPES LLTPRLAPSK PFTGCIRHFV IDGHPVSFSK
1801  AALVSGAVSI NSCPAA
```

Structural and functional sites

Signal peptide: 1–24
EGF (8C) repeats: 82–129, 132–184, 187–238, 241–265 (partial)
G repeats: 835–1028, 1029–1190, 1191–1450, 1451–1633, 1634–1816
Potential N-linked glycosylation sites: 104, 215, 308, 458, 524, 550, 571, 574, 631, 639, 735, 751, 754, 780, 803, 1086, 1281, 1330, 1359, 1411.

Accession number

U37501

Primary structure: α5 chain (mouse; partial)

Ala	A	278	Cys	C	215	Asp	D	165	Glu	E	189
Phe	F	126	Gly	G	325	His	H	119	Ile	I	81
Lys	K	79	Leu	L	351	Met	M	49	Asn	N	121
Pro	P	272	Gln	Q	221	Arg	R	227	Ser	S	250
Thr	T	190	Val	V	224	Trp	W	34	Tyr	Y	94

Mol. wt (calc.) = 393 038 Residues = 3610

```
1     DLYCKLVGGP VAGGDPNQTI QGQYCDICTA ANSNKAHPVS NAIDGTERWW
51    QSPPLSRGLE YNEVNVTLDL GQVFHVAYVL IKFANSPRPD LWVLERSTDF
101   GHTYQPWQFF ASSKRDCLER FGPRTLERIT QDDDVICTTE YSRIVPLENG
151   EIVVSLVNGR PGALNFSYSP LLRDFTKATN IRLRFLRTNT LLGHLMGKAL
201   RDPTVTRRYY YSIKDISIGG RCVCHGHADV CDAKDPLDPF RLQCACQHNT
251   CGGSCDRCCP GFNQQPWKPA TTDSANECQS CNCHGHAYDC YYDPEVDRRN
301   ASQNQDNVYQ GGGVCLDCQH HTTGINCERC LPGFFRAPDQ PLDSPHVCRP
351   CDCESDFTDG TCEDLTGRCY CRPNFTGELC AACAEGYTDF PHCYPLPSFP
401   HNDTREQVLP AGQIVNCDCN AAGTQGNACR KDPRLGRCVC KPNFRGAHCE
451   LCAPGFHGPS CHPCQCSSPG VANSLCDPES GQCMCRTGFE GDRCDHCALG
501   YFHFPLCQLC GCSPAGTLPE GCDEAGRCQC RPGFDGPHCD RCLPGYHGYP
551   DCHACACDPR GALDQQCGVG GLCHCRPGNT GATCQECSPG FYGFPSCIPC
601   HCSADGSLHT TCDPTTGQCR CRPRVTGLHC DMCVPGAYNF PYCEAGSCHP
651   AGLAPANPAL PETQAPCMCR AHVEGPSCDR CKPGYWGLSA SNPEGCTRCS
701   CDPRGTLGGV TECQGNGQCF CKAHVCGKTC AACKDGFFGL DYADYFGCRS
751   CRCDVGGALG QGCEPKTGAC RCRPNTQGPT CSEPAKDHYL PDLHHMRLEL
801   EEAATPEGHA VRFGFNPLEF ENFSWRGYAH MMAIQPRIVA RLNVTSPDLF
851   RLVFRYVNRG STSVNGQISV REEGKLSSCT NCTEQSQPVA FPPSTEPAFV
901   TVPQRGFGEP FVLNPGIWAL LVEAEGVLLD YVVLLPSTYY EAALLQHRVT
951   EACTYRPSAL HSTENCLVYA HLPLDGFPSA AGTEALCRHD NSLPRPCPTE
1001  QLSPSHPPLA TCFGSDVDIQ LEMAVPQPGQ YVLVVEYVGE DSHQEMGVAV
1051  HTPQRAPQQG VLNLHPCPYS SLCRSPARDT QHHLAIFHLD SEASIRLTAE
1101  QAHFFLHSVT LVPVEEFSTE FVEPRVFCVS SHGTFNPSSA ACLASRFPKP
1151  PQPIILKDCQ VLPLPPDLPL TQSQELSPGA PPEGPQPRPP TAVDPNAEPT
1201  LLRHPQGTVV FTTQVPTLGR YAFLLHGYQP VHPSFPVEVL INGGRIWQGH
1251  ANASFCPHGY GCRTLVLCEG QTMLDVTDNE LTVTVRVPEG RWLWLDYVLI
1301  VPEDAYSSSY LQEEPLDKSY DFISHCATQG YHISPSSSSP FCRNAATSLS
1351  LFYNNGALPC GCHEVGAVSP TCEPFGGQCP CRGHVIGRDC SRCATGYWGF
1401  PNCRPCDCGA RLCDELTGQC ICPPRTVPPD CLVCQPQSFG CHPLVGCEEC
1451  NCSGPGVQEL TDPTCDMDSG QCRCRPNVAG RRCDTCAPGF YGYPSCRPCD
1501  CHEAGTMASV CDPLTGQCHC KENVQGSRCD QCRVGTFSLD AANPKGCTRC
1551  FCFGATERCG NSNLARHEFV DMEGWVLLSS DRQVVPHEHR PEIELLHADL
1601  RSVADTFSEL YWQAPPSYLG DRVSSYGGTL HYELHSETQR GDIFIPYESR
1651  PDVVLQGNQM SIAFLELAYP PPGQVHRGQL QLVEGNFRHL ETHNPVSREE
1701  LMMVLAGLEQ LQIRALFSQT SSSVSLRRVV LEVASEAGRG PPASNVELCM
1751  CPANYRGDSC QECAPGYYRD TKGLFLGRCV PCQCHGHSDR CLPGSGICVG
1801  CQHNTEGDQC ERCRPGFVSS DPSNPASPCV SCPCPLAVPS NNFADGCVLR
1851  NGRTQCLCRP GYAGASCERC APGFFGNPLV LGSSCQPCDC SGNGDPNMIF
1901  SDCDPLTGAC RGCLRHTTGP HCERCAPGFY GNALLPGNCT RCDCSPCGTE
1951  TCDPQSGRCL CKAGVTGQRC DRCLEGYFGF EQCQGCRPCA CGPAAKGSEC
2001  HPQSGQCHCQ PGTTGPQCLE CAPGYWGLPE KGCRRCQCPR GHCDPHTGHC
2051  TCPPGLSGER CDTCSQQHQV PVPGKPGGHG IHCEVCDHCV VLLLDDLERA
2101  GALLPAIREQ LQGINASSAA WARLHRLNAS IADLQSKLRR PPGPRYQAAQ
2151  QLQTLEQQSI SLQQDTERLG SQATGVQGQA GQLLDTTEST LGRAQKLLES
2201  VRAVGRALNE LASRMGQGSP GDALVPSGEQ LRWALAEVER LLWDMRTRDL
2251  GAQGAVAEAE LAEAQRLMAR VQEQLTSFWE ENQSLATHIR DQLAQYESGL
2301  MDLREALNQA VNTTREAEEL NSRNQERVKE ALQWKQELSQ DNATLKATLQ
2351  AASLILGHVS ELLQGIDQAK EDLEHLAASL DGAWTPLLKR MQAFSPASSK
2401  VDLVEAAEAH AQKLNQLAIN LSGIILGINQ DRFIQRAVEA SNAYSSILQA
2451  VQAAEDAAGQ ALRQASRTWE MVVQRGLAAG ARQLLANSSA LEETILGHQG
2501  RLGLAQGRLQ AAGIQLHNVW ARKNQLAAQI QEAQAMLAMD TSETSEKIAH
2551  AKAVAAEALS TATHVQSQLQ GMQKNVERWQ SQLGGLQGQD LSQVERDASS
2601  SVSTLEKTLP QLLAKLSRLE NRGVHNASLA LSANIGRVRK LIAQARSAAS
2651  KVKVSMKFNG RSGVRLRPPR DLADLAAYTA LKFHIQSPVP APEPGKNTGD
```

```
2701 HFVLYMGSRQ ATGDYMGVSL RNQKVHWVYR LGKAGPTTLS IDENIGEQFA
2751 AVSIDRTLQF GHMSVTVEKQ MVHEIKGDTV APGSEGLLNL HPDDFVFYVG
2801 GYPSNFTPPE PLRFPGYLGC IEMETLNEEV VSLYNFEQTF MLDTAVDKPC
2851 ARSKATGDPW LTDGSYLDGS GFARISFEKQ FSNTKRFDQE LRLVSYNGII
2901 FFLKQESQFL CLAVQEGTLV LFYDFGSGLK KADPLQPPQA LTAASKAIQV
2951 FLLAGNRKRV LVRVERATVF SVDQDNMLEM ADAYYLGGVP PEQLPLSLRQ
3001 LFPSGGSVRG CIKGIKALGK YVDLKRLNTT GISFGCTADL LVGRTMTFHG
3051 HGFLPLALPN VAPITEVVYS GFGFRGTQDN NLLYYRTSPD GPYQVSLREG
3101 HVTLRFMNQE VETQRVFADG APHYVAFYSN VTGVWLYVDD QLQLVKSHER
3151 TTPMLQLQPE EPSRLLLGGL PVSGTFHNFS GCISNVFVQR LRGPQRVFDL
3201 HQNMGSVNVS VGCTPAQLIE TSRATAQKVS RRSRQPSQDL ACTTPWLPGT
3251 IQDAYQFGGP LPSYLQFVGI SPSHRNRLHL SMLVRPHAAS QGLLLSTAPM
3301 SGRSPSLVLF LNHGHFVAQT EGPGPRLQVQ SRQHSRAGQW HRVSVRWGMQ
3351 QIQLVVDGSQ TWSQKALHHR VPRAERPQPY TLSVGGLPAS SYSSKLPVSV
3401 GFSGCLKKLQ LDKQPLRTPT QMVGVTPCVS GPLEDGLFFP GSEGVVTLEL
3451 PKAKMPYVSL ELEMRPLAAA GLIFHLGQAL ATPYMQLKVL TEQVLLQAND
3501 GAGEFSTWVT YPKLCDGRWH RVAVIMGRDT LRLEVDTQSN HTTGRLPESL
3551 AGSPALLHLG SLPKSSTARP ELPAYRGCWR KLLINGPPVN VTASVQIQGA
3601 VGMRGCPSGT
```

Structural and functional sites

Signal peptide: Not defined
G repeats: 2679–3610 (5 repeats)
EGF (8C) repeats: 211–780 (11 repeats), 1359–1558 (4 repeats), 1748–2084 (7 repeats)
Putative RGD cell adhesion sites: 1640–1642, 1756–1758
Putative LRE cell adhesion sites: 2302–2304, 3097–3099
Potential N-linked glycosylation sites: 17, 65, 165, 300, 374, 402, 843, 881, 1134, 1252, 1451, 1938, 2115, 2128, 2282, 2312, 2342, 2420, 2487, 2626, 2805, 3028, 3130, 3178, 3208, 3240, 3590

Accession number

P07942

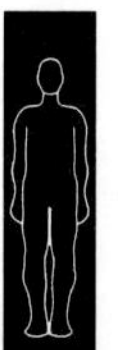

Primary structure: β1 chain

Ala	A	119	Cys	C	127	Asp	D	118	Glu	E	153
Phe	F	55	Gly	G	126	His	H	39	Ile	I	70
Lys	K	85	Leu	L	137	Met	M	32	Asn	N	80
Pro	P	87	Gln	Q	89	Arg	R	89	Ser	S	120
Thr	T	93	Val	V	98	Trp	W	13	Tyr	Y	56

Mol. wt (calc.) = 197 844 Residues = 1786

```
1    MGLLQLLAFS FLALCRARVR AQEPEFSYGC AEGSCYPATG DLLIGRAQKL
51   SVTSTCGLHK PEPYCIVSHL QEDKKCFICN SQDPYHETLN PDSHLIENVV
```

```
101   TTFAPNRLKI  WWQSENGVEN  VTIQLDLEAE  FHFTHLIMTF  KTFRPAAMLI
151   ERSSDFGKTW  GVYRYFAYDC  EASFPGISTG  PMKKVDDIIC  DSRYSDIEPS
201   TEGEVIFRAL  DPAFKIEDPY  SPRIQNLLKI  TNLRIKFVKL  HTLGDNLLDS
251   RMEIREKYYY  AVYDMVVRGN  CFCYGHASEC  APVDGFNEEV  EGMVHGHCMC
301   RHNTKGLNCE  LCMDFYHDLP  WRPAEGRNSN  ACKKCNCNEH  SISCHFDMAV
351   YLATGNVSGG  VCDDCQHNTM  GRNCEQCKPF  YYQHPERDIR  DPNFCERCTC
401   DPAGSQNEGI  CDSYTDFSTG  LIAGQCRCKL  NVEGEHCDVC  KEGFYDLSSE
451   DPFGCKSCAC  NPLGTIPGGN  PCDSETGHCY  CKRLVTGQHC  DQCLPEHWGL
501   SNDLDGCRPC  DCDLGGALNN  SCFAESGQCS  CRPHMIGRQC  NEVEPGYYFA
551   TLDHYLYEAE  EANLGPGVSI  VERQYIQDRI  PSWTGAGFVR  VPEGAYLEFF
601   IDNIPYSMEY  DILIRYEPQL  PDHWEKAVIT  VQRPGRIPTS  SRCGNTIPDD
651   DNQVVSLSPG  SRYVVLPRPV  CFEKGTNYTV  RLELPQYTSS  DSDVESPYTL
701   IDSLVLMPYC  KSLDIFTVGG  SGDGVVTNSA  WETFQRYRCL  ENSRSVVKTP
751   MTDVCRNIIF  SISALLHQTG  LACECDPQGS  LSSVCDPNGG  QCQCRPNVVG
801   RTCNRCAPGT  FGFGPSGCKP  CECHLQGSVN  AFCNPVTGQC  HCFQGVYARQ
851   CDRCLPGHWG  FPSCQPCQCN  GHADDCDPVT  GECLNCQDYT  MGHNCERCLA
901   GYYGDPIIGS  GDHCRPCPCP  DGPDSGRQFA  RSCYQDPVTL  QLACVCDPGY
951   IGSRCDDCAS  GYFGNPSEVG  GSCQPCQCHN  NIDTTDPEAC  DKETGRCLKC
1001  LYHTEGEHCQ  FCRFGYYGDA  LRQDCRKCVC  NYLGTVQEHC  NGSDCQCDKA
1051  TGQCLCLPNV  IGQNCDRCAP  NTWQLASGTG  CDPCNCNAAH  SFGPSCNEFT
1101  GQCQCMPGFG  GRTCSECQEL  FWGDPDVECR  ACDCDPRGIE  TPQCDQSTGQ
1151  CVCVEGVEGP  RCDKCTRGYS  GVFPDCTPCH  QCFALWDVII  AELTNRTHRF
1201  LEKAKALKIS  GVIGPYRETV  DSVERKVSEI  KDILAQSPAA  EPLKNIGNLF
1251  EEAEKLIKDV  TEMMAQVEVK  LSDTTSQSNS  TAKELDSLQT  EAESLDNTVK
1301  ELAEQLEFIK  NSDIRGALDS  ITKYFQMSLE  AEERVNASTT  EPNSTVEQSA
1351  LMRDRVEDVM  MERESQFKEK  QEEQARLLDE  LAGKLQSLDL  SAAAEMTCGT
1401  PPGASCSETE  CGGPNCRTDE  GERKCGGPGC  GGLVTVAHNA  WQKAMDLDQD
1451  VLSALAEVEQ  LSKMVSEAKL  RADEAKQSAE  DILLKTNATK  EKMDKSNEEL
1501  RNLIKQIRNF  LTQDSADLDS  IEAVANEVLK  MEMPSTPQQL  QNLTEDIRER
1551  VESLSQVEVI  LQHSAADIAR  AEMLLEEAKR  ASKSATDVKV  TADMVKEALE
1601  EAEKAQVAAE  KAIKQADEDI  QGTQNLLTSI  ESETAASEET  LFNASQRISE
1651  LERNVEELKR  KAAQNSGEAE  YIEKVVYTVK  QSAEDVKKTL  DGELDEKYKK
1701  VENLIAKKTE  ESADARRKAE  MLQNEAKTLL  AQANSKLQLL  KDLERKYEDN
1751  QRYLEDKAQE  LARLEGEVRS  LLKDISQKVA  VYSTCL
```

Structural and functional sites

Signal peptide: 1–21
EGF (8C) repeats: 271–334, 335–397, 398–457, 458–509, 510–540 (partial), 773–820, 821–866, 867–916, 917–975, 976–1027, 1028–1087, 1088–1121 (partial), 1122–1178
Potential N-linked glycosylation sites: 120, 356, 519, 677, 1041, 1195, 1279, 1336, 1343, 1487, 1542
LGTIPG cell adhesion site: 463–468
RYVVLPR (F9) cell adhesion site: 662–668
PDSGR cell adhesion site: 923–927
YIGSR cell adhesion site: 950–954

Accession number

P55268

Primary structure: β2 chain

Ala	A	177	Cys	C	125	Asp	D	101	Glu	E	113
Phe	F	44	Gly	G	161	His	H	73	Ile	I	43
Lys	K	33	Leu	L	170	Met	M	16	Asn	N	47
Pro	P	108	Gln	Q	120	Arg	R	146	Ser	S	109
Thr	T	82	Val	V	80	Trp	W	16	Tyr	Y	34

Mol. wt (calc.) = 196 080 Residues = 1798

```
1     MELTSRERGR GQPLPWELRL GLLLSVLAAT LAQAPAPDVP GCSRGSCYPA
51    TGDLLVGRAD RLTASSTCGL NGPQPYCIVS HLQDEKKCFL CDSRRPFSAR
101   DNPHSHRIQN VVTSFAPQRR AAWWQSENGI PAVTIQLDLE AEFHFTHLIM
151   TFKTFRPAAM LVERSADFGR TWHVYRYFSY DCGADFPGVP LAPPRHWDDV
201   VCESRYSEIE PSTEGEVIYR VLDPAIPIPD PYSSRIQNLL KITNLRVNLT
251   RLHTLGDNLL DPRREIREKY YYALYELVVR GNCFCYGHAS ECAPAPGAPA
301   HAEGMVHGAC ICKHNTRGLN CEQCQDFYRD LPWRPAEDGH SHACRKCECH
351   GHTHSCHFDM AVYLASGNVS GGVCDGCQHN TAGRHCELCR PFFYRDPTKD
401   LRDPAVCRSC DCDPMGSQDG GRCDSHDDPA LGLVSGQCRC KEHVVGTRCQ
451   QCRDGFFGLS ISDRLGCRRC QCNARGTVPG STPCDPNSGS CYCKRLVTGR
501   GCDRCLPGHW GLSHDLLGCR PCDCDVGGAL DPQCDEGTGQ CHCRQHMVGR
551   RCEQVQPGYF RPFLDHLIWE AEDTRGQVLD VVERLVTPGE TPSWTGSGFV
601   RLQEGQTLEF LVASVPKAMD YDLLLRLEPQ VPEQWAELEL IVQRPGPVPA
651   HSLCGHLVPK DDRIQGTLQP HARYLIFPNP VCLEPGISYK LHLKLVRTGG
701   SAQPETPYSG PGLLIDSLVL LPRVLVLEMF SGGDAAALER QATFERYQCH
751   EEGLVPSKTS PSEACAPLLI SLSTLIYNGA LPCQCNPQGS LSSECNPHGG
801   QCLCKPGVVG RRCDLCAPGY YGFGPTGCQA CQCSHEGALS SLCEKTSGQC
851   LCRTGAFGLR CDRCQRGQWG FPSCRPCVCN GHADECNTHT GACLGCRDHT
901   GGEHCERCIA GFHRDPRLPY GGQCRPCPCP EGPGSQRHFA TSCHQDEYSQ
951   QIVCHCRAGY TGLRCEACAP GHFGDPSRPG GRCQLCECSG NIDPMDPDAC
1001  DPHTGQCLRC LHHTEGPHCA HCKPGFHGQA ARQSCHRCTC NLLGTNPQQC
1051  PSPDQCHCDP SSGQCPCLPN VQGPSCDRCA PNFWNLTSGH GCQPCACHPS
1101  RARGPTCNEF TGQCHCRAGF GGRTCSECQE LHWGDPGLQC HACDCDSRGI
1151  DTPQCHRFTG HCSCRPGVSG VRCDQCARGF SGIFPACHPC HACFGDWDRV
1201  VQDLAARTQR LEQRAQELQQ TGVLGAFESS FWHMQEKLGI VQGIVGARNT
1251  SAASTAQLVE ATEELRREIG EATEHLTQLE ADLTDVQDEN FNANHALSGL
1301  ERDRLALNLT LRQLDQHLDL LKHSNFLGAY DSIRHAHSQS AEAERRANTS
1351  ALAVPSPVSN SASARHRTEA LMDAQKEDFN SKHMANQRAL GKLSAHTHTL
1401  SLTDINELVC GAPGDAPCAT SPCGGAGCRD EDGQPRCGGL SCNGAAATAD
1451  LALGRARHTQ AELQRALAEG GSILSRVAET RRQASEAQQR AQAALDKANA
1501  SRGQVEQANQ ELQELIQSVK DFLNQEGADP DSIEMVATRV LELSIPASAE
1551  QIQHLAGAIA ERVRSLADVD AILARTVGDV RRAEQLLQDA RRARSWAEDE
1601  KQKAETVQAA LEEAQRAQGI AQGAIRGAVA DTRDTEQTLY QVQERMAGAE
1651  RALSSAGERA RQLDALLEAL KLKRAGNSLA ASTAEETAGS AQGRAQEAEQ
1701  LLRGPLGDQY QTVKALAERK AQGVLAAQAR AEQLRDEARD LLQAAQDKLQ
1751  RLQELEGTYE ENERALESKA AQLDGLEARM RSVLQAINLQ VQIYNTCQ
```

Structural and functional sites

Signal peptide: 1–32

EGF (8C) repeats: 251–312, 315–378, 380–438, 440–490, 492–520 (partial), 751–796, 799–842, 845–892, 895–950, 953–1002 (imperfect), 1005–1046, 1062–1107 (imperfect), 1110–1154.

Potential N-linked glycosylation sites: 216, 336, 1052, 1216, 1275, 1315, 1466

Accession number

A53612

Primary structure: β3 chain

Ala	A	108	Cys	C	68	Asp	D	60	Glu	E	75
Phe	F	30	Gly	G	88	His	H	21	Ile	I	27
Lys	K	30	Leu	L	105	Met	M	28	Asn	N	33
Pro	P	63	Gln	Q	98	Arg	R	106	Ser	S	80
Thr	T	55	Val	V	61	Trp	W	10	Tyr	Y	24

Mol. wt (calc.) = 129 417 Residues = 1170

```
1    MRPFFLLCFA LPGLLHAQQA CSRGACYPPV GDLLVGRTRF LRASSTCGLT
51   KPETYCTQYG EWQMKCCKCD SRQPHNYYSH RVENVASSSG PMRWWQSQND
101  VNPVSLQLDL DRRFQLQEVM MEFRGPMPAG MLIERSSDFG KTWRVYQYLA
151  ADCTSTFPRV RQGRPQSWQD VRCQSLPQRP NARLNGGKVQ LNLMDLVSGI
201  PATQSQKIQE VGEITNLRVN FTRLAPVPQR GYHPPSAYYA VSQLRLQGSC
251  FCHGHADRCA PKPGASAGST AVQVHDVCVC QHNTAGPNCE RCAPFYNNRP
301  WRPAEGQDAH ECQRCDCNGH SETCHFDPAV FAASQGAYGG VCDNCRDHTE
351  GKNCERCQLH YFRNRRPGAS IQETCISCEC DPDGAVAGAP CDPVTGQCVC
401  KEHVQGERCD LCKPGFTGLT YANPRRCHRC DCNILGSREM PCDEESGRCL
451  CLPNVVGPKC DQCAPYHWKL ASGQGCEPCA CDPHNSLSPQ CNQFTGQCPC
501  REGFGGLMCS AAAIRQCPDR TYGDVATGCR ACDCDFRGTE GPGCDKASGR
551  CLCRPGLTGP RCDQCQRGYC NRYPVCVACH PCFQTYDADL REQALRFGRL
601  PNATASLWSG PGLEDRGLAS RILDAKSKIE QIRAVLSSPA VTEQEVAQVA
651  SAILSLRRTL QGLQLDLPLE EETLSLPRDL ESLDRSFNGL LTMYQRKREQ
701  FEKISSADPS GAFRMLSTAY EQSAQAAQQV SDSSRLLDQL RDSRREAERL
751  VRQAGGGGGT GSPKLVALRL EMSSLPDLTP TFNKLCGNSR QMACTPISCP
801  GELCPQDNGT ACASRCRGVL PRAGGAFLMA GQVAEQLRGF NAQLQRTRQM
851  IRAAEESASQ IQSSAQRLET QVSASRSQME EDVRRTRLLI QQVRDFLTDP
901  DTDAATIQEV SEAVLALWLP TDSATVLQKM NEIQAIAARL PNVDLVLSQT
951  KQDIARARRL QAEAEEARSR AHAVEGQVED VVGNLRQGTV ALQEAQDTMQ
1001 GTSRSLRLIQ DRVAEVQQVL RPAEKLVTSM TKQLGDFWTR MEELRHQARQ
1051 QGAEAVQAQQ LAEGASEQAL SAQEGFERIK QKYAELKDRL GQSSMLGEQG
1101 ARIQSVKTEA EELFGETMEM MDRMKDMELE LLRGSQAIML RSADLTGLEK
1151 RVEQIRDHIN GRVLYYATCK
```

Structural and functional sites

Signal peptide: 1–17
EGF (8C) repeats: 233–295, 298–358, 361–410, 413–459, 462–512, 515–553
Potential N-linked glycosylation sites: 203, 585, 791

Accession number

P11047

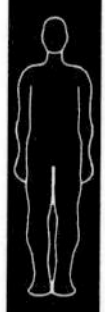

Primary structure: γ1 chain

Ala	A	139	Cys	C	100	Asp	D	106	Glu	E	135
Phe	F	50	Gly	G	113	His	H	27	Ile	I	47
Lys	K	81	Leu	L	117	Met	M	22	Asn	N	100
Pro	P	72	Gln	Q	74	Arg	R	96	Ser	S	99
Thr	T	97	Val	V	78	Trp	W	11	Tyr	Y	45

Mol. wt (calc.) = 177 409 Residues = 1609

```
1     MRGSHRAAPA LRPRGRLWPV LAVLAAAAAA GCAQAAMDEC TDEGGRPQRC
51    MPEFVNAAFN VTVVATNTCG TPPEEYCVQT GVTGVTKSCH LCDAGQPHLQ
101   HGAAFLTDYN NQADTTWWQS QTMLAGVQYP SSINLTLHLG KAFDITYVRL
151   KFHTSRPESF AIYKRTREDG PWIPYQYYSG SCENTYSKAN RGFIRTGGDE
201   QQALCTDEFS DFSPLTGGNV AFSTLEGRPS AYNFDNSPVL QEWVTATDIR
251   VTLNRLNTFG DEVFNDPKVL KSYYYAISDF AVGGRCKCNG HASECMKNEF
301   DKLVCNCKHN TYGVDCEKCL PFFNDRPWRR ATAESASECL PCDCNGRSQE
351   CYFDPELYRS TGHGGHCTNC QDNTDGAHCE RCRENFFRLG NNEACSSCHC
401   SPVGSLSTQC DSYGRCSCKP GVMGDKCDRC QPGFHSLTEA GCRPCSCDPS
451   GSIDECNVET GRCVCKDNVE GFNCERCKPG FFNLESSNPR GCTPCFCFGH
501   SSVCTNAVGY SVYSISSTFQ IDEDGWRAEQ RDGSEASLEW SSERQDIAVI
551   SDSYFPRYFI APAKFLGKQV LSYGQNLSFS FRVDRRDTRL SAEDLVLEGA
601   GLRVSVPLIA QGNSYPSETT VKYVFRLHEA TDYPWRPALT PFEFQKLLNN
651   LTSIKIRGTY SERSAGYLDD VTLASARPGP GVPATWVESC TCPVGYGGQF
701   CEMCLSGYRR ETPNLGPYSP CVLCACNGHS ETCDPETGVC NCRDNTAGPH
751   CEKCSDGYYG DSTAGTSSDC QPCPCPGGSS CAVVPKTKEV VCTNCPTGTT
801   GKRCELCDDG YFGDPLGRNG PVRLCRLCQC SDNIDPNAVG NCNRLTGECL
851   KCIYNTAGFY CDRCKDGFFG NPLAPNPADK CKACNCNPYG TMKQQSSCNP
901   VTGQCECLPH VTGQDCGACD PGFYNLQSGQ GCERCDCHAL GSTNGQCDIR
951   TGQCECQPGI TGQHCERCEV NHFGFGPEGC KPCDCHPEGS LSLQCKDDGR
1001  CECREGFVGN RCDQCEENYF YNRSWPGCQE CPACYRLVKD KVADHRVKLQ
1051  ELESLIANLG TGDEMVTDQA FEDRLKEAER EVMDLLREAQ DVKDVDQNLM
1101  DRLQRVNNTL SSQISRLQNI RNTIEETGNL AEQARAHVEN TERLIEIASR
1151  ELEKAKVAAA NVSVTQPEST GDPNNMTLLA EEARKLAERH KQEADDIVRV
1201  AKTANDTSTE AYNLLLRTLA GENQTAFEIE ELNRKYEQAK NISQDLEKQA
1251  ARVHEEAKRA GDKAVEIYAS VAQLSPLDSE TLENEANNIK MEAENLEQLI
1301  DQKLKDYEDL REDMRGKELE VKNLLEKGKT EQQTADQLLA RADAAKALAE
1351  EAAKKGRDTL QEANDILNNL KDFDRRVNDN KTAAEEALRK IPAINQTITE
1401  ANEKTREAQQ ALGSAAADAT EAKNKAHEAE RIASAVQKNA TSTKAEAERT
1451  FAEVTDLDNE VNNMLKQLQE AEKELKRKQD DADQDMMMAG MASQAAQEAE
1501  INARKAKNSV TSLLSIINDL LEQLGQLDTV DLNKLNEIEG TLNKAKDEMK
1551  VSDLDRKVSD LENEAKKQEA AIMDYNRDIE EIMKDIRNLE DIRKTLPSGC
1601  FNTPSIEKP
```

Structural and functional sites

Signal peptide: 1–33

EGF (8C) repeats: 286–341, 342–397, 398–444, 445–494, 495–509 (partial), 690–723 (partial), 724–772, 773–827, 828–883, 884–934, 935–982, 983–1030

Potential N-linked glycosylation sites: 60, 134, 576, 650, 1022, 1107, 1161, 1175, 1205, 1223, 1241, 1380, 1395, 1439

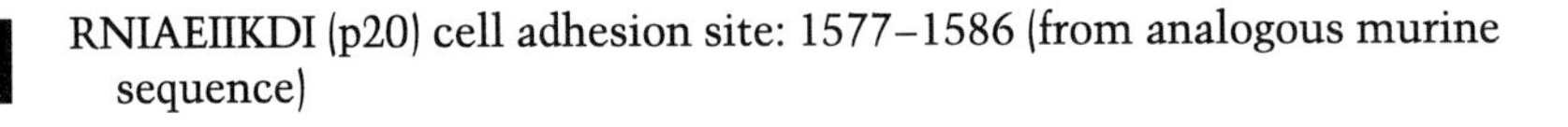

RNIAEIIKDI (p20) cell adhesion site: 1577–1586 (from analogous murine sequence)

Accession number

Z15008

Primary structure: γ2 chain

Ala	A	96	Cys	C	68	Asp	D	74	Glu	E	80
Phe	F	29	Gly	G	98	His	H	25	Ile	I	35
Lys	K	53	Leu	L	107	Met	M	21	Asn	N	57
Pro	P	56	Gln	Q	77	Arg	R	81	Ser	S	94
Thr	T	51	Val	V	55	Trp	W	6	Tyr	Y	30

Mol. wt (calc.) = 130 725 Residues = 1193

```
1      MPALWLGCCL  CFSLLLPAAR  ATSRREVCDC  NGKSRQCIFD  RELHRQTGNG
51     FRCLNCNDNT  DGIHCEKCKN  GFYRHRERDR  CLPCNCNSKG  SLSARCDNSG
101    RCSCKPGVTG  ARCDRCLPGF  HMLTDAGCTQ  DQRLLDSKCD  CDPAGIAGPC
151    DAGRCVCKPA  VTGERCDRCR  SGYYNLDGGN  PEGCTQCFCY  GHSASCRSSA
201    EYSVHKITST  FHQDVDGWKA  VQRNGSPAKL  QWSQRHQDVF  SSAQRLDPVY
251    FVAPAKFLGN  QQVSYGQSLS  FDYRVDRGGR  HPSAHDVILE  GAGLRITAPL
301    MPLGKTLPCG  LTKTYTFRLN  EHPSNNWSPQ  LSYFEYRRLL  RNLTALRIRA
351    TYGEYSTGYI  DNVTLISARP  VSGAPAPWVE  QCICPVGYKG  QFCQDCASGY
401    KRDSARLGPF  GTCIPCNCQG  GGACDPDTGD  CYSGDENPDI  ECADCPIGFY
451    NDPHDPRSCK  PCPCHNGFSC  SVIPETEEVV  CNNCPPGVTG  ARCELCADGY
501    FGDPFGEHGP  VRPCQPCQCN  SNVDPSASGN  CDRLTGRCLK  CIHNTAGIYC
551    DQCKAGYFGD  PLAPNPADKC  RACNCNPMGS  EPVGCRSDGT  CVCKPGFGGP
601    NCEHGAFSCP  ACYNQVKIQM  DQFMQQLQRM  EALISKAQGG  DGVVPDTELE
651    GRMQQAEQAL  QDILRDAQIS  EGASRSLGLQ  LAKVRSQENS  YQSRLDDLKM
701    TVERVRALGS  QYQNRVRDTH  RLITQMQLSL  AESEASLGNT  NIPASDHYVG
751    PNGFKSLAQE  ATRLAESHVE  SASNMEQLTR  ETEDYSKQAL  SLVRKALHEG
801    VGSGSGSPDG  AVVQGLVEKL  EKTKSLAQQL  TREATQAEIE  ADRSYQHSLR
851    LLDSVSPLQG  VSDQSFQVEE  AKRIKQKADS  LSSLVTRHMD  EFKRTQKNLG
901    NWKEEAQQLL  QNGKSGREKS  DQLLSRANLA  KSRAQEALSM  GNATFYEVES
951    ILKNLREFDL  QVDNRKAEAE  EAMKRLSYIS  QKVSDASDKT  QQAERALGSA
1001   AADAQRAKNG  AGEALEISSE  IEQEIGSLNL  EANVTADGAL  AMEKGLASLK
1051   SEMREVEGEL  ERKELEFDTN  MDAVQMVITE  AQKVDTRAKN  AGVTIQDTLN
1101   TLDGLLHLMD  QPLSVDEEGL  VLLEQKLSRA  KTQINSQLRP  MMSELEERAR
1151   QQRGHLHLLE  TSIDGILADV  KNLENIRDNL  PPGCYNTQAL  EQQ
```

Variant is identical up to 1109 then:

```
1110  GM
```

Structural and functional sites

Signal peptide: 1–21

EGF (8C) repeats: 28–83, 84–130, 139–185, 186–196 (partial), 382–415 (partial), 416–461, 462–516, 517–572, 573–608

Potential N-linked glycosylation sites: 224, 326, 342, 362, 942, 1033

Gene structure

The LAMA1 gene is localized to human chromosome 18p11.3 [22]. The LAMA2 gene contains 64 exons varying in size from 6 to 270 bp in a gene of >260 kb localized to chromosome 6q22–23 [23]. Mutations in the LAMA2 gene are associated with congenital muscular dystrophy [23,24]. The LAMA3 gene which is thought to encode both the $a3_a$ and $_b$ chains has been localized to chromosome 18q11.2 [10,14]. Mutations in this gene (as well as the genes for the other chains that associate to form laminin 5, namely LAMB3 and LAMC2) underlie the junctional forms of epidermolysis bullosa (JEB) [10,25]. The LAMA4 gene has been localized to chromosome 6q.21 [26]. The LAMB1 gene contains 34 exons varying in size from 64 to 370 bp in a gene of >80 kb localized to chromosome 7q22 [27]. The LAMB2 gene contains 33 exons varying in size from 64 to 373 bp in a gene of approximately 12 kb localized to chromosome 3p21 [28]. Knockout of the LAMB2 gene in mice leads to aberrant differentiation of neuromuscular junctions and renal nephrosis within 2 weeks of birth [10]. The LAMB3 gene contains 23 exons varying in length from 64 to 379 bp in a gene of approximately 29 kb localized to chromosome 1q25–q31 [29]. Mutations in this gene (as well as the genes for the other chains that associate to form laminin 5, namely LAMA3 and LAMC2) underlie the junctional forms of epidermolysis bullosa (JEB). Both the LAMC1 and LAMC2 are localized to chromosome 1q25–q31 [30,31]. The LAMC1 gene is >58 kb and contains 28 exons varying in size from 80 to 3051 bp [30]. The LAMC2 gene is 55 kb and contains 23 exons varying in size from 80 to >2005 bp [31]. Mutations in this gene (as well as the genes for the other chains that associate to form laminin 5, namely LAMA3 and LAMB3) underlie the junctional forms of epidermolysis bullosa (JEB).

References

1 Timpl, R. and Brown, J.C. (1994) The laminins. Matrix Biol. 14: 275–281.

2 Burgeson, R.E. et al. (1994) A new nomenclature for the laminins. Matrix Biol. 14: 209–211.

3 Timpl, R. et al. (1979) Laminin – a glycoprotein from basement membranes. J. Biol. Chem. 254: 9933–9937.

4 Graf, J. et al. (1987) Identification of an amino acid sequence in laminin mediating cell attachment, chemotaxis, and receptor binding. Cell 48: 989–996.

5 Hunter, D.D. et al. (1989) Primary sequence of a motor neuron-selective adhesive site in the synaptic basal lamina protein S-laminin. Cell 59: 905–913.

6 Kleinman, H.K. et al. (1989) Identification of a second active site in laminin for promotion of cell adhesion, migration, and inhibition of *in vivo* melanoma lung colonization. Arch. Biochem. Biophys. 272: 39–45.

7 Liesi, P. et al. (1989) Identification of a neurite outgrowth-promoting domain of laminin using synthetic peptides. FEBS Lett. 244: 141–148.

8 Tashiro, K.-I. et al. (1989) A synthetic peptide containing the IKVAV sequence from the A chain of laminin mediates cell attachment, migration, and neurite outgrowth. J. Biol. Chem. 264: 16174–16182.

[9] Gehlsen, K. R. et al. (1992) A synthetic peptide derived from the carboxy terminus of the laminin A chain represents a binding site for the $\alpha3\beta1$ integrin. J. Cell Biol. 117: 449–459.
[10] Ryan, M.C. et al. (1996) The functions of laminins: Lessons from *in vitro* studies. Matrix Biol. 15: 369–381.
[11] Paulsson, M. et al. (1987) Laminin–nidogen complex. Extraction with chelating agents and structural characterization. Eur. J. Biochem. 166: 467–478.
[12] Nissinen, M. et al. (1991) Primary structure of the human laminin A chain. Limited expression in human tissues. Biochem. J. 276: 369–379 ($\alpha1$ chain).
[13] Vuolteenaho, R. et al. (1994) Human laminin M chain (Merosin): Complete primary structure, chromosomal assignment, and expression of the M and A chain in human fetal tissues. J. Cell Biol. 124: 381–394 ($\alpha2$ chain).
[14] Ryan, M.C. et al. (1994) Cloning of the LamA3 gene encoding the $\alpha3$ chain of the adhesive ligand epiligrin. J. Biol. Chem. 269: 22779–22787.
[15] Iivanainen, A. et al. (1995) Primary structure and expression of a novel human laminin alpha 4 chain. FEBS Lett. 365: 183–188.
[16] Miner, J.H. et al. (1995) Molecular cloning of a novel laminin chain, alpha 5, and widespread expression in adult mouse tissues. J. Biol. Chem. 270: 28523–28526.
[17] Pikkarainen, T. et al. (1987) Human laminin B1 chain. J. Biol. Chem. 262: 10454–10462 ($\beta1$ chain).
[18] Wewer, U.M. et al. (1994) Human beta 2 chain of laminin (formerly S chain): cDNA cloning, chromosomal localization, and expression in carcinomas. Genomics 24: 243–252.
[19] Gerecke, D.R. et al. (1994) The complete primary structure for a novel laminin chain, the laminin B1k chain. J. Biol. Chem. 269: 11073–11080 ($\beta3$ chain).
[20] Pikkarainen, T. et al. (1988) Human laminin B2 chain. J. Biol. Chem. 263: 6751–6758 ($\gamma1$ chain).
[21] Kallunki, P. et al. (1992) A truncated laminin chain homologous to the B2 chain: Structure, spatial expression and chromosomal assignment. J. Cell Biol. 119: 679–693 ($\gamma2$ chain).
[22] Nagayoshi, T. et al. (1989) Human laminin A chain (LAMA) gene: Chromosomal mapping to locus 18p11.3. Genomics 5: 932–935.
[23] Zhang, X. et al. (1996) Structure of the human laminin alpha 2-chain gene (LAMA2), which is affected in congenital muscular dystrophy. J. Biol. Chem. 44: 27664–27669.
[24] Helbling-Leclerc, A. et al. (1996) Mutations in the laminin $\alpha2$-chain gene (LAMA2) cause merosin-deficient congenital muscular dystrophy. Nat. Genet. 11: 216–218.
[25] Vidal, F. et al. (1995) Cloning of the laminin $\alpha3$ chain gene (LAMA3) and identification of a homozygous deletion in a patient with Herlitz junctional epidermolysis bullosa. Genomics 30: 273–280.
[26] Richards, A.J. et al. (1994) Localisation of the gene LAMA4 to chromosome 6q.21 and isolation of a partial cDNA encoding a varient laminin A chain. Genomics 22: 237–239.
[27] Vuolteenaho, R. et al. (1990) Structure of the human laminin B1 gene. J. Biol. Chem. 265: 15611–15616.

[28] Durkin, M.E. et al. (1996) Structural organisation of the human and mouse laminin $\beta 2$ chain genes, and alternative splicing at the 5′ end of the human transcript. J. Biol. Chem. 271: 13407–13416.
[29] Pulkkinen, L. et al. (1995) Cloning of the b3 chain gene (LAMB3) of human laminin 5, a candidate gene in junctional epidermolysis bullosa. Genomics 25: 192–198.
[30] Kallunki, T. et al. (1991) Structure of the laminin B2 gene reveals extensive divergence from the laminin B1 chain gene. J. Biol. Chem. 266: 221–228.
[31] Airenne, T. et al. (1996) Structure of the human laminin $\gamma 2$ chain gene (LAMC2): Alternative splicing with different tissue distribution of two transcripts. Genomics 32: 54–64.

Latent transforming growth factor-β binding protein-1

LTBP1

Latent transforming growth factor-β binding protein-1 (LTBP1) belongs to a family of transforming growth factor-β (TGFβ) binding proteins. The functions of LTBP are not fully understood, but TGFβs are secreted by virtually all cells as latent high molecular weight complexes, and LTBPs are components of these complexes. LTBP does not appear to be involved in the latency of TGFβ, but facilitates secretion and may be involved in activation of TGFβ by targeting to cell surface. In addition LTBP can be immobilized in the extracellular matrix, and bound TGFβ released by proteolytic cleavage. The product from platelets is smaller than that produced by fibroblasts, suggesting either alternative splicing or specific proteolysis.

Molecular structure

LTBP1 is observed as a 190–205 kDa protein under reducing conditions with the majority of the sequence being accounted for by eight cysteine motifs and epidermal growth factor (EGF)-like motifs. These motifs are similar to those found in the fibrillin family of proteins and are arranged in the molecule in a similar way. The central portion of the molecule is characterized by 12 EGF-like repeats and there are four copies of the eight-cysteine repeat. The RGD cell binding domain as well as an eight amino acid motif are identical to the cell binding domain of the β2-laminin[1–6].

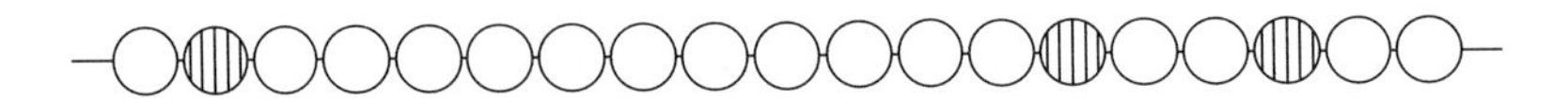

Isolation

Conditioned medium was passed through siliconized glass wool and protein precipitated with 95% ammonium sulphate, dissolved in 50 mM NaCl, 10 mM phosphate pH 7.4 and applied to Q-Sepharose[2].

Accession number

P22064

Primary structure

Ala	A	12	Cys	C	13	Asp	D	8	Glu	E	16
Phe	F	6	Gly	G	8	His	H	3	Ile	I	4
Lys	K	5	Leu	L	18	Met	M	2	Asn	N	3
Pro	P	18	Gln	Q	16	Arg	R	10	Ser	S	10
Thr	T	8	Val	V	12	Trp	W	0	Tyr	Y	11

Mol. wt. (calc.) = 152 791   Residues = 1394

```
1    MDTKLMCLLF FFSLPPLLVS NHTGRIKVVF TPSICKVTCT KGSCQNSCEK
51   GNTTTLISEN GHAADTLTAT NFRVVICHLP CMNGGQCSSR DKCQCPPNFT
101  GKLCQIPVHG ASVPKLYQHS QQPGKALGTH VIHSTHTLPL TVTSQQGVKV
151  KFPPNIVNIH VKHPPEASVQ IHQVSRIDGP TGQKTKEAQP GQSQVSYQGL
201  PVQKTQTIHS TYSHQQVIPH VYPVAAKTQL GRCFQETIGS QCGKALPGLS
251  KQEDCCGTVG TSWGFNKCQK CPKKPSYHGY NQMMECLPGY KRVNNTFCQD
```

```
301   INECQLQGVC PNGECLNTMG SYRCTCKIGF GPDPTFSSCV PDPPVISEEK
351   GPCYRLVSSG RQCMYPLSVH LTKQLCCCSV GKAGPHCEKC PLPGTAAFKE
401   ICPGGMGYTV SGVHRRRPIH HHVGKGPVFV KPKNTQPVAK STHPPPLPAK
451   EEPVEALTFS REHGARSAEP EVATAPPEKE IPSLDQEKTK LEPGQPQLSP
501   GISAIHLHPQ FPVVIEKTSP PVPVEVAPEA STSSASQVIA PTQVTEINEC
551   TVNPDICGAG HCINLPVRYT CICYEGYRFS EQQRKCVDID ECTQVQHLCS
601   QGRCENTEGS FLCICPAGFM ASEEGTNCID VDECLRPDVC GEGHCVNTVG
651   AFRCEYCDSG YRMTQRGRCE DIDECLNPST CPDEQCVNSP GSYQCVPCTE
701   GFRGWNGQCL DVDECLEPNV CANGDCSNLE GSYMCSCHKG YTRTPDHKHC
751   RDIDECQQGN LCVNGQCKNT EGSFRCTCGQ GYQLSAAKDQ CEDIDECQHR
801   HLCAHGQCRN TEGSFQCVCD QGYRASGLGD HCEDINECLE DKSVCQRGDC
851   INTAGSYDCT CPDGFQLDDN KTCQDINECE HPGLCGPQGE CLNTEGSFHC
901   VCQQGFSISA DGRTCEDIDE CVNNTVCDSH GFCDNTAGSF RCLCYQGFQA
951   PQDGQGCVDV NECELLSGVC GEAFCENVEG SFLCVCADEN QEYSPMTGQC
1001  RSRTSTDLDV DVDQPKEEKK ECYYNLNDAS LCDNVLAPNV TKQECCCTSG
1051  AGWGDNCEIF PCPVLGTAEF TEMCPKGKGF VPAGESSSEA GGENYKDADE
1101  CLLFGQEICK NGFCLNTRPG YECYCKQGTY YDPVKLQCFD MDECQDPSSC
1151  IDGQCVNTEG SYNCFCTHPM VLDASEKRCI RPAESNEQIE ETDVYQDLCW
1201  EHLSDEYVCS RPLVGKQTTY TECCCLYGEA WGMQCALCPL KDSDDYAQLC
1251  NIPVTGRRQP YGRDALVDFS EQYTPEADPY FIQDRFLNSF EELQAEECGI
1301  LNGCENGRCV RVQEGYTCDC LDGYHLDTAK MTCFDVNECD ELNNRMSLCK
1351  NAKCINTDGS YKCLCLPGYV PSDKPNYCTP LNTALNLEKD SDLE
```

Structural and functional sites

Signal peptide: 1–20
EGF (6C) repeats: 300–340, 546–587, 588–629, 630–670, 671–710, 711–751, 752–792, 793–833, 834–874, 875–916, 917–958, 959–1001, 1097–1139, 1140–1180, 1294–1334, 1335–1379
TGFβ1 receptor repeats: 348–412, 1017–1084, 1190–1262
Putative RGD cell attachment site: 847–849
Potential N-linked glycosylation sites: 21, 52, 98, 294, 870, 923, 1039

Gene structure

Gene is found on human chromosome 2 region p12–q22 [7].

References

[1] Miyazono, K. et al. (1988) Latent high molecular weight complex of transforming growth factor beta. J. Biol. Chem. 263: 6407–6415.

[2] Olofsson, A. et al. (1992) Transforming growth factor β1-,β2-,β3 secreted by human glioblastoma cells. J. Biol. Chem. 267: 19482–19488.

[3] Flaumenhaft, A. et al. (1993) Role of the latent TGF-beta binding protein in the activation of latent TGF-beta by co-culture of endothelial and smooth muscle cells. J. Cell Biol. 120: 995–1002.

[4] Yamazaki, M. et al. (1994) Expression of transforming growth factor-β1 and its relation to endomysial fibrosis in progressive muscular dystrophy. Amer. J. Pathol. 144: 221–226.

[5] Taipale, J. et al. (1994) Latent transforming growth factor-β associates to fibroblast extracellular matrix via latent TGF-β binding protein. J. Cell Biol. 124: 171–181.

[6] Dallas, S.L. et al. (1995) Dual role for the latent transforming growth factor-beta binding protein in storage of latent TGF-beta in the extracellular matrix and as structural matrix protein. J. Cell Biol. 131: 539–549.
[7] Stenman, G. et al. (1994) Assignment of the gene encoding the latent TGF-beta 1-binding protein (LTBP1) to human chromosome 2, region p12 → q22. Cytogenet. Cell Genet. 66: 117–119.

Latent transforming growth factor-β binding protein-2

LTBP2

Latent transforming growth factor-β binding proteins (LTBPs) are a family of structurally related molecules that facilitate the normal assembly and secretion of latent transforming growth factor-β (TGFβ) complexes. At present three members of this family have been described, and evidence is increasing to show that these complexes are associated with the extracellular matrix and that this is an important site for growth factor modulation. Immunolocalization has demonstrated an extracellular association of LTBP2 with elastic-associated microfibrils.

Molecular structure

LTBP2 is synthesized as a 240 kDa protein by human foreskin fibroblasts and by COS cells transfected with cDNA. It is composed of 20 epidermal growth factor (EGF)-like repeats and four copies of the eight-cysteine repeat. The amino acid sequence of LTBP2 is 41% identical to LTBP1 and 25% to that of fibrillin-1. LTBP2 has been shown to form a complex with TGFβ1 precursor in transfected fibroblast cell lines [1–4].

Isolation

Immunoprecipitates of medium and cell layers of COS cells cotransfected with TGFβ cDNA and LTBP2 cDNA revealed a 255 kDa protein band, probably a complex of LTBP2 and TGFβ1 precursor [1].

Accession number

A55494

Primary structure

Ala	A	12	Cys	C	13	Asp	D	8	Glu	E	16
Phe	F	6	Gly	G	8	His	H	3	Ile	I	4
Lys	K	5	Leu	L	18	Met	M	2	Asn	N	3
Pro	P	18	Gln	Q	16	Arg	R	10	Ser	S	10
Thr	T	8	Val	V	12	Trp	W	0	Tyr	Y	11

Mol. wt (calc.) = 195 128 Residues = 1820

```
1    MRPRTKARSP GRALRNPWRG FLPLTLALFV GAGHAQRDPV GRYEPAGGDA
51   NRLRRPGGSY PAAAAAKVYS LFREQDAPVA GLQPVERAQP GWGSPRRPTE
101  AEARRPSRAQ QSRRVQPPAQ TRRSTPLGQQ QPAPRTRAAP ALPRLGTPQR
151  SGAAPPTPPR GRLTGRNVCG GQCCPGWTTA NSTNHCIKPV CEPPCQNRGS
201  CSRPQLCVCR SGFRGARCEE VIPDEEFDPQ NSRLAPRRWA ERSPNLRRSS
251  AAGEGTLARA QPPAPQSPPA PQSPPAGTLS GLSQTHPSQQ HVGLSRTVRL
301  HPTATASSQL SSNALPPGPG LEQRDGTQQA VPLEHPSSPW GLNLTEKIKK
351  IKIVFTPTIC KQTCARGHCA NSCERGDTTT LYSQGGHGHD PKSGFRIYFC
401  QIPCLNGGRC IGRDECWCPA NSTGKFCHLP IPQPDREPPG RGSRPRALLE
451  APLKQSTFTL PLSNQLASVN PSLVKVHIHH PPEASVQIHQ VAQVRGGVEE
```

```
501   ALVENSVETR  PPPWLPASPG  HSLWDSNNIP  ARSGEPPRPL  PPAAPRPRGL
551   LGRCYLNTVN  GQCANPLLEL  TTQEDCCGSV  GAFWGVTLCA  PCPPRPASPV
601   IENGQLECPQ  GYKRLNLTHC  QDINECLTLG  LCKDAECVNT  RGSYLCTCRP
651   GLMLDPSRSR  CVSDKAISML  QGLCYRSLGP  GTCTLPLAQR  ITKQICCCSR
701   VGKAWGSECE  KCPLPGTEAF  REICPAGHGY  TYASSDIRLS  MRKAEEEELA
751   RPPREQGQRS  SGALPGPAER  QPLRVVTDTW  LEAGTIPDKG  DSQAGQVTTS
801   VTHAPAWVTG  NATTPPMPEQ  GIAEIQEEQV  TPSTDVLVTL  STPGIDRCAA
851   GATNVCGPGT  CVNLPDGYRC  VCSPGYQLHP  SQAYCTDDNE  CLRDPCQGKG
901   RCINRVGSYS  CFCYPGYTLA  TSGATQECQD  INECEQPGVC  SGGQCTNTEG
951   SYHCECDQGY  IMVRKGHCQD  INECRHPGTC  PDGRCVNSPG  SYTCLACEEG
1001  YRGQSGSCVD  VNECLTPGVC  AHGKCTNLEG  SFRCSCEQGY  EVTSDEKGCQ
1051  DVDECASRAS  CPTGLCLNTE  GSFACSACEN  GYWVNEDGTA  CEDLDECAFP
1101  GVCPSGVCTN  TAGSFSCKDC  DGGYRPSPLG  DSCEDVDECE  DPQSSCLGGE
1151  CKNTVGSYQC  LCPQGFQLAN  GTVCEDVNEC  MGEEHCAPHG  ECLNSHGSFF
1201  CLCAPGFVSA  EGGTSCQDVD  ECATTDPCVG  GHCVNTEGSF  NCLCETGFQP
1251  SPESGECVDI  DECEDYGDPV  CGTWKCENSP  GSYRCVLGCQ  PGFHMAPNGD
1301  CIDIDECAND  TMCGSHGFCD  NTDGSFRCLC  DQGFEISPSG  WDCVDVNECE
1351  LMLAVCGAAL  CENVEGSFLC  LCASDLEEYD  AQEGTAAHGG  LEVRVCLRPQ
1401  RDHAPAPTRM  DCYSGQKGHA  PCSSVLGRNT  TQAECCCTQG  ATWGDACDLC
1451  PSEDSAEFSE  ICPSGKGYIP  VEGAWTFGQT  MYTDADECVI  FGPGLCPNGR
1501  CLNTVPGYVC  LCNPGFHYDA  SHKKCEDHDE  CQDLACENGE  CVNTEGSFHC
1551  FCSPPLTLDL  SQQRCMNSTS  STEDLPDHDI  HMDICWKKVT  NDVCSEPLRG
1601  HRTTYTECCC  QDGKAWSQQC  ALCPPRSSEV  YAQLCNVARI  EAEREAGVHF
1651  RPGYEYGPGP  DDLHYSIYGP  DGAPFYNYLG  PEDTVPEPAF  PNTAGHSADR
1701  TPILESPLQP  SELQPHYVAS  HPEPPAGFEG  LQAEECGILN  GCENGRCVRV
1751  REGYTCDCFE  GFQLDAAHMA  CVDVNECDDL  NGPAVLCVHG  YCENTEGSYR
1801  CHCSPGYVAE  AGPPHCTAKE
```

Structural and functional sites

Signal peptide: Not defined

EGF (6C) repeats: 187–219, 396–429, 622–662, 844–886, 887–929, 930–969, 970–1009, 1010–1050, 1051–1092, 1093–1134, 1135–1175, 1176–1217, 1218–1258, 1259–1302, 1303–1344, 1345–1385, 1484–1526, 1527–1566, 1732–1772, 1773–1817

TGFβ1 receptor repeats: 554–609, 674–725, 1412–1463, 1585–1636

Potential N-linked glycosylation sites: 181, 343, 422, 471, 616, 811, 1170, 1309, 1439, 1567

Gene structure

The LTBP2 gene has been assigned to human chromosome 14q24.3. Northern blot analysis revealed two transcripts of 7.5 and 9.0 kb, which like LTBP1 may undergo alternative splicing [1,2,4].

References

[1] Moren, A. et al. (1994) Identification and characterisation of LTBP-2, a novel latent transforming growth factor-β-binding protein. J. Biol. Chem. 269: 32469–32478.

[2] Li, X. et al. (1995) Mapping of the human and murine gene for the latent TGF-beta binding protein-2 (LTBP2). Mamm. Genome 6: 42–45.

[3] Gibson, M.A. et al. (1995) Bovine latent transforming growth factor beta-1-binding protein2 – molecular cloning, identification of tissue isoforms, and immunolocalisation to elastin-associated microfibrils. Molec. Cell. Biol. 15: 6932–6942.
[4] Bashir, M.M. et al. (1996) Analysis of the human gene encoding latent transforming growth factor-beta-binding protein-2. Int. J. Biochem. Cell Biol. 28: 531–542.

Latent transforming growth factor-β binding protein-3

LTBP3

Latent transforming growth factor-β binding proteins (LTBPs) are a family of related proteins that bind latent transforming growth factor-β (TGFβ) complexes. At present three members of the family have been described and, while the function of these molecules has not been resolved, it is clear that they are involved in both intracellular and extracellular organization and availability of TGFβ.

Molecular structure

Sequence analysis of cDNA clones suggested a protein of 1251 amino acids with an estimated pI of 5.92, a predicted molecular mass of 134 710 Da and five potential N-linked glycosylation sites. The N-terminal portion contains 28 amino acids with a net basic charge (estimated pI of 12.36) and, apart from a central proline-rich region (21%) and a region near the N-terminal which is rich in proline (20.7%) and glycine (11.8%), the remainder of the molecule contains epidermal growth factor (EGF)-like and eight-cysteine motifs. The average identity to LTBP1 is 38.4% [1–3].

Isolation

The protein only isolated as an immunoprecipitate from cell medium and lysate [1].

Accession number

A57293, A60487

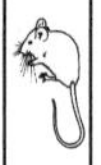

Primary structure (mouse)

Ala	A	12	Cys	C	13	Asp	D	8	Glu	E	16
Phe	F	6	Gly	G	8	His	H	3	Ile	I	4
Lys	K	5	Leu	L	18	Met	M	2	Asn	N	3
Pro	P	18	Gln	Q	16	Arg	R	10	Ser	S	10
Thr	T	8	Val	V	12	Trp	W	0	Tyr	Y	11

Mol. wt (calc.) = 134 646 Residues = 1251

```
1    MRQAALGLLA LLLLALLGPG GRGVGRPGSG AQAGAGRWAQ RFKVVFAPVI
51   CKRTCLKGQC RDSCQQGSNM TLIGENGHST DTLTGSAFRV VVCPLPCMNG
101  GQCSSRNQCL CPPDFTGRFC QVPAAGTGAG TGSSGPGWPD RAMSTGPLPP
151  LAPEGESVAS KHAIYAVQVI ADPPGPGEGP PAQHAAFLVP LGPGQISAEV
201  QAPPPVVNVR VHHPPEASVQ VHRIEGPNAE GPASSQHLLP HPKPQHPRPP
251  TQKPLGRCFQ DTLPKQPCGS NPLPGLTKQE DCCGSIGTAW GQSKCHKCPQ
301  LQYTGVQKPV PVRGEVGADC PQGYKRLNST HCQDINECAM PGNVCHGDCL
351  NNPGSYRCVC PPGHSLGPLA AQCIADKPEE KSLCFRLVST EHQCQHPLTT
401  RLTRQLCCCS VGKAWGARCQ RCPADGTAAF KEICPGWERV PYPHLPPDAH
451  HPGGKRLLPL PAPDGPPKPQ QLPESPSRAP PLEDTEEERG VTMDPPVSEE
501  RSVQQSHPTT TTSPPRPYPE LISRPSPPTF HRFLPDLPPS RSAVEIAPTQ
```

```
551   VTETDECRLN QNICGHGQCV PGPSDYSCHC NAGYRSHPQH RYCVDVNECE
601   AEPCGPGKGI CMNTGGSYNC HCNRGYRLHV GAGGRSCVDL NECAKPHLCG
651   DGGFCINFPG HYKCNCYPGY RLKASRPPIC EDIDECRDPS TCPDGKCENK
701   PGSFKCIACQ PGYRSQGGGA CRDVNECSEG TPCSPGWCEK LPGSYRCTCA
751   QGIRTRTGRL SCIDVDDCEA GKVCQDGICT NTPGSFQCQC LSGYHLSRDR
801   SRCEDIDECD FPAACIGGDC INTNGSYRCL CPLGHRLVGG RKCKKDIDEC
851   SQDPGLCLPH ACENLQGSYV CVCDEGFTLT QDQHGCEEVE QPHHKKECYL
901   NFDDTVFCDS VLATNVTQQE CCCSLGAGWG DHCEIYPCPV YSSAEFHSLV
951   PDGKRLHSGQ QHCELCIPAH RDIDECILFG AEICKEGKCV NSQPGYECYC
1001  KQGFYYDGNL LECVDVDECL DESNCRNGVC ENTWRLPCAC TPPAEYSPAQ
1051  AQCLSPEEME HAPERREVCW GQRGEDGMCM GPLAGPALTF DDCCCRQPRL
1101  GYQCRPCPPR GTGSQCPTSQ SESNSFWDTS PLLLGKSPRD EDSSEEDSDE
1151  CRCVSGPCVP RPGGAVCECP GGFQLDASRA RCVDIDECRE LNQRGLLCKS
1201  ERCVNTSGSF RCVCKAGFTR SRPHGPACLS AAADDAAIAH TSVIDHRGYF
1251  H
```

Structural and functional sites

Signal peptide: 1–21
Potential N-linked glycosylation sites: 69, 328, 824, 915, 1205
EGF (6C) repeats: 258–298, 556–593, 594–637, 638–680, 681–721, 722–762, 763–803, 804–843, 844–886, 975–1012, 1013–1053, 1151–1182, 1187–1228
TGFβ1 receptor repeats: 380–439, 897–963, 1069–1118

Gene structure

Human chromosome 11q12. Northern blot analysis of mouse embryo RNA identified a transcript band of 4.6 kb [1,2].

References

[1] Yin, W. et al. (1995) Isolation of a novel latent transforming growth factor-beta binding protein gene (LTBP-3). J. Biol. Chem. 270: 10147–10160.

[2] Li, X. et al. (1995) Assignment of the gene for LTBP3 to human chromosome 11q12. Mamm. Genome 6: 42–45.

[3] Reichelt, K.L. et al. (1990) Isolation of a growth and mitosis inhibitory peptide from mouse liver. Virchows Arch. B Cell Pathol. 59: 137–142.

Link protein

Link protein (LP) binds to both cartilage proteoglycan (aggrecan) and hyaluronan in cartilage extracellular matrix, thereby stabilizing their aggregation and producing supramolecular assemblies. It is a glycoprotein of approximate molecular weight 45 000. Three forms have been identified: LP1 and LP2 differ in glycosylation at their NH_2-termini; and LP3 is a biologically active proteolytic cleavage product that lacks 13–16 amino acids at its NH_2-terminus. Link protein has been detected in other connective tissues and may also interact with proteoglycans (such as versican and neurocan) and hyaluronan in these locations.

Molecular structure

Link protein exhibits homology to the G1 domain of aggrecan and contains both an immunoglobulin repeat and a two-loop structure stabilized by disulphide bonds. The NH_2-terminal portion of the molecule binds to the proteoglycan and has homology to an immunoglobulin variable repeat [1–6].

Aggrecan Hyaluronan

Isolation

A 4 M guanidine–HCl extract of cartilage is reassociated and separated by isopycnic caesium chloride density gradient centrifugation. The bottom fraction containing proteoglycan aggregates is taken, dissociated, recentrifuged and the top fraction containing link protein isolated. Further purification is achieved by gel filtration [3].

Accession number

P10915

Primary structure

Ala	A	25	Cys	C	11	Asp	D	29	Glu	E	12
Phe	F	18	Gly	G	31	His	H	10	Ile	I	16
Lys	K	20	Leu	L	31	Met	M	2	Asn	N	12
Pro	P	14	Gln	Q	13	Arg	R	21	Ser	S	18
Thr	T	15	Val	V	26	Trp	W	8	Tyr	Y	22

Mol. wt (calc.) = 40 120 Residues = 354

```
1     MKSLLLLVLI  SICWADHLSD  NYTLDHDRAI  HIQAENGPHL  LVEAEQAKVF
51    SHRGGNVTLP  CKFYRDPTAF  GSGIHKIRIK  WTKLTSDYLK  EVDVFVSMGY
101   HKKTYGGYQG  RVFLKGGSDS  DASLVITDLT  LEDYGRYKCE  VIEGLEDDTV
151   VVALDLQGVV  FPYFPRLGRY  NLNFHEAQQA  CLDQDAVIAS  FDQLYDAWRG
201   GLDWCNAGWL  SDGSVQYPIT  KPREPCGGQN  TVPGVRNYGF  WDKDKSRYDV
251   FCFTSNFNGR  FYYLIHPTKL  TYDEAVQACL  NDGAQIAKVG  QIFAAWKILG
```

```
301   YDRCDAGWLA DGSVRYPISR PRRRCSPTEA AVRFVGFPDK KHKLYGVYCF
351   RAYN
```

Structural and functional sites

Signal peptide: 1–15
Immunoglobulin repeat: 47–142
Link repeats: 176–253, 274–350
Potential N-linked glycosylation sites: 21, 56
Disulphide bonds: 61–139, 181–252, 205–226, 279–349, 304–325

Gene structure

A single proteoglycan link protein gene is found on chromosome 5 at locus q13–14[5].

References

1. Caterson, B. et al. (1985) Monoclonal antibodies as probes for determining the microheterogeneity of the link proteins of cartilage proteoglycans. J. Biol. Chem. 260: 11348–11356.
2. Doege, K. et al. (1986) Link protein cDNA sequence reveals a tandemly repeated protein structure. Proc. Natl Acad. Sci. USA 83: 3761–3765.
3. Neame, P.J. et al. (1986) The primary structure of link protein from rat chondrosarcoma proteoglycan aggregate. J. Biol. Chem. 261: 3519–3535.
4. Goetinck, P.F. et al. (1987) The tandemly repeated sequences of cartilage link protein contain the sites for interaction with hyaluronic acid. J. Cell Biol. 105: 2403–2408.
5. Byers, M.G. et al. (1990) Complete amino acid sequence of human cartilage link protein (CRTL1) deduced from cDNA clones and chromosomal assignment of the gene. Genomics 8: 562–567.
6. Dudhia, J. and Hardingham, T.E. (1990) Primary structure of human cartilage-link protein. Nucleic Acids Res. 18: 1292.

Lumican

corneal keratan sulphate proteoglycan, 37B

Lumican is a small keratan sulphate proteoglycan whose core protein is homologous to the leucine-rich proteoglycans decorin, biglycan and fibromodulin. It is expressed in a number of tissues including cornea (it was originally found in chicken cornea), muscle, aorta and intestine. Several lines of evidence suggest that it is important in corneal transparency: it is present early in the development of the chick cornea, it is absent from the opaque cornea of healing rabbit corneal wounds, and the return of transparency is accompanied by the presence of lumican. The abundance of low sulphate lumican in many tissues indicates that this protein occurs predominantly as a glycoprotein rather than the highly sulphated proteoglycan present in cornea.

Molecular structure

Lumican contains 320 amino acids and has a central leucine-rich domain that accounts for 62% of the protein. The central domain contains nine repeats of the sequence LXXLXLXXNXL/I where X is any amino acid. There are three variations of this sequence arranged in tandem to form three repeating units[1–4].

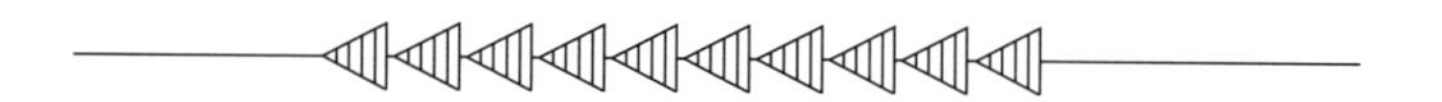

Isolation

Lumican can be purified from extracts of aorta using DEAE ion exchange chromatography, gel filtration, hydrophobic interaction and reverse-phase chromatography[5].

Accession number

A41748

Primary structure

Ala	A	11	Cys	C	7	Asp	D	19	Glu	E	13
Phe	F	15	Gly	G	17	His	H	8	Ile	I	22
Lys	K	22	Leu	L	55	Met	M	5	Asn	N	34
Pro	P	20	Gln	Q	12	Arg	R	6	Ser	S	24
Thr	T	20	Val	V	16	Trp	W	1	Tyr	Y	16

Mol. wt (calc.) = 38 598 Residues = 343

```
1     MTLNSLPIFL  VLISGIFCQY  DYGPADDYGY  DPFGPSTAVC  APECNCPLSY
51    PTAMYCDNLK  LKTIPIVPSG  IKYLYLRNNM  IEAIEENTFD  NVTDLQWLIL
101   DHNHLENSKI  KGRVFSKLKN  LKKLHINYNN  LTEAVGPLPK  TLDDLQLSHN
151   KITKVNPGAL  EGLVNLTVIH  LQNNQLKTDS  ISGAFKGLNS  LLYLDLSFNQ
201   LTKLPTGLPH  SLLMLYFDNN  QISNIPDEYF  QGFKTLQYLR  LSHNKLTDSG
251   IPGNVFNITS  LVELDLSFNQ  LKSIPTVSEN  LENFYLQVNK  INKFPLSSFC
301   KVVGPLTYSK  ITHLRLDGNN  LTRADLPQEM  YNCLRVAADI  SLE
```

Structural and functional sites

Signal peptide: 1–18
Leucine-rich repeats: 68–78, 92–102, 118–128, 139–149, 161–171, 186–196, 207–217, 231–241, 256–266, 276–280
Potential N-linked glycosylation sites: 88, 127, 160, 252

Gene structure

Human lumican gene is about 7.5 kb of genomic DNA and located on chromosome 12q21.3–q22. The gene consists of three exons. RNA from adult chondrocytes showed a single species of 1.8 kb [6,7].

References

1 Cintron, C. et al. (1978) Biochemical and ultrastructural changes in collagen during corneal wound healing. J. Ultrastruct. Res. 65: 13–22.
2 Hassell, J.R. et al. (1983) Proteoglycan changes during restoration of transparency in corneal scars. Arch. Biochem. Biophys. 222: 362–369.
3 Funderburgh, J.L. et al. (1986) Keratan sulfate proteoglycan during embryonic development of the chicken cornea. Dev. Biol. 116: 267–277.
4 Blochberger, T.C. et al. (1992) cDNA to chick lumican (corneal keratan sulfate proteoglycan) reveals homology to the small interstitial proteoglycan gene family and expression in muscle and intestine. J. Biol. Chem. 267: 347–352.
5 Funderburgh, J.L et al. (1991) Unique glycosylation of 3 keratan sulfate proteoglycan isoforms. J. Biol. Chem. 266: 14226–14231.
6 Grover, J. et al. (1995) The human lumican gene. Organisation, chromosomal location and expression in cartilage. J. Biol. Chem. 270: 21942–21949.
7 Chakravati, S. et al. (1995) Primary structure of the human lumican (keratan sulphate proteoglycan) and localisation of the gene (LUM) to chromosome 12q21.3–q22. Genomics 27: 481–488.

Lysyl oxidase protein-lysine 6-oxidase

Lysyl oxidase is a copper-dependent amine oxidase which catalyses the oxidative deamination of specific lysyl and hydroxylysyl residues in collagen and elastin. The resultant peptidyl lysyl and hydroxylysyl residues can spontaneously condense with the ε-amino group of other lysyl residues to form inter-chain covalent cross-links in collagen and elastin. The cross-links stabilize the collagen fibrillar structure of tissues, and the elastin structure of elastic tissues, and therefore play a fundamental role in the mechanical properties of tissues. Defects in lysyl oxidase are the cause of type VI variant of the Ehlers–Danlos syndrome (EDS-VI), a heritable connective tissue disorder. Lysyl oxidase is homologous to the *ras* recision gene product (rrg) which has been shown to counteract the transforming activity of *ras* [1].

Molecular structure

Lysyl oxidase is a glycoprotein which is synthesized as a single polypeptide. Several isoforms of the enzyme may exist. The molecule is secreted in a precursor form [2] which is converted to the active enzyme by removal of the N-terminal prodomain by BMP-1 (and possibly mammalian tolloid). Recent studies suggest that the lysyl oxidase produced by smooth muscle cells is translated as a 46 kDa preproenzyme. Following removal of the signal peptide, the proenzyme undergoes *N*-glycosylation (apparently within the propeptide region) and the resultant 50 kDa molecule is secreted. The secreted form then undergoes proteolytic cleavage to the active enzyme, which ranges in different tissues in size from 28 to 32 kDa [3]. The enzyme responsible for extracellular activation has been shown to be BMP-1 [4].

Isolation

Lysyl oxidase has been purified from bovine aorta and cultured fibroblasts, in addition to other sources including smooth muscle cells making recombinant lysyl oxidase [5]. It is apparently difficult to purify because of its low solubility and its instability in neural salt buffers. The inclusion of 4–6 M urea in solutions markedly increases the yields from cultured cells and tissues. All procedures now use anion exchange chromatography as an initial purification step [6], and others include size chromatography [7] or the use of Cibacron blue dye and hydroxyapatite chromatography.

Accession number

P28300

Primary structure

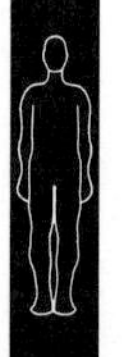

Ala	A	42	Cys	C	12	Asp	D	26	Glu	E	13
Phe	F	9	Gly	G	30	His	H	13	Ile	I	10
Lys	K	6	Leu	L	26	Met	M	5	Asn	N	17
Pro	P	35	Gln	Q	24	Arg	R	37	Ser	S	29
Thr	T	23	Val	V	20	Trp	W	7	Tyr	Y	33

Mol. wt (calc.) = 46 944 Residues = 417

```
1     MRFAWTVLLL GPLQLCALVH CAPPAAGQQQ PPREPPAAPG AWRQQIQWEN
51    NGQVFSLLSL GSQYQPQRRR DPGAAVPGAA NASAQQPRTP ILLIRDNRTA
101   AARTRTAGSS GVTAGRPRPT ARHWFQAGYS TSRAREAGAS RAENQTAPGE
151   VPALSNLRPP SRVDGMVGDD PYNPYKYSDD NPYYNYYDTY ERPRPGGRYR
201   PGYGTGYFQY GLPDLVADPY YIQASTYVQK MSMYNLRCAA EENCLASTAY
251   RADVRDYDHR VLLRFPQRVK NQGTSDFLPS RPRYSWEWHS CHQHYHSMDE
301   FSHYDLLDAN TQRRVAEGHK ASFCLEDTSC DYGYHRRFAC TAHTQGLSPG
351   CYDTYGADID CQWIDITDVK PGNYILKVSV NPSYLVPESD YTNNVVRCDI
401   RYTGHHAYAS GCTISPY
```

Structural and functional sites

Signal peptide: 1–19
Potential N-linked glycosylation sites: 81, 381

Gene structure

The gene for lysyl oxidase is located at human chromosome 5q23.3–31.2 [8]. The gene is 15 kb in size and consists of seven exons. Transcription is initiated at one major site and four minor sites. The 5′-flanking region contains a TATA-like sequence at –30 relative to the major transcription initiation site and a CCAAT motif at –109. The 5′-flanking region and the downstream sequences present in the first exon and first intron contain altogether five possible binding sequences for Sp1, six for AP-2, one for AP-1, three for PEA3, three for MEP-1, and three CCCTCCC motifs, all of which may be involved in the regulation of the expression of the gene *in vivo* [8].

References

[1] Kenyon, K. et al. (1991) Lysyl oxidase and rrg messenger RNA. Science 253: 802.

[2] Kagan, H.M. (1986) Characterisation and regulation of lysyl oxidase: In: Regulation of Matrix Accumulation, ed. Mecham, R.P., Academic Press, Orlando, FL, pp. 321–389.

[3] Trackman, P.C. et al. (1992) Posttranslational glycosylation and proteolytic processing of a lysyl oxidase precursor. J. Biol. Chem. 267: 8666–8671.

[4] Panchenko, M.V. et al. (1996) Metalloproteinase activity secreted by fibrogenic cells in the processing of prolysyl oxidase. Potential role of procollagen C-proteinase. J. Biol. Chem. 271: 7113–7119.

[5] Narayanan, A.S. et al. (1974) Stability and purification of lysyl oxidase. Arch. Biochem. Biophys. 162: 231–237.

[6] Kagan, H.M. et al. (1979) Purification and properties of four species of lysyl oxidase from bovine aorta. Biochem. J. 177: 203–204.

[7] Hamalainen, E.R. et al. (1991) Molecular cloning of human lysyl oxidase and assignment of the gene to chromosome 5q23.3–31.2. Genomics 11: 508–516.

[8] Hamalainen, E.R. et al. (1993) Structure of the human lysyl oxidase gene. Genomics 17: 544–548.

Matrix Gla protein

Matrix Gla protein is a vitamin K-dependent protein initially isolated from bovine bone and associated with the organic matrix. The protein was subsequently shown to be expressed in many tissues including cartilage and most visceral organs. In each tissue it is restricted to a discrete set of tissue-specific cells. In bone, but not in kidney, 1,25-dihydroxyvitamin D3 up-regulates matrix Gla protein expression. Matrix Gla protein is the only known vitamin K-dependent protein that lacks a propeptide. Mice that lack matrix Gla protein develop to term, but die within 2 months as a result of arterial calcification which leads to blood vessel rupture [1]. These studies indicate that matrix Gla protein may act as an inhibitor of calcification in arteries.

Molecular structure

Matrix Gla protein is a single-chain polypeptide which contains five Gla residues and is stabilized by one intra-chain disulphide bond. Despite its high content of hydrophilic amino acids, it is exceptionally water-insoluble and requires 4 M guanidine–HCl for extraction. The two forms of matrix Gla protein isolated so far contain 79 and 83 residues, and lack 5 and 1 amino acids from the predicted COOH-terminus. Residues 2–12 encode an α-helical portion which shares no homology with other proteins. This is followed by a putative γ-carboxylation recognition site and a Gla-containing region. This region shows some identity to osteocalcin and other vitamin K-dependent proteins [2–7].

Isolation

Matrix Gla protein can be isolated by extracting demineralized bone matrix with denaturing agents and fractionating the extract by gel filtration and ion-exchange chromatography [2].

Accession number

P08493

Primary structure

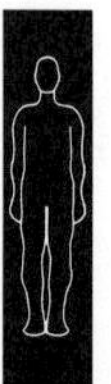

Ala	A	11	Cys	C	3	Asp	D	2	Glu	E	9
Phe	F	3	Gly	G	2	His	H	2	Ile	I	5
Lys	K	5	Leu	L	9	Met	M	3	Asn	N	7
Pro	P	3	Gln	Q	3	Arg	R	14	Ser	S	6
Thr	T	2	Val	V	5	Trp	W	1	Tyr	Y	8

Mol. wt (calc.) = 12 309 Residues = 103

```
1     MKSLILLAIL AALAVVTLCY ESHESMESYE LNPFINRRNA NTFISPQQRW
51    RAKVQERIRE RSKPVHELNR EACDDYRLCE RYAMVYGYNA AYNRYFRKRR
101   GAK
```

Structural and functional sites

Signal peptide: 1–19
Intrachain disulphide bond: 73, 79
γ-Carboxyglutamic acid residues: 21, 56, 60, 67, 71

Gene structure

The human matrix Gla protein gene spans 3.9 kb of chromosomal DNA and consists of four exons separated by intervening sequences which account for more than 80% of the gene. The gene has been assigned to the short arm (p) of human chromosome 12. The four exon organization is similar to that of osteocalcin, but different from the two exon organization found in other vitamin K-dependent proteins[6].

References

[1] Luo, G. et al. (1997) Spontaneous calcification of arteries and cartilage in mice lacking matrix GLA protein. Nature 386: 78–81.

[2] Price, P.A. and Williamson, M.K. (1985) Primary structure of bovine matrix Gla protein, a new vitamin K-dependent bone protein. J. Biol. Chem. 260: 14971–14975.

[3] Price, P.A. et al. (1987) Molecular cloning of matrix Gla protein. Implications for substrate recognition by the vitamin K-dependent γ carboxylase. Proc. Natl Acad. Sci. USA 84: 8335–8339.

[4] Fraser, J.D. and Price, P.A. (1988) Lung, heart, and kidney express high levels of mRNA for the vitamin K-dependent matrix Gla protein. J. Biol. Chem. 263: 11033–11036.

[5] Kiefer, M.C. et al. (1988) The cDNA and derived amino acid sequences of human and bovine matrix Gla protein. Nucleic Acids Res. 16: 5213.

[6] Cancela, L. et al. (1990) Molecular structure, chromosome assignment, and promoter organisation of the human matrix Gla protein gene. J. Biol. Chem. 265: 15040–15048.

[7] Chen, L. et al. (1990) Overexpression of matrix Gla protein mRNA in malignant human breast cells: Isolation by differential cDNA hybridization. Oncogene 5: 1391–1395.

Microfibril-associated glycoprotein-1

MAGP, MFAP2

Microfibril-associated glycoprotein-1 (MAGP1) is a component of 12 nm microfibrils that are found in a variety of elastic and non-elastic tissues. In elastic tissues these 'elastin-associated microfibrils' become incorporated into elastic fibres and act as elastic fibre determinants. It has been shown to bind to the carboxyl-terminus of tropoelastin and to be a substrate for transglutaminase [1]. MAGP1 has been shown to localize on the beaded microfibrils which contain fibrillin as a major component. Two other members of this family have been identified, MAGP2 and MAGP3, although the relationship between one another and to the fibrillin family of proteins is only just beginning to be assessed.

Molecular structure

Microfibril-associated glycoprotein is an acidic glycoprotein which is synthesized as a 21-kDa polypeptide. At least two structural regions have been described: an NH_2-terminal domain containing high levels of glutamine, proline and acidic amino acids and a COOH-terminal domain which contains all 13 cysteine residues and most of the basic amino acids. The presence of high levels of proline and glutamic acid has been suggested to account for the anomalous electrophoretic behaviour of microfibril-associated glycoprotein. There appears to be no *N*-glycosylation, and O-linked sites to serine or threonine have not been delineated [2–6].

Isolation

Glycoproteins from elastin-associated microfibrils have been solubilized from fetal bovine nuchal ligament using treatment with reductive saline extracts. Microfibril-associated glycoprotein can be purified in low yield from this extract by DEAE chromatography followed by gel filtration on Sephacryl S-300 [5]. Microfibril-associated glycoprotein was originally prepared from reductive guanidinium–HCl extracts.

Accession number

P55001

Primary structure

Ala	A	12	Cys	C	13	Asp	D	10	Glu	E	14
Phe	F	7	Gly	G	7	His	H	4	Ile	I	5
Lys	K	4	Leu	L	17	Met	M	2	Asn	N	4
Pro	P	17	Gln	Q	16	Arg	R	11	Ser	S	12
Thr	T	6	Val	V	11	Trp	W	0	Tyr	Y	11

Mol. wt (calc.) = 20 826 Residues = 183

```
1     MRAAYLFLLF  LPAGLLAQGQ  YDLDPLPPFP  DHVQYTHYSD  QIDNPDYYDY
51    QEVTPRPSEE  QFQFQSQQQV  QQEVIPAPTP  EPGNAELEPT  EPGPLDCREE
101   QYPCTRLYSI  HRPCKQCLNE  VCFYSLRRVY  VINKEICVRT  VCAHEELLRA
151   DLCRDKFSKC  GVMASSGLCQ  SVAASCARSC  GSC
```

Structural and functional sites

Signal peptide: 1–17 (possibly 1–19)

Gene structure

Northern blot hybridization of poly-A RNA has identified a single mRNA species of approximately 1.1 kb. The gene is located on human chromosome 1p36.1–p35, contains eight exons and occupies about 4 kb [7].

References

1. Brown-Augsberger, P. et al. (1994) Microfibril-associated glycoprotein binds to carboxyl-terminal domain of tropoelastin and is a substrate for transglutaminase. J. Biol. Chem. 269: 28443–28449.
2. Gibson, M.A. et al. (1986) The major antigen of elastin-associated microfibrils is a 31-kDa glycoprotein. J. Biol. Chem. 261: 11429–11436.
3. Cleary, E.G. (1987) The microfibrillar component of elastic fibres: Morphology and biochemistry. In: Connective Tissue Disease. Molecular Pathology of the Extracellular Matrix, ed. Uitto, J. and Perjeda, A.J., Marcel Dekker, New York, pp. 55–81.
4. Gibson, M.A. and Cleary, E.G. (1987) The immunohistochemical localisation of microfibril-associated glycoprotein (MAGP) in elastic and non-elastic tissues. Immunol. Cell Biol. 65: 345–356.
5. Gibson, M.A. et al. (1989) The protein components of the 12 nm microfibrils of elastic and non-elastic tissues. J. Biol. Chem. 264: 4590–4598.
6. Gibson, M.A. et al. (1991) Complementary DNA cloning establishes microfibril-associated glycoprotein (MAGP) to be a discrete component of the elastin-associated microfibrils. J. Biol. Chem. 266: 7596–7601.
7. Faraco J. et al. (1995) Characterisation of the human gene for microfibril-associated glycoprotein (MFAP2), assignment to chromosome 1p36.1–p35, and linkage to D1S170. Genomics 25: 630–637.

Microfibril-associated glycoprotein-2

MAGP2, MP25

Microfibril-associated glycoprotein-2 (MAGP2) is a glycoprotein component, distinct from MAGP1, of 12 nm microfibrils found in a variety of elastic and non-elastic tissues. MAGP2 has been shown to localize on the beaded microfibrils which contain fibrillin as a major component, although it has a more restricted distribution than MAGP1. There is some indication that MAGP2 gene expression is more closely linked than that of MAGP1 to the switching on of elastin gene expression in nuchal ligament.

Molecular structure

MAGP2 is a highly hydrophilic protein containing two distinct regions, a basic cysteine-rich carboxy-terminal half and an acidic cysteine-free amino-terminal half. The amino-terminal region is rich in serine and threonine, contains a RGD putative integrin-binding motif and a consensus sequence for *N*-glycosylation. Overall MAGP2 has an isoelectric point of 6.6. Close structural similarity to MAGP1 is confined to 60-amino acid region in the centre of the molecule where the first seven cysteines in MAGP2 align with residues in MAGP1. A polypeptide of around 17 kDa is predicted, smaller than the apparent size seen by SDS–PAGE [1–4].

Isolation

Glycoproteins from elastin-associated microfibrils have been solubilized from fetal bovine nuchal ligament using treatment with reductive saline extracts. MAGP2 was purified in low yield from this extract by DEAE chromatography followed by gel filtration on Sephacryl S-300 [3]. It was originally prepared from reductive guanidinium chloride extracts.

Accession number

U37283

Primary structure

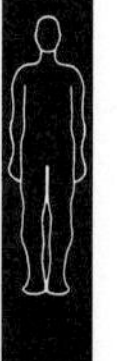

Ala	A	11	Cys	C	8	Asp	D	12	Glu	E	11
Phe	F	6	Gly	G	5	His	H	3	Ile	I	7
Lys	K	7	Leu	L	21	Met	M	4	Asn	N	8
Pro	P	12	Gln	Q	6	Arg	R	13	Ser	S	10
Thr	T	13	Val	V	11	Trp	W	2	Tyr	Y	3

Mol. wt (calc.) = 19 612 Residues = 173

```
1     MSLLGPKVLL  FLAAFIITSD  WIPLGVNSQR  GDDVTQATPE  TFTEDPNLVN
51    DPATDETVLA  VLADIAPSTD  DLASLSEKNT  TAECWDEKFT  CTRLYSVHRP
101   VKQCIHQLCF  TSLRRMYIVN  KEICSRLVCK  EHEAMKDELC  RQMAGLPPRR
151   LRRSNYFRLP  PCENVDLQRP  NGL
```

Structural and functional sites

Signal peptide: 1–18
Putative RGD cell binding motif: 30–32
Potential N-linked glycosylation site: 79

Gene structure

Located on human chromosome 12 in region of 12p12.3–12p13.1 [4].

References
1 Cleary, E.G. and Gibson, M.A. (1995) In: Structurc and Function of Extracellular Components, ed. Comper, W.D., Gordon Breach Science Publishers, New York, pp. 95–140.
2 Gibson, M.A. et al. (1986) The major antigen of elastin-associated microfibrils is a 31-kDa glycoprotein. J. Biol. Chem. 261: 11429–11436.
3 Gibson, M.A. et al. (1989) The protein components of the 12 nm microfibrils of elastic and non-elastic tissues. J. Biol. Chem. 264: 4590–4598.
4 Gibson, M.A. et al. (1996) Further characterisation of proteins associated with elastic fiber microfibrils including the cloning of MAGP-2 (MP25). J. Biol. Chem. 271, 1096–1103.

Microfibril-associated protein-1

AMP, MFAP1

A variety of proteins/glycoproteins have been found to be associated with elastic-fibre microfibrils, the best characterized of which are microfibril-associated glycoprotein (MAGP) 1 and 2. Extraction difficulties have hampered isolation and characterization of many of the associated proteins, and associated microfibrillar protein (AMP) was found through cloning strategies from chick tissues. Recombinant protein and peptides were shown to be associated with elastic-fibre microfibrils by immunolocalization.

Molecular structure

MFAP1 contains 439 amino acids, including a large amount of glutamic acid (22%), some of which are clustered at the N-terminal, and a cluster of lysine residues at the C-terminal. Human microfibril-associated protein-1 lacks cysteine entirely. It does not share homology or domain motifs with any other proteins [1–4].

Isolation

A 58 kDa protein could be isolated from the medium and mild salt extracts of cultured chick embryonic aorta, whereas harsh denaturants extracted a 32 kDa C-terminal derived protein. This processing was confirmed by pulse chase experiments [3].

Accession number

P55081

Primary structure

Ala	A	29	Cys	C	0	Asp	D	29	Glu	E	86
Phe	F	13	Gly	G	16	His	H	4	Ile	I	13
Lys	K	41	Leu	L	20	Met	M	14	Asn	N	12
Pro	P	15	Gln	Q	19	Arg	R	44	Ser	S	27
Thr	T	17	Val	V	26	Trp	W	3	Tyr	Y	11

Mol. wt (calc.) = 51 855 Residues = 439

```
1     MSVPSALMKQ  PPIQSTAGAV  PVRNEKGEIS  MEKVKVKRYV  SGKRPDYAPM
51    ESSDEEDEEF  QFIKKAKEQE  AEPEEQEEDS  SSDPRLRRLQ  NRISEDVEER
101   LARHRKIVEP  EVVGESDSEV  EGDAWRMERE  DSSEEEEEEI  DDEEIERRRG
151   MMRQRAQERK  NEEMEVMEVE  DEGRSGEESE  SESEYEEYTD  SEDEMEPRLK
201   PVFIRKKDRV  TVQEREAEAL  KQKELEQEAK  RMAEERRQYT  LQIVGEETPK
251   ELEENKRSLA  ALDALNTDDE  NDEEEYEAWK  VRELKRIKRD  REDREALEKE
301   KAEIERMRNL  TEEERRAELR  ANGKVITNKA  VKGKYKFLQK  YYHRGAFFMD
351   EDEEVYKRDF  SAPTLEDHFN  KTILPKVMQV  KNFGRSGRTK  YTHLVDQDTT
401   SFDSAWGQES  AQNTKFFKQK  AAGVRDVFER  PSAKKRKTT
```

Structural and functional sites

Signal peptide: Not defined
No known sequence homologies

Gene structure

Northern blot hybridization of total RNA from a fibroblast cell line has identified a single mRNA species of approximately 2.4 kb. The coding portion is contained in nine exons and is localized on human chromosome 15q15–q21, close to the region encoding FBN1 [4].

References

[1] Cleary, E.G. (1987) The microfibrillar component of elastic fibres: Morphology and biochemistry. In: Connective Tissue Disease. Molecular Pathology of the Extracellular Matrix, ed. Uitto, J. and Perjeda, A.J., Marcel Dekker, New York, pp. 55–81.

[2] Gibson, M.A. et al. (1989) The protein components of the 12 nm microfibrils of elastic and non-elastic tissues. J. Biol. Chem. 264: 4590–4598.

[3] Horrigan, S.K. et al. (1992) Characterization of an associated microfibril protein through recombinant DNA approach. J. Biol. Chem. 267: 10087–10095.

[4] Yeh, H. et al. (1994) Structure of the gene encoding the associated microfibrillar protein (MFAP1) and localization to chromosome 15q15q–21. Genomics 23: 443–449.

Microfibril-associated protein-3

MFAP3

A variety of proteins/glycoproteins have been found to be associated with elastic-fibre microfibrils. Many of these are present in low amounts and are difficult to solubilize. Therefore, characterization has been achieved largely through a number of cloning strategies. Microfibril-associated protein-3 (MFAP3) has been cloned and characterized as a microfibrillar protein and recombinantly expressed proteins shown to be reactive to ocular zonules.

Molecular structure

MFAP3 contains 362 amino acids, and extraction of nuchal ligament by 4 M guanidine–HCl solutions revealed a 41 kDa component, in close agreement to predicted protein size and suggesting glycosylation of protein occurs *in vivo*. It does not share homology or domain motifs with any other proteins [1–3].

Isolation

A 41 kDa protein has been shown to be present in 4 M guanidine–HCl extracts of fetal bovine nuchal ligament [3].

Accession number

L35251

Primary structure

Ala	A	26	Cys	C	9	Asp	D	23	Glu	E	27
Phe	F	18	Gly	G	19	His	H	8	Ile	I	25
Lys	K	15	Leu	L	30	Met	M	9	Asn	N	16
Pro	P	13	Gln	Q	12	Arg	R	17	Ser	S	39
Thr	T	19	Val	V	25	Trp	W	2	Tyr	Y	10

Mol. wt (calc.) = 40 165 Residues = 362

```
1     MKLHCCLFTL  VASIIVPAAF  VLEDVDFDQM  VSLEANRSSY  NASFPSSFEL
51    SASSHSDDDV  IIAKEGTSVS  IECLLTASHY  EDVHWHNSKG  QQLDGRSRGG
101   KWLVSDNFLN  ITNVAFDDRG  LYTCFVTSPI  RASYSVTLRV  IFTSGDMSVY
151   YMIVCLIAFT  ITLILNVTRL  CMMSSHLRKT  EKAINEFFRT  EGAEKLQKAF
201   EIAKRIPIIT  SAKTLELAKV  TQFKTMEFAR  YIEELARSVP  LPPLILNCRA
251   FVEEMFEAVR  VDDPDDLGER  IKERPALNAQ  GGIYVINPEM  GRSNSPGGDS
301   DDGSLNEQGQ  EIAVQVSVHL  QSETKSIDTE  SQGSSHFSPP  DDIGSAESNC
351   NYKDGAYENC  QL
```

Structural and functional sites

Signal peptide: Not defined
No known sequence homologies

Gene structure

Northern blot hybridization of total RNA from fetal bovine aorta has identified a single mRNA species of approximately 4.8 kb. The coding portion is contained in two exons and is localized on human chromosome 5q32–q33.2, close to the locus for FBN2 reported at 5q21–q31 [4].

References

[1] Cleary, E.G. (1987) The microfibrillar component of elastic fibres: Morphology and biochemistry. In: Connective Tissue Disease. Molecular Pathology of the Extracellular Matrix, ed. Uitto, J. and Perjeda, A.J., Marcel Dekker, New York, pp. 55–81.

[2] Gibson, M.A. et al. (1989) The protein components of the 12 nm microfibrils of elastic and non-elastic tissues. J. Biol. Chem. 264: 4590–4598.

[3] Abrams, V.R. et al. (1995) Molecular cloning of the microfibrillar protein MFAP3 and assignment of the gene to human chromosome 5q32–q33.2. Genomics 26: 47–54.

[4] Yeh, H. et al. (1994) Structure of the gene encoding the associated microfibrillar protein (MFAP1) and localization to chromosome 15q15q–21. Genomics 23: 443–449.

MMP1 fibroblast collagenase, FIB-CL, interstitial collagenase, CLG

MMP1 was the first member of the family of matrix metalloproteinases to be identified. It takes its long name from the interstitial fibrillar collagens I, II, III which it cleaves. It also cleaves collagens type VII, VIII and X, in addition to gelatine (non-triple helical collagen) and proteoglycan. MMP1 is an endopeptidase which catalyses the hydrolytic cleavage of the triple helical portions of collagens I, II and III at the one-quarter/three-quarter site, at the Gly-Ile or Gly-Leu residues numbered 775–776.

Molecular structure

MMP1, like all MMPs, is a neutral metalloproteinase which belongs to the M10A family of matrixin. The enzyme requires zinc and calcium cations for activity. An atomic zinc is coordinated by specific histidine, tyrosine and glutamic acid residues within the active site of the enzyme. Removal of the zinc inactivates the enzyme. MMP1 is secreted as an inactive proenzyme containing an amino-terminal prodomain. The prodomain maintains the molecule in a catalytically inactive form. The enzyme is activated by the cysteine-switch mechanism that requires the proteolytic cleavage or a conformational change in the propeptide to liberate active enzyme. Possible important activators of MMP1 include cellular proteases, plasmin and plasminogen activator. The central portions of the enzyme contain the catalytic domain. The carboxyl portion contains the hemopexin domain, which is essential for collagenase activity. The hemopexin-like domain contains four units of four-stranded anti-parallel beta-sheet stabilized on its fourfold axis by a cation, which is probably calcium. The domain constitutes a four-bladed beta-propeller structure in which the blades are scarcely twisted[1]. Loss of the hemopexin domain causes the enzyme to lose its ability to hydrolyse types I, II and III collagens, but not its ability to hydrolyse peptide substrates[2–5]. The structure of the catalytic domain has been solved by X-ray crystallography[6,7].

Propeptide

Isolation

MMP1 can be purified from a wide variety of sources including animal tissues, cultured cells and cell culture medium. In addition, cDNA clones are available which has made recombinant protein expression an attractive alternative to purification of the native enzyme. For a review of the sources and the purification methods, refer to references 8–10.

Accession number

P03956

Primary structure

Ala	A	25	Cys	C	5	Asp	D	34	Glu	E	27
Phe	F	34	Gly	G	36	His	H	11	Ile	I	19
Lys	K	27	Leu	L	35	Met	M	9	Asn	N	25
Pro	P	26	Gln	Q	24	Arg	R	22	Ser	S	24
Thr	T	26	Val	V	30	Trp	W	7	Tyr	Y	23

Mol. wt (calc.) = 53 666 Residues = 469

```
1    MHSFPPLLLL LFWGVVSHSF PATLETQEQD VDLVQKYLEK YYNLKNDGRQ
51   VEKRRNSGPV VEKLKQMQEF FGLKVTGKPD AETLKVMKQP RCGVPDVAQF
101  VLTEGNPRWE QTHLTYRIEN YTPDLPRADV DHAIEKAFQL WSNVTPLTFT
151  KVSEGQADIM ISFVRGDHRD NSPFDGPGGN LAHAFQPGPG IGGDAHFDED
201  ERWTNNFREY NLHRVAAHEL GHSLGLSHST DIGALMYPSY TFSGDVQLAQ
251  DDIDGIQAIY GRSQNPVQPI GPQTPKACDS KLTFDAITTI RGEVMFFKDR
301  FYMRTNPFYP EVELNFISVF WPQLPNGLEA AYEFADRDEV RFFKGNKYWA
351  VQGQNVLHGY PKDIYSSFGF PRTVKHIDAA LSEENTGKTY FFVANKYWRY
401  DEYKRSMDPG YPKMIAHDFP GIGHKVDAVF MKDGFFYFFH GTRQYKFDPK
451  TKRILTLQKA NSWFNCRKN
```

Structural and functional sites

Signal peptide: 1–19
Propeptide: 20–99
Auto-inhibitor region: 90–97
Catalytic domain: 8–275
Hemopexin module: 276–469
Zinc binding residues: 218, 222, 228
Potential N-linked glycosylation sites: 120

Gene structure

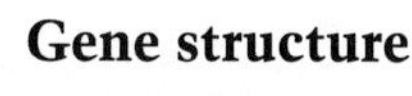

The human gene is located at 11q22–23 and is 17 kb in length[11,12]. Further information and original citations can be found by pointing a web browser at the GeneCard for MMP1 (http://bioinfo.weizmann.ac.il/cgi-bin/lvrebhan/carddisp?MMP1).

References

1 Li, J. et al. (1995) Structure of full-length porcine synovial collagenase reveals a C-terminal domain containing a calcium-linked, four-bladed beta-propeller. Structure 3: 541–549.

2 Clark, I.M. and Cawston, T.E. (1989) Fragments of human fibroblast collagenase. Purification and characterization. Biochem. J. 263: 201–206.

3 Birkedal-Hansen, B. et al. (1988) Monoclonal antibodies to human fibroblast procollagenase. Inhibition of enzymatic activity, affinity purification of the enzyme, and evidence for clustering of epitopes in the NH_2-terminal end of the activated enzyme. Biochemistry 27: 6751–6758.

4 Knauper, V. et al. (1993) Fragmentation of human polymorphonuclear-leucocyte collagenase. Biochem. J. 291: 847–854.

[5] Murphy, G. and Knauper, V. (1997) Relating matrix metalloproteinase structure to function: why the 'hemopexin' domain? Matrix Biol. 15: 511–518.
[6] Lovejoy B. et al. (1994) Structure of the catalytic domain of fibroblast collagenase complexed with an inhibitor. Science 263: 375–377.
[7] Lovejoy B. et al. (1994) Crystal structures of recombinant 19-kDa human fibroblast collagenase complexed to itself. Biochemistry 33: 8207–8217.
[8] Dioszegi, M. et al. (1995) Vertebrate collagenases. Methods Enzymol. 248: 413–431.
[9] Murphy, G. et al. (1987) Stromelysin is an activator of procollagenase. A study with natural and recombinant enzymes. Biochem. J. 248: 265–268.
[10] Murphy, G. et al. (1992) The role of the C-terminal domain in collagenase and stromelysin specificity. J. Biol. Chem. 267: 9612–9618.
[11] Church, R.L. et al. (1983) Human skin collagenase: Assignment of the structural gene to chromosome 11 in both normal and recessive dystrophic epidermolysis bullosa cells using human–mouse somatic cell hybrids. Collagen Rel. Res. 3: 115–124.
[12] Gerhard, D.S. et al. (1987) Human collagenase gene is localised to 11q. (Abstract) Cytogenet. Cell Genet. 46: 619.

MMP2 gelatinase A, 72 kDa type IV collagenase, 72 kDa gelatinase, TBE-1, CLG4A

Gelatinase A (EC 3.4.24.24) as well as gelatinase B are active in cleaving denatured collagens, native type V collagen, elastin and other matrix proteins, and also carry out specific cleavage of native type IV collagen chains (reviewed by Murphy and Crabbe, 1995 [1]). Matrix metalloproteinase-2 (MMP2) is widely distributed and occurs in many types of cultured cells and has been detected in the extracellular matrix (secreted by mesenchymal cells), in polymorphonuclear leukocytes and in plasma. The major role of gelatinase A appears to be in basement membrane turnover. This activity means that gelatinase A plays a key step in metastasis and tumour invasion. Consequently, activation of progelatinase A is a critically important step in tumour biology. Current evidence suggests that gelatinase A can be activated by membrane-type I MMP, and probably involves the formation of a complex between MT1-MMP, tissue inhibitor of metalloproteinase-2 (TIMP2) and progelatinase A [2,3] and by MT2-MMP [4]. Activation of progelatinase A occurs in the presence of some cell types or membrane preparations at two specific sites [5].

Molecular structure

Analysis of the primary structure of gelatinase A shows that it is a modular molecule containing a prodomain, catalytic domain, collagen-like hinge and hemopexin-like domain [6–9]. Proteolytic removal of the amino-terminal prodomain generates the active enzyme forms. The catalytic domain contains the active site zinc and contains three fibronectin type II repeats involved in substrate binding. The type II repeats define a collagen-binding domain (CBD123) [10] which possesses at least two collagen-binding sites that can each be simultaneously occupied. Recombinant CBD123 avidly binds elastin and denatured types IV and V collagens, but neither native types IV and V collagens nor fibronectin, all of which are substrates of the enzyme. The hemopexin-like domain confers specific substrate binding properties, cell-binding properties and TIMP interactions. The C-terminal domain binds fibronectin and heparin, and requires calcium for this interaction [11].

Propeptide

Isolation

Progelatinase A is usually prepared from the conditioned medium of cultured skin, gingival or synovial fibroblasts, and tumour cells. The enzyme is prone to autolytic activation and is best maintained as the latent proenzyme during purification. For a detailed description of the purification of gelatinase A, taking into account the potential contamination of preparations with gelatinase B, see reference 1.

Accession number

P08253

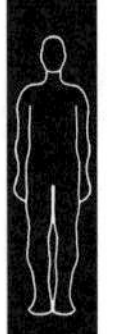

Primary structure

Ala	A	47	Cys	C	19	Asp	D	54	Glu	E	33
Phe	F	47	Gly	G	66	His	H	10	Ile	I	25
Lys	K	45	Leu	L	47	Met	M	12	Asn	N	24
Pro	P	44	Gln	Q	19	Arg	R	26	Ser	S	29
Thr	T	42	Val	V	26	Trp	W	15	Tyr	Y	30

Mol. wt (calc.) = 73 882 Residues = 660

```
1     MEALMARGAL  TGPLRALCLL  GCLLSHAAAA  PSPIIKFPGD  VAPKTDKELA
51    VQYLNTFYGC  PKESCNLFVL  KDTLKKMQKF  FGLPQTGDLD  QNTIETMRKP
101   RCGNPDVANY  NFFPRKPKWD  KNQITYRIIG  YTPDLDPETV  DDAFARAFQV
151   WSDVTPLRFS  RIHDGEADIM  INFGRWEHGD  GYPFDGKDGL  LAHAFAPGTG
201   VGGDSHFDDD  ELWTLGEGQV  VRVKYGNADG  EYCKFPFLFN  GKEYNSCTDT
251   GRSDGFLWCS  TTYNFEKDGK  YGFCPHEALF  TMGGNAEGQP  CKFPFRFQGT
301   SYDSCTTEGR  TDGYRWCGTT  EDYDRDKKYG  FCPETAMSTV  GGNSEGAPCV
351   FPFTFLGNKY  ESCTSAGRSD  GKMWCATTAN  YDDDRKWGFC  PDQGYSLFLV
401   AAHEFGHAMG  LEHSQDPGAL  MAPIYTYTKN  FRLSQDDIKG  IQELYGASPD
451   IDLGTGPTPT  LGPVTPEICK  QDIVFDGIAQ  IRGEIFFFKD  RFIWRTVTPR
501   DKPMGPLLVA  TFWPELPEKI  DAVYEAPQEE  KAVFFAGNEY  WIYSASTLER
551   GYPKPLTSLG  LPPDVQRVDA  AFNWSKNKKT  YIFAGDKFWR  YNEVKKKMDP
601   GFPKLIADAW  NAIPDNLDAV  VDLQGGGHSY  FFKGAYYLKL  ENQSLKSVKF
651   GSIKSDWLGC
```

Structural and functional sites

Signal peptide: 1–29
Propeptide: 30–109
MT1-MMP cleavage site: 66–67
APMA and MT1-MMP cleavage site: 109–110
Catalytic domain: 110–660
Autoinhibitor region: 100–107
Collagenase-like region: 110–221
Collagen-binding region: 222–396
Catalytic domain: 397–465
Hemopexin-like domain: 466–660
Fibronectin type II repeats: 226–283, 284–341, 342–399
Catalytic zinc binding residues: 403, 407, 413
Active site glutamic acid residue: 404
Potential N-linked glycosylation sites: 573, 642

Gene structure

The human gene is located at 16q13. The gene is 17 kb in length and contains 13 exons varying in size from 110 to 901 bp and 12 introns ranging in size from 175 to 4350 bp. Sequence alignments show a close structural relationship with MMP1 and stromelysin. Further information and original citations can be found by pointing a web browser at the GeneCard for MMP2 (http://bioinfo.weizmann.ac.il/cgi-bin/lvrebhan/carddisp?MMP2).

References

[1] Murphy, G. and Crabbe, T. (1995) Gelatinases A and B. Methods Enzymol. 248: 470–484.
[2] Ueno, H. et al. (1997) Expression and tissue localization of membrane-types 1, 2, and 3 matrix metalloproteinases in human invasive breast carcinomas. Cancer Res. 57: 2055–2060.
[3] Lee, A.Y. et al. (1997) Intracellular activation of gelatinase A (72-kDa type IV collagenase) by normal fibroblasts. Proc. Natl Acad. Sci. USA 94: 4424–4429.
[4] Butler, G.S. et al. (1997) Membrane-type-2 matrix metalloproteinase can initiate the processing of progelatinase A and is regulated by the tissue inhibitors of metalloproteinases. Eur. J. Biochem. 244: 653–657.
[5] Strongin, A.Y. et al. (1993) Plasma membrane-dependent activation of the 72-kDa type IV collagenase is prevented by complex formation with TIMP-2. J. Biol. Chem. 268: 14033–14039.
[6] Collier, I.E. et al. (1988) H-ras oncogene-transformed human bronchial epithelial cells (TBE-1) secrete a single metalloprotease capable of degrading basement membrane collagen. J. Biol. Chem. 263: 6579–6587.
[7] Collier, I.E. et al. (1991) On the structure and chromosome location of the 72- and 92-kDa human type IV collagenase genes. Genomics 9: 429–434.
[8] Huhtala, P. et al. (1990) Structure of the human type IV collagenase gene. J. Biol. Chem. 265: 11077–11082.
[9] Huhtala, P. et al. (1990) Completion of the primary structure of the human type IV collagenase preproenzyme and assignment of the gene (CLG4) to the q21 region of chromosome 16. Genomics 6: 554–559.
[10] Steffensen, B. et al. (1995) Extracellular matrix binding properties of recombinant fibronectin type II-like modules of human 72-kDa gelatinase/ type IV collagenase. High affinity binding to native type I collagen but not native type IV collagen. J. Biol. Chem. 270: 11555–11566.
[11] Wallon, U.M. and Overall, C.M. (1997) The hemopexin-like domain (C domain) of human gelatinase A (matrix metalloproteinase-2) requires Ca^{2+} for fibronectin and heparin binding. Binding properties of recombinant gelatinase A C domain to extracellular matrix and basement membrane components. J. Biol. Chem. 272: 7473–7481.

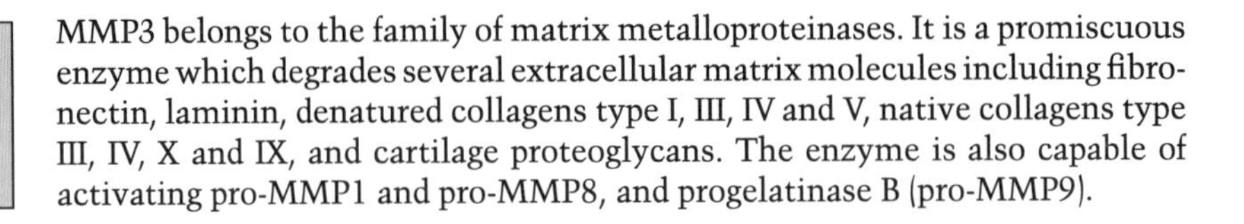

MMP3 stromelysin-1, transin-1, SL-1, STMY1

MMP3 belongs to the family of matrix metalloproteinases. It is a promiscuous enzyme which degrades several extracellular matrix molecules including fibronectin, laminin, denatured collagens type I, III, IV and V, native collagens type III, IV, X and IX, and cartilage proteoglycans. The enzyme is also capable of activating pro-MMP1 and pro-MMP8, and progelatinase B (pro-MMP9).

Molecular structure

The enzyme has a typical MMP structure in having a propeptide, catalytic domain and a hemopexin-like domain. Sequence similarities suggest that the enzyme belongs to the family M10A, also known as the matrixin subfamily, of zincin neutral metalloproteinases. The enzyme is activated by dislocation of the prodomain by the cysteine switch mechanism. The three-dimensional structure of the catalytic domain is known[1-3]. Pro-MMP3 is activated *in vitro* by mercurial compounds, endopeptidases (which remove the prodomain), by treatment with SDS and by elevated temperatures. Activation converts the 57–59 kDa proenzyme to the 45 and 28 kDa active enzymes. The 45 kDa and 28 kDa forms lack the N-terminus of the proenzyme, and the 28 kDa form lacks a portion of the C-terminus of the molecule[4].

Propeptide

Isolation

MMP3 and pro-MMP3 are relatively abundant in dermal fibroblasts and are readily purified from cultured cells, especially after treatment of the cells with interleukin-1, tumour necrosis factor-α, growth factors, and phorbol myristate acetate. Levels of MMP3 mRNA are suppressed by retinoic acid. The proenzyme is purified from the conditioned media of mesenchymal cells or tissue by chromatography on columns of DEAE–cellulose, Green A Matrex gel resin, gelatin–Sepharose and Sephacryl S-200. Alternatively, pro-MMP3 can be purified by affinity chromatography using anti-stromelysin antibodies[5].

Accession number

P08254

Primary structure

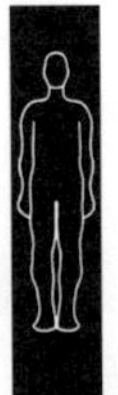

Ala	A	29	Cys	C	5	Asp	D	36	Glu	E	30
Phe	F	32	Gly	G	32	His	H	13	Ile	I	21
Lys	K	33	Leu	L	43	Met	M	7	Asn	N	16
Pro	P	37	Gln	Q	9	Arg	R	23	Ser	S	29
Thr	T	27	Val	V	30	Trp	W	8	Tyr	Y	17

Mol. wt (calc.) = 53 977 Residues = 477

```
1     MKSLPILLLL CVAVCSAYPL DGAARGEDTS MNLVQKYLEN YYDLKKDVKQ
51    FVRRKDSGPV VKKIREMQKF LGLEVTGKLD SDTLEVMRKP RCGVPDVGHF
101   RTFPGIPKWR KTHLTYRIVN YTPDLPKDAV DSAVEKALKV WEEVTPLTFS
151   RLYEGEADIM ISFAVREHGD FYPFDGPGNV LAHAYAPGPG INGDAHFDDD
201   EQWTKDTTGT NLFLVAAHEI GHSLGLFHSA NTEALMYPLY HSLTDLTRFR
251   LSQDDINGIQ SLYGPPPDSP ETPLVPTEPV PPEPGTPANC DPALSFDAVS
301   TLRGEILIFK DRHFWRKSLR KLEPELHLIS SFWPSLPSGV DAAYEVTSKD
351   LVFIFKGNQF WAIRGNEVRA GYPRGIHTLG FPPTVRKIDA AISDKEKNKT
401   YFFVEDKYWR FDEKRNSMEP GFPKQIAEDF PGIDSKIDAV FEEFGFFYFF
451   TGSSQLEFDP NAKKVTHTLK SNSWLNC
```

Structural and functional sites

Signal peptide: 1–17
Propeptide: 18–99
Autoinhibitor region (cysteine switch): 90–97
Catalytic domain: 100–286
Hemopexin-like domain: 287–477
Catalytic residues: 218, 219, 222, 228
Potential N-linked glycosylation site: 120

Gene structure

The human gene is located at 11q23 [6]. Further information may be obtained from GeneCard (http://bioinfo.weizmann.ac.il/cgi-bin/lvrebhan/carddisp?MMP3).

References

[1] Gooley, P.R. (1994) The NMR structure of the inhibited catalytic domain of human stromelysin-1. Nat. Struct. Biol. 1: 111–118.

[2] Becker, J.W. et al. (1995) Stomelysin-1: Three-dimensional structure of the inhibited catalytic domain and of the C-truncated proenzyme. Protein Sci. 4: 1966–1976.

[3] Dhanaraj, V. et al. (1996) X-ray structure of a hydroxamate complex of stromelysin catalytic domain and its comparison with members of the zinc metalloproteinase superfamily. Structure 4: 375–386.

[4] Nagase, H. et al. (1990) Stepwise activation mechanisms of the precursor of matrix metalloproteinase 3 (stromelysin) by proteinases and (4-aminophenyl)mercuric acetate. Biochemistry 29: 5783–5789.

[5] Nagase, H. (1995) Human stromelysins 1 and 2. Methods Enzymol. 248: 449–470.

[6] Formstone, C.J. et al. (1993) The order and orientation of a cluster of metalloproteinase genes, stromelysin 2, collagenase, and stromelysin, together with D11S385, on chromosome 11q22–q23. Genomics 16: 289–291.

MMP7 matrilysin, PUMP

Matrilysin is a member of the matrix metalloproteinase (MMP) family of metalloendopeptidases that degrade the extracellular matrix. Like other MMPs it requires calcium and zinc for activity, and is a member of the family M10A, also known as matrixins, of zincin metalloproteinases. This enzyme degrades casein, denatured collagen types I, III, IV and V, fibronectin, decorin and insulin. Matrilysin notably cleaves aggrecan[1]. MMP7 has no action on native collagen types I, II, IV and V. A major function of this enzyme is to activate procollagenases.

Molecular structure

Matrilysin contains a signal peptide, a propeptide, which is important in activation of the enzyme, and a catalytic domain that lacks a hemopexin-like domain found in other MMP molecules. It exists as a monomer of 30 kDa.

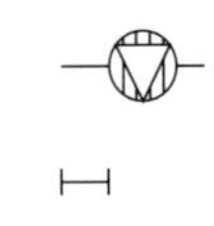

Propeptide

Isolation

Human MMP7 was overproduced in *E. coli* as a histidine-tagged recombinant proenzyme[2,3]. The recombinant protein was refolded into a form that was activatable by *p*-amino-phenylmercuric acetate. The activated enzyme cleaved a synthetic peptide substrate at the reported site for MMP7.

Accession number

P09237

Primary structure

Ala	A	18	Cys	C	3	Asp	D	16	Glu	E	12
Phe	F	11	Gly	G	26	His	H	8	Ile	I	11
Lys	K	17	Leu	L	27	Met	M	9	Asn	N	11
Pro	P	16	Gln	Q	9	Arg	R	12	Ser	S	19
Thr	T	13	Val	V	13	Trp	W	5	Tyr	Y	11

Mol. wt (calc.) = 29 677 Residues = 267

```
1     MRLTVLCAVC  LLPGSLALPL  PQEAGGMSEL  QWEQAQDYLK  RFYLYDSETK
51    NANSLEAKLK  EMQKFFGLPI  TGMLNSRVIE  IMQKPRCGVP  DVAEYSLFPN
101   SPKWTSKVVT  YRIVSYTRDL  PHITVDRLVS  KALNMWGKEI  PLHFRKVVWG
151   TADIMIGFAR  GAHGDSYPFD  GPGNTLAHAF  APGTGLGGDA  HFDEDERWTD
201   GSSLGINFLY  AATHELGHSL  GMGHSSDPNA  VMYPTYGNGD  PQNFKLSQDD
251   IKGIQKLYGK  RSNSRKK
```

Structural and functional sites

Signal peptide: 1–17
Propeptide: 18–94
Catalytic domain: 95–267
Autoinhibitor region (cysteine-switch): 85–92
Zinc (catalytic residues): 214, 215, 218, 224

Gene structure

The human gene is located at 11q21–q22[4], where a cluster of MMP genes is found. Further information may be obtained from GeneCard (http://bioinfo.weizmann.ac.il/cgi-bin/lvrebhan/carddisp?MMP7).

References

[1] Fosang, A. J. et al. (1992) The interglobular domain of cartilage aggrecan is cleaved by PUMP, gelatinases, and cathepsin B. J. Biol. Chem. 267: 19470–19474.

[2] Imai, K. et al. (1995) Matrix metalloproteinase 7 (matrilysin) from human rectal carcinoma cells. Activation of the precursor, interaction with other matrix metalloproteinases and enzymic properties. J. Biol. Chem. 270: 6691–6697.

[3] Itoh, M. et al. (1996) Purification and refolding of recombinant human proMMP-7 (pro-matrilysin) expressed in *Escherichia coli* and its characterization. J. Biochem. (Tokyo) 119: 667–673.

[4] Knox, J. D. et al. (1996) Mapping of the metalloproteinase gene matrilysin (MMP7) to human chromosome 11q21–q22. Cytogenet. Cell Genet. 72: 179–182.

Matrix metalloproteinase-8 (MMP8) degrades fibrillar collagens type I, II and III. Type III collagen is cleaved more slowly than type I collagen. Calcium and zinc are required for this activity. MMP8 cannot be activated without removal of its prodomain at the N-terminus of the molecule. The enzyme appears to be stored in the granules of polymorphonuclear leukocytes, but is also made by other cell types including chondrocytes[1]. MMP8 cleaves the triple helical domain of fibrillar collagens. It is synthesized by polymorphonuclear lymphocyte MMP8 cleaves aggrecan at a site distinct from the site cleaved by aggrecanase[2].

Molecular structure

MMP8 is a neutral metalloproteinase that belongs to the M10A family of matrixin, within the larger zincin superfamily of endometalloproteinases. The enzyme requires zinc and calcium cations for activity. Specific histidine, tyrosine and glutamic acid residues within the active site of the enzyme co-ordinate an atom of zinc. Removal of the zinc inactivates the enzyme. MMP8 is secreted as an inactive proenzyme containing an amino-terminal prodomain. The prodomain maintains the molecule in a catalytically inactive form. The enzyme is activated by the cysteine-switch mechanism that requires the proteolytic cleavage or a conformational change in the propeptide to liberate active enzyme. Pro-MMP8 can be activated by stromelysin-2 by cleavage of the Gly78–Phe79 peptide bond at the end of the propeptide domain[3]. The X-ray crystal structure of the metalloprotease domain is known to a resolution of 2.1 angstroms.

Propeptide

Isolation

The purification of MMP8 from whole human blood has been described[4]. Briefly, whole blood is used as a source of leukocytes from which granules are prepared. The PMNL procollagenase is purified in five chromatographic steps, which include zinc-chelate Sepharose CL-6B, DEAE–Sepharose, Q-Sepharose and Orange-Sepharose, and affinity chromatography on hydroxymate–Sepharose[5].

Accession number

P22894

Primary structure

Ala	A	30	Cys	C	3	Asp	D	27	Glu	E	22
Phe	F	37	Gly	G	33	His	H	10	Ile	I	25
Lys	K	22	Leu	L	36	Met	M	8	Asn	N	25
Pro	P	31	Gln	Q	26	Arg	R	24	Ser	S	33
Thr	T	24	Val	V	19	Trp	W	8	Tyr	Y	24

Mol. wt (calc.) = 53 412 Residues = 467

```
1     MFSLKTLPFL LLLHVQISKA FPVSSKEKNT KTVQDYLEKF YQLPSNQYQS
51    TRKNGTNVIV EKLKEMQRFF GLNVTGKPNE ETLDMMKKPR CGVPDSGGFM
101   LTPGNPKWER TNLTYRIRNY TPQLSEAEVE RAIKDAFELW SVASPLIFTR
151   ISQGEADINI AFYQRDHGDN SPFDGPNGIL AHAFQPGQGI GGDAHFDAEE
201   TWTNTSANYN LFLVAAHEFG HSLGLAHSSD PGALMYPNYA FRETSNYSLP
251   QDDIDGIQAI YGLSSNPIQP TGPSTPKPCD PSLTFDAITT LRGEILFFKD
301   RYFWRRHPQL QRVEMNFISL FWPSLPTGIQ AAYEDFDRDL IFLFKGNQYW
351   ALSGYDILQG YPKDISNYGF PSSVQAIDAA VFYRSKTYFF VNDQFWRYDN
401   QRQFMEPGYP KSISGAFPGI ESKVDAVFQQ EHFFHVFSGP RYYAFDLIAQ
451   RVTRVARGNK WLNCRYG
```

Structural and functional sites

Signal peptide: 1–20
Prodomain: 21–100
Autoinhibitor region (cysteine switch): 89–96
Catalytic domain: 101–467
Hemopexin-like domain: 276–467
Catalytic zinc binding residues: 217, 218, 221, 227
Potential N-linked glycosylation sites: 54, 73, 112, 204, 246

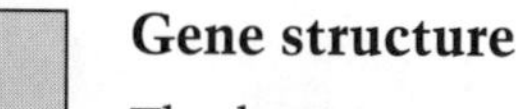

Gene structure

The human gene is located at 11q21–q22[6]. Further information may be available from GeneCard (http://bioinfo.weizmann.ac.il/cgi-bin/lvrebhan/carddisp?MMP8).

References

[1] Cole, A.A. et al. (1996) Chondrocyte matrix metalloproteinase-8. Human articular chondrocytes express neutrophil collagenase. J. Biol. Chem. 271: 11023–11026.

[2] Arner, E.C. et al. (1997) Cleavage of native cartilage aggrecan by neutrophil collagenase (MMP-8) is distinct from endogenous cleavage by aggrecanase. J. Biol. Chem. 272: 9294–9299.

[3] Knauper, V. et al. (1996) Activation of human neutrophil procollagenase by stromelysin 2. Eur. J. Biochem. 235: 187–191.

[4] Tschesche, H. (1995) Human neutrophil collagenase. Methods Enzymol. 248: 431–449.

[5] Knauper, V. et al. (1990) Characterisation and activation of procollagenase from human polymorphonuclear leucocytes. N-terminal sequence

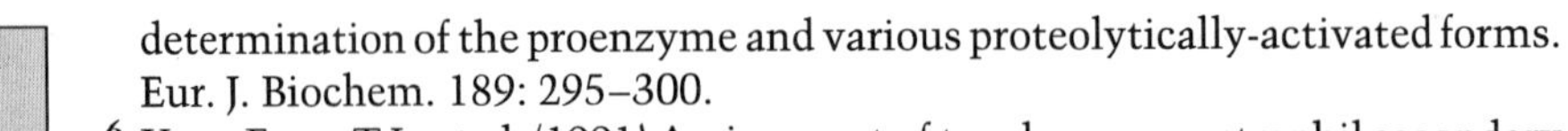

determination of the proenzyme and various proteolytically-activated forms. Eur. J. Biochem. 189: 295–300.

[6] Yang-Feng, T.L. et al. (1991) Assignment of two human neutrophil secondary granule protein genes, transcobalamin I and neutrophil collagenase to chromosome 11. (Abstract) Cytogenet. Cell Genet. 58: 1974.

Gelatinase B[1–4], which is part of the family M10A of zincin metalloproteinases, is a major degradative proteinase in the extracellular matrix. It is produced by stimulated mesenchymal cells in culture, by normal alveolar macrophages and granulocytes, and is a major product of tumour cells. In addition to gelatinase A, gelatinase B most probably plays a major role in basement membrane turnover and tumour metastasis. Gelatinase B could play a further role in osteoclastic resorption and is active on gelatin types I and V.

Molecular structure

Matrix metalloproteinase-9 (MMP9) requires zinc and calcium for activity. It exists as a monomer, a disulphide-linked homodimer, and a heterodimer with a 25 kDa protein. Macrophages and transformed cell lines produce only the monomeric form. The central domain of the molecule differs from most other MMP molecules in that it shows similarity to the fibronectin type-II repeats of the collagen-binding domain of fibronectin. This domain binds gelatin. Gelatinase B is a modular molecule and contains a prodomain, a catalytic domain, a collagen-like hinge domain and a C-terminal domain that contains the hemopexin-like repeats. The molecule is, therefore, structurally related to gelatinase A.

Progelatinase B undergoes cleavage in its N-terminal prodomain during activation and processing to the final mature enzyme. Progelatinase B is also autolytically processed in the presence of mercuric chloride and 4-aminophenylmercuric acetate (APMA)[5,6]. Stromelysin cleaves at two sites within the propeptide to yield active enzyme[7,8].

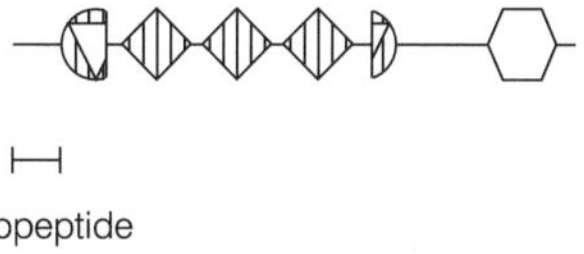

Propeptide

Isolation

Gelatinase B is secreted by mesenchymal cells in culture after induction by cytokines, and is a major product of monocytes and tumour cells. For a detailed description of purification methods, and comparison with the sources and purification of gelatinase A, see Murphy and Crabbe (1995)[9].

Accession number

P14780

Primary structure

Ala	A	51	Cys	C	19	Asp	D	54	Glu	E	29
Phe	F	44	Gly	G	67	His	H	14	Ile	I	14
Lys	K	22	Leu	L	62	Met	M	9	Asn	N	14
Pro	P	60	Gln	Q	25	Arg	R	49	Ser	S	44
Thr	T	50	Val	V	39	Trp	W	14	Tyr	Y	27

Mol. wt (calc.) = 78 427 Residues = 707

```
1     MSLWQPLVLV  LLVLGCCFAA  PRQRQSTLVL  FPGDLRTNLT  DRQLAEEYLY
51    RYGYTRVAEM  RGESKSLGPA  LLLLQKQLSL  PETGELDSAT  LKAMRTPRCG
101   VPDLGRFQTF  EGDLKWHHHN  ITYWIQNYSE  DLPRAVIDDA  FARAFALWSA
151   VTPLTFTRVY  SRDADIVIQF  GVAEHGDGYP  FDGKDGLLAH  AFPPGPGIQG
201   DAHFDDDELW  SLGKGVVVPT  RFGNADGAAC  HFPFIFEGRS  YSACTTDGRS
251   DGLPWCSTTA  NYDTDDRFGF  CPSERLYTRD  GNADGKPCQF  PFIFQGQSYS
301   ACTTDGRSDG  YRWCATTANY  DRDKLFGFCP  TRADSTVMGG  NSAGELCVFP
351   FTFLGKEYST  CTSEGRGDGR  LWCATTSNFD  SDKKWGFCPD  QGYSLFLVAA
401   HEFGHALGLD  HSSVPEALMY  PMYRFTEGPP  LHKDDVNGIR  HLYGPRPEPE
451   PRPPTTTTPQ  PTAPPTVCPT  GPPTVHPSER  PTAGPTGPPS  AGPTGPPTAG
501   PSTATTVPLS  PVDDACNVNI  FDAIAEIGNQ  LYLFKDGKYW  RFSEGRGSRP
551   QGPFLIADKW  PALPRKLDSV  FEEPLSKKLF  FFSGRQVWVY  TGASVLGPRR
601   LDKLGLGADV  AQVTGALRSG  RGKMLLFSGR  RLWRFDVKAQ  MVDPRSASEV
651   DRMFPGVPLD  THDVFQYREK  AYFCQDRFYW  RVSSRSELNQ  VDQVGYVTYD
701   ILQCPED
```

Structural and functional sites

Signal peptide: 1–19
Propeptide: 20–106
Autoinhibitor region: 97–104
Catalytic domain: 107–511
Stromelysin cleavage sites: 59–60, 106–107
$HgCl_2$ cleavage sites: 34–35, 59–60, 71–72, 93–94
APMA cleavage site: 93–94
Fibronectin type II repeats: 223–280, 281–339, 340–397
Hemopexin-like domain: 511–707
Zinc-binding catalytic residues: 401, 402, 405, 411
Potential N-linked glycosylation sites: 38, 120, 127

Gene structure

The human gene contain 13 exons [10] and therefore three more than other members of this gene family. The extra exons encode the amino acids of the fibronectin-like domain. The gene is located at 20q11.2–13.1 [11,12]. Further information and original citations can be found by pointing a web browser at the GeneCard for MMP9 (http://bioinfo.weizmann.ac.il/cgi-bin/lvrebhan/carddisp?MMP9).

References

[1] Wilhelm, S.M. et al. (1989) SV40-transformed human lung fibroblasts secrete a 92-kDa type IV collagenase which is identical to that secreted by normal human macrophages [published erratum appears in J. Biol. Chem. 1990 Dec 25; 265(36): 22570] J. Biol. Chem. 264: 17213–17221.

[2] Sato H. and Seiki M. (1993) Regulatory mechanism of 92 kDa type IV collagenase gene expression which is associated with invasiveness of tumor cells. Oncogene 8: 395–405.

[3] Van Ranst, M. et al. (1991) The cytokine–protease connection: Identification of a 96-kD THP-1 gelatinase and regulation by interleukin-1 and cytokine inducers. Cytokine 3: 231–239.

[4] Masure, S. et al. (1991) Purification and identification of 91-kDa neutrophil gelatinase. Release by the activating peptide interleukin-8. Eur. J. Biochem. 198: 391–398.

[5] Okada, Y. et al. (1992) Matrix metalloproteinase 9 (92-kDa gelatinase/type IV collagenase) from HT 1080 human fibrosarcoma cells. Purification and activation of the precursor and enzymic properties. J. Biol. Chem. 267: 21712–21719.

[6] Triebel, S. et al. (1992) Mercurial activation of human PMN leucocyte type IV procollagenase (gelatinase). FEBS Lett. 298: 280–284.

[7] Ogata, Y. et al. (1992) Matrix metalloproteinase 3 (stromelysin) activates the precursor for the human matrix metalloproteinase 9. J. Biol. Chem. 267: 3581–3584.

[8] O'Connell, J.P. et al. (1994) Analysis of the role of the COOH-terminal domain in the activation, proteolytic activity, and tissue inhibitor of metalloproteinase interactions of gelatinase B. J. Biol. Chem. 269: 14967–14973.

[9] Murphy, G. and Crabbe, T. (1995) Gelatinases A and B. Methods Enzymol. 248: 470–484.

[10] Huhtala, P. et al. (1991) Complete structure of the human gene for 92-kDa type IV collagenase: Divergent regulation of expression for the 92- and 72-kilodalton enzyme genes in HT-1080 cells. J. Biol. Chem. 266: 16485–16490.

[11] St Jean, P.L. et al. (1995) Characterization of a dinucleotide repeat in the 92 kDa type IV collagenase gene (CLG4B), localization of CLG4B to chromosome 20 and the role of CLG4B in aortic aneurysmal disease. Ann. Hum. Genet. 59: 17–24.

[12] Linn, R. et al. (1996) Reassignment of the 92-kDa type IV collagenase gene (CLG4B) to human chromosome 20. Cytogenet. Cell Genet. 72: 159–161.

MMP10 transin-2, stromelysin-2, SL-2, STMY2

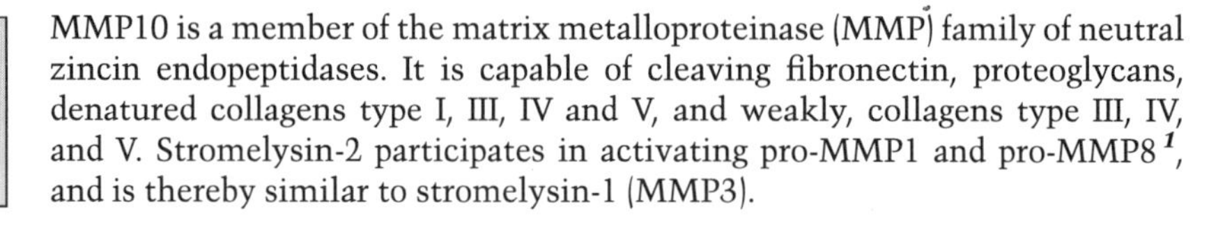

MMP10 is a member of the matrix metalloproteinase (MMP) family of neutral zincin endopeptidases. It is capable of cleaving fibronectin, proteoglycans, denatured collagens type I, III, IV and V, and weakly, collagens type III, IV, and V. Stromelysin-2 participates in activating pro-MMP1 and pro-MMP8 [1], and is thereby similar to stromelysin-1 (MMP3).

Molecular structure

Stromelysin-2 has the same domain structure as stromelysin-1 and several other MMPs in having a prodomain (which functions as a latency peptide), a catalytic domain containing a highly conserved and extended HEXXH motif seen in all zincin metallopeptidases, and a hemopexin-like domain at the C-terminus of the molecule. Activation of pro-MMP10 occurs by proteolytic removal of the prodomain or by dislocation of the domain by mercuric compounds, as for several other MMPs.

Propeptide

Isolation

Cultured human keratinocytes are a suitable source of stromelysin-2. The proenzyme can be purified using conventional liquid chromatography methods developed for stromelysin-1 [2].

Accession number

P09238

Primary structure

Ala	A	35	Cys	C	5	Asp	D	33	Glu	E	31
Phe	F	36	Gly	G	30	His	H	15	Ile	I	17
Lys	K	29	Leu	L	43	Met	M	10	Asn	N	15
Pro	P	33	Gln	Q	13	Arg	R	20	Ser	S	36
Thr	T	20	Val	V	25	Trp	W	9	Tyr	Y	21

Mol. wt (calc.) = 54 151 Residues = 476

```
1     MMHLAFLVLL  CLPVCSAYPL  SGAAKEEDSN  KDLAQQYLEK  YYNLEKDVKQ
51    FRRKDSNLIV  KKIQGMQKFL  GLEVTGKLDT  DTLEVMRKPR  CGVPDVGHFS
101   SFPGMPKWRK  THLTYRIVNY  TPDLPRDAVD  SAIEKALKVW  EEVTPLTFSR
151   LYEGEADIMI  SFAVKEHGDF  YSFDGPGHSL  AHAYPPGPGL  YGDIHFDDDE
201   KWTEDASGTN  LFLVAAHELG  HSLGLFHSAN  TEALMYPLYN  SFTELAQFRL
251   SQDDVNGIQS  LYGPPPASTE  EPLVPTKSVP  SGSEMPAKCD  PALSFDAIST
301   LRGEYLFFKD  RYFWRRSHWN  PEPEFHLISA  FWPSLPSYLD  AAYEVNSRDT
351   VFIFKGNEFW  AIRGNEVQAG  YPRGIHTLGF  PPTIRKIDAA  VSDKEKKKTY
401   FFAADKYWRF  DENSQSMEQG  FPRLIADDFP  GVEPKVDAVL  QAFGFFYFFS
451   GSSQFEFDPN  ARMVTHILKS  NSWLHC
```

Structural and functional sites

Signal peptide: 1–17
Prodomain: 18–98
Autoinhibitor region (cysteine switch): 89–96
Catalytic domain: 99–285
Hemopexin-like domain: 286–476
Catalytic residues: 217, 218, 221, 227

Gene structure

The human gene is located at 11q22.3–23[3]. Further information can be obtained from GeneCard (http://bioinfo.weizmann.ac.il/cgi-bin/lvrebhan/carddisp?MMP10).

References

1. Knauper, V. et al. (1996) Activation of human neutrophil procollagenase by stromelysin 2. Eur. J. Biochem. 235: 187–191.
2. Nagase, H. (1995) Human stromelysins 1 and 2. Methods Enzymol. 248: 449–470.
3. Jung, J.Y. et al. (1990) Localization of stromelysin 2 gene to the q22.3–23 region of chromosome 11 by in situ hybridization. Ann. Genet. 33: 21–23.

MMP11 stromelysin-3, ST3, SL-3, STMY3

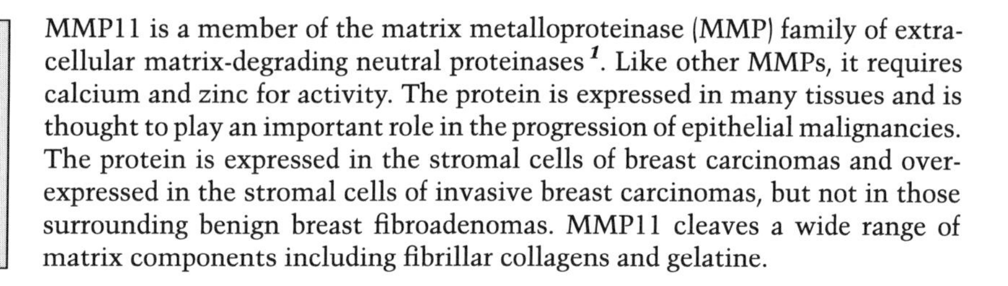

MMP11 is a member of the matrix metalloproteinase (MMP) family of extracellular matrix-degrading neutral proteinases [1]. Like other MMPs, it requires calcium and zinc for activity. The protein is expressed in many tissues and is thought to play an important role in the progression of epithelial malignancies. The protein is expressed in the stromal cells of breast carcinomas and overexpressed in the stromal cells of invasive breast carcinomas, but not in those surrounding benign breast fibroadenomas. MMP11 cleaves a wide range of matrix components including fibrillar collagens and gelatine.

Molecular structure

MMP11 is a single-chain polypeptide of molecular weight 54 500. The protein comprises a signal peptide, a prodomain, catalytic domain and a hemopexin-like domain, and is therefore like other members of the MMP superfamily. MMP11 belongs to the family of M10A (matrixin) zinc-dependent metalloproteinases. Stromelysin-3 can be processed directly to its enzymatically active form by an obligate intracellular proteolytic event that occurs within the constitutive secretory pathway. Intracellular activation is regulated by an unusual 10-amino-acid insert sandwiched between the pro- and catalytic-domains of stromelysin-3, which is encrypted with an Arg-X-Arg-X-Lys-Arg recognition motif for the Golgi-associated proteinase, furin. A furin–stromelysin-3 processing axis distinguishes the regulation of this enzyme from other MMPs [2].

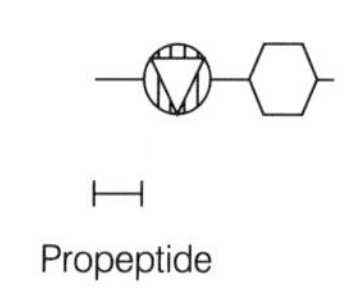

Isolation

The availability of cDNA clones for stromelysin-3 has made it possible to obtain purified recombinant enzyme from bacterial cells and from a mouse myeloma cell line [3].

Accession number

P24347

Primary structure

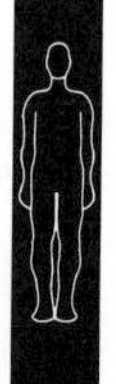

Ala	A	55	Cys	C	4	Asp	D	34	Glu	E	21
Phe	F	30	Gly	G	38	His	H	16	Ile	I	11
Lys	K	12	Leu	L	48	Met	M	5	Asn	N	4
Pro	P	52	Gln	Q	21	Arg	R	37	Ser	S	20
Thr	T	23	Val	V	27	Trp	W	16	Tyr	Y	14

Mol. wt (calc.) = 54 596 Residues = 488

```
1     MAPAAWLRSA AARALLPPML LLLLQPPPLL ARALPPDVHH LHAERRGPQP
51    WHAALPSSPA PAPATQEAPR PASSLRPPRC GVPDPSDGLS ARNRQKRFVL
101   SGGRWEKTDL TYRILRFPWQ LVQEQVRQTM AEALKVWSDV TPLTFTEVHE
151   GRADIMIDFA RYWDGDDLPF DGPGGILAHA FFPKTHREGD VHFDYDETWT
201   IGDDQGTDLL QVAAHEFGHV LGLQHTTAAK ALMSAFYTFR YPLSLSPDDC
251   RGVQHLYGQP WPTVTSRTPA LGPQAGIDTN EIAPLEPDAP PDACEASFDA
301   VSTIRGELFF FKAGFVWRLR GGQLQPGYPA LASRHWQGLP SPVDAAFEDA
351   QGHIWFFQGA QYWVYDGEKP VLGPAPLTEL GLVRFPVHAA LVWGPEKNKI
401   YFFRGRDYWR FHPSTRRVDS PVPRRATDWR GVPSEIDAAF QDADGYAYFL
451   RGRLYWKFDP VKVKALEGFP RLVGPDFFGC AEPANTFL
```

Structural and functional sites

Signal peptide: 1–31 (R33, potential)
Prodomain: 32–97
Autoinhibitor region (cysteine-switch): 78–85
Catalytic domain: 98–286
Hemopexin-like: 287–488
Catalytic zinc-binding residues: 215, 216, 219, 225

Gene structure

The human gene is located at 22q11.2 [4]. Further information can be obtained from GeneCard (http://bioinfo.weizmann.ac.il/cgi-bin/lvrebhan/carddisp?MMP11).

References

[1] Basset, P. et al. (1990) A novel metalloproteinase gene specifically expressed in stromal cells of breast carcinomas. Nature 348: 699–704.

[2] Pei, D. and Weiss, S.J. (1995) Furin-dependent intracellular activation of the human stromelysin-3 zymogen. Nature 375: 244–247.

[3] Murphy, G. et al. (1993) The 28-kDa N-terminal domain of mouse stromelysin-3 has the general properties of a weak metalloproteinase. J. Biol. Chem. 268: 15435–15441.

[4] Levy, A. et al. (1992) Assignment of the human stromelysin 3 (STMY3) gene to the q11.2 region of chromosome 22. Genomics 13: 881–883.

MMP12 macrophage metalloelastase, HME

Matrix metalloproteinase-12 (MMP12) appears to play a role in tissue remodelling in that it has significant elastolytic activity. It is capable of hydrolysing soluble and insoluble elastin. The enzyme is present in alveolar macrophages but is absent from peripheral blood monocytes. The activity of the enzyme is induced by exposure to lipopolysaccharide and is inhibited by dexamethasone.

Molecular structure

The enzyme possesses a signal peptide, a prodomain, a catalytic domain and a substrate-binding hemopexin-like domain.

Propeptide

Isolation

The enzyme is available in recombinant form, which possesses elastolytic activity [1].

Primary structure

Ala	A	24	Cys	C	3	Asp	D	28	Glu	E	21
Phe	F	38	Gly	G	35	His	H	14	Ile	I	25
Lys	K	33	Leu	L	40	Met	M	12	Asn	N	28
Pro	P	27	Gln	Q	14	Arg	R	21	Ser	S	29
Thr	T	24	Val	V	23	Trp	W	8	Tyr	Y	23

Mol. wt (calc.) = 54 001 Residues = 470

```
1    MKFLLILLLQ ATASGALPLN SSTSLEKNNV LFGERYLEKF YGLEINKLPV
51   TKMKYSGNLM KEKIQEMQHF LGLKVTGQLD TSTLEMMHAP RCGVPDVHHF
101  REMPGGPVWR KHYITYRINN YTPDMNREDV DYAIRKAFQV WSNVTPLKFS
151  KINTGMADIL VVFARGAHGD FHAFDGKGGI LAHAFGPGSG IGGDAHFDED
201  EFWTTHSGGT NLFLTAVHEI GHSLGLGHSS DPKAVMFPTY KYVDINTFRL
251  SADDIRGIQS LYGDPKENQR LPNPDNSEPA LCDPNLSFDA VTTVGNKIFF
301  FKDRFFWLKV SERPKTSVNL ISSLWPTLPS GIEAAYEIEA RNQVFLFKDD
351  KYWLISNLRP EPNYPKSIHS FGFPNFVKKI DAAVFNPRFY RTYFFVDNQY
401  WRYDERRQMM DPGYPKLITK NFQGIGPKID AVFYSKNKYY YFFQGSNQFE
451  YDFLLQRITK TLKSNSWFGC
```

Structural and functional sites

Signal peptide: 1–16
Prodomain: 17–105
Autoinhibitor region (cysteine-switch): 90–97
Catalytic domain: 106–278
Hemopexin-like domain: 279–470
Catalytic residues: 218, 219, 222, 228
Potential N-linked glycosylation sites: 20, 285

Gene structure

The 13-kb human gene is composed of ten exons and is located at 11q22.2–22.3 [2]. Further information can be obtained from GeneCard (http://bioinfo.weizmann.ac.il/cgi-bin/lvrebhan/carddisp?MMP12).

References

[1] Shapiro, S.D. et al. (1993) Cloning and characterization of a unique elastolytic metalloproteinase produced by human alveolar macrophages. J. Biol. Chem. 268: 23824–23829.

[2] Belaaouaj, A. et al. (1995) Human macrophage metalloelastase: Genomic organization, chromosomal location, gene linkage, and tissue-specific expression. J. Biol. Chem. 270: 14568–14575.

The major function of this enzyme is to degrade collagens type I and II and has no action on gelatin or casein. Matrix metalloproteinase-13 (MMP13) cleaves the collagens into the characteristic three-quarters and one-quarter fragments. Notably, the enzyme has telopeptidase activity and efficiently removes the extrahelical telopeptides at the ends of collagen molecules. The hemopexin-like domain of the molecule is only essential for the triple helicase activity. MMP13 also cleaves large tenascin C isoform, fibronectin, recombinant fibronectin fragments, and type IV, IX, X and XIV collagens. The presence of the enzyme in breast carcinoma suggests that it may have a role in tumorigenesis.

Molecular structure

The molecule contains a signal peptide, a prodomain, a catalytic domain and a hemopexin-like domain at the C-terminus. The primary structure of the enzyme shows greater similarity to the collagenases than to the stromelysins. The X-ray crystal structure of the domain enclosed by residues 265–471 has been obtained[1].

Propeptide

Isolation

Recombinant enzyme is available from a vaccinia virus system of protein overexpression[2]. Recombinant human procollagenase-3 and a C-terminal truncated form have been stably expressed in myeloma cells and purified[3].

Accession number

P45452

Primary structure

Ala	A	25	Cys	C	4	Asp	D	41	Glu	E	27
Phe	F	32	Gly	G	35	His	H	18	Ile	I	24
Lys	K	29	Leu	L	43	Met	M	12	Asn	N	18
Pro	P	33	Gln	Q	8	Arg	R	19	Ser	S	30
Thr	T	22	Val	V	20	Trp	W	10	Tyr	Y	21

Mol. wt (calc.) = 53 820 Residues = 471

```
1     MHPGVLAAFL  FLSWTHCRAL  PLPSGGDEDD  LSEEDLQFAE  RYLRSYYHPT
51    NLAGILKENA  ASSMTERLRE  MQSFFGLEVT  GKLDDNTLDV  MKKPRCGVPD
101   VGEYNVFPRT  LKWSKMNLTY  RIVNYTPDMT  HSEVEKAFKK  AFKVWSDVTP
151   LNFTRLHDGI  ADIMISFGIK  EHGDFYPFDG  PSGLLAHAFP  PGPNYGGDAH
201   FDDDETWTSS  SKGYNLFLVA  AHEFGHSLGL  DHSKDPGALM  FPIYTYTGKS
```

```
251  HFMLPDDDVQ GIQSLYGPGD EDPNPKHPKT PDKCDPSLSL DAITSLRGET
301  MIFKDRFFWR LHPQQVDAEL FLTKSFWPEL PNRIDAAYEH PSHDLIFIFR
351  GRKFWALNGY DILEGYPKKI SELGLPKEVK KISAAVHFED TGKTLLFSGN
401  QVWRYDDTNH IMDKDYPRLI EEDFPGIGDK VDAVYEKNGY IYFFNGPIQF
451  EYSIWSNRIV RVMPANSILW C
```

Structural and functional sites

Signal peptide: 1–19
Prodomain: 20–103
Autoinhibitor region: 94–101
Catalytic domain: 104–280
Hemopexin-like domain: 281–471
Catalytic residues: 222, 223, 226, 232
Potential N-linked glycosylation sites: 117, 152

Gene structure

The human gene is located at 11q22.3[4]. Further information can be obtained from GeneCard (http://bioinfo.weizmann.ac.il/cgi-bin/lvrebhan/carddisp?MMP13).

References

1. Gomis-Ruth, F.X. et al. (1996) The helping hand of collagenase-3 (MMP-13): 2.7 A crystal structure of its C-terminal hemopexin-like domain. J. Mol. Biol. 264: 556–566.
2. Freije, J.M.P. et al. (1994) Molecular cloning and expression of collagenase-3, a novel human matrix metalloproteinase produced by breast carcinomas. J. Biol. Chem. 269: 16766–16773.
3. Knauper, V. et al. (1997) The role of the C-terminal domain of human collagenase-3 (MMP-13) in the activation of procollagenase-3, substrate specificity, and tissue inhibitor of metalloproteinase interaction. J. Biol. Chem. 272: 7608–7616.
4. Pendas, A.M. et al. (1995) The human collagenase-3 (CLG3) gene is located on chromosome 11q22.3 clustered to other members of the matrix metalloproteinase gene family. Genomics 26: 615–618.

Neurocan is a developmentally regulated chondroitin sulphate proteoglycan found in brain. In addition to binding with hyaluronan the core protein appears to undergo a developmentally regulated *in vivo* proteolytic processing. Concentration decreases during development from 20% of the total soluble chondroitin sulphate proteoglycan protein at 7 days postnatal to 6% in adult brain. In both young and old animals the proteoglycan contains about 40% carbohydrate by weight, comprising equal proportions of chondroitin sulphate and oligosaccharides. Neurocan seems to facilitate axonal adhesion and may be important in defining migration.

Molecular structure

The deduced amino acid composition shows a protein of molecular mass of 136 kDa with 1257 amino acids. The N-terminal region shows an immunoglobulin domain followed by two link protein-like tandem repeats, closely followed by an RGDS sequence. The central portion of the molecule has no known homology and has a high concentration (35%) of serine, threonine and proline. The C-terminal portion contains two epidermal growth factor (EGF)-like repeats, followed by a lectin-like domain and a complement regulatory-like protein. There are five potential N-linked sites, seven chondroitin sulphate attachment sites and 25 threonine *O*-glycosylation sites. In the adult the protein appears to be proteolytically cleaved near residue 638 to produce a shortened protein which contains only one chondroitin sulphate chain and does not bind hyaluronan [1–4].

Isolation

Radiolabelled extracts have been extracted in 50 mM PBS (pH 7.2) containing 0.1% CHAPS and purified on monoclonal affinity columns [2].

Accession number

P55066

Primary structure (mouse)

Ala	A	110	Cys	C	34	Asp	D	65	Glu	E	90
Phe	F	29	Gly	G	111	His	H	42	Ile	I	33
Lys	K	31	Leu	L	99	Met	M	23	Asn	N	31
Pro	P	118	Gln	Q	48	Arg	R	81	Ser	S	117
Thr	T	75	Val	V	80	Trp	W	28	Tyr	Y	23

Mol. wt (calc.) = 137 200   Residues = 1268

```
1     MGAGSVWASG  LLLLWLLLLV  AGDQDTQDTT  ATEKGLRMLK  SGSGPVRAAL
51    AELVALPCFF  TLQPRLSSLR  DIPRIKWTKV  QTASGQRQDL  PILVAKDNVV
101   RVAKGWQGRV  SLPAYPRHRA  NATLLLGPLR  ASDSGLYRCQ  VVKGIEDEQD
```

```
151   LVTLEVTGVV FHYRAARDRY ALTFAEAQEA CRLSSATIAA PRHLQAAFED
201   GFDNCDAGWL SDRTVRYPIT QSRPGCYGDR SSLPGVRSYG RRDPQELYDV
251   YCFARELGGE VFYVGPARRL TLAGARAQCQ RQGAALASVG QLHLAWHEGL
301   DQCDPGWLAD GSVRYPIQTP RRRCGGPAPG VRTVYRFANR TGFPAPGARF
351   DAYCFRAHHH TAQHGDSEIP SSGDEGEIVS AEGPPGRELK PSLGEQEVIA
401   PDFQEPLMSS GEGEPPDLTW TQAPEETLGS TPGGPTLASW PSSEKWLFTG
451   APSSMGVSSP SDMGVDMEAT TPLGTQVAPT PTMRRGRFKG LNGRHFQQQG
501   PEDQLPEVAE PSAQPPTLGA TANHMRPSAA TEASESDQSH SPWAILTNEV
551   DEPGAGSLGS RSLPESLMWS PSLISPSVPS TESTPSPKPG AAEAPSVKSA
601   IPHLPRLPSE PPAPSPGPSE ALSAVSLQAS SADGSPDFPI VAMLRAPKLW
651   LLPRSTLVPN MTPVPLSPAS PLPSWVPEEQ AVRPVSLGAE DLETPFQTTI
701   AAPVEASHRS PDADSIEIEG TSSMRATKHP ISGPWASLDS SNVTMNPVPS
751   DAGILGTESG VLDLPGSPTS GGQATVEKVL ATWLPLPGQG LDPGSQSTPM
801   EAHGVAVSME PTVALEGGAT EGPMEATREV VPSTADATWE SESRSAISST
851   HIAVTMARAQ GMPTLTSTSS EGHPEPKGQM VAQESLEPLN TLPSHPWSSL
901   VVPMDEVASV SSGEPTGLWD IPSTLIPVSL GLDESVLNVV AESPSVEGFW
951   EEVASGQEDP TDPCENNPCL HGGTCHTNGT VYGCSCDQGY AGENCEIDID
1001  DCLCSPCENG GTCIDEVNGF ICLCLPSYGG SLCEKDTEGC DRGWHKFQGH
1051  CYRYFAHRRA WEDAERDCRR RAGHLTSVHS PEEHKFINSF GHENSWIGLN
1101  DRTVERDFQW TDNTGLQYEN WREKQPDNFF AGGEDCVVMV AHESGRWNDV
1151  PCNYNLPYVC KKGTVLCGPP PAVENASLVG VRKIKYNVHA TVRYQCDEGF
1201  SQHRVATIRC RNNGKWDRPQ IMCIKPRRSH RMRRHHHPH  RHHKPRKEHR
1251  KHKRHPAEDW EKDEGDFC
```

Structural and functional sites

Signal peptide: 1–22
Hyaluronic acid binding domain: 48–346
  Immunoglobulin repeat: 36–160
  Link protein tandem repeats: 161–355
Putative RGD cell adhesion site: 364–367
Potential N-linked glycosylation sites: 121, 339, 742, 978, 1175
EGF (6C) repeats: 998–1035, 1036–1073
Lectin repeat: 1089–1215
CCP repeat: 1226–1268

Gene structure

Northern blot analysis revealed a single band 7.5 kb in bovine brain mRNA. The murine gene has a size of approximately 25 kb and contains 15 exons, and is located on mouse chromosome 8 [4].

References

[1] Rauch, U. et al. (1992) Cloning and primary structure of neurocan, a developmentally regulated, aggregating chondroitin sulphate proteoglycan of brain. J. Biol. Chem. 267: 19536–19547.

[2] Rauch, U. et al. (1991) Isolation and characterization of developmentally regulated chondroitin/keratan sulfate proteoglycan of brain identified with monoclonal antibodies. J. Biol. Chem. 266: 14785–14801.

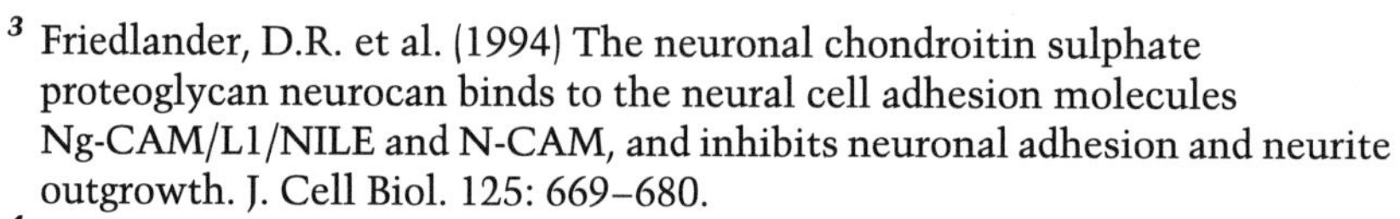

[3] Friedlander, D.R. et al. (1994) The neuronal chondroitin sulphate proteoglycan neurocan binds to the neural cell adhesion molecules Ng-CAM/L1/NILE and N-CAM, and inhibits neuronal adhesion and neurite outgrowth. J. Cell Biol. 125: 669–680.

[4] Rauch, U. et al. (1995) Structure and chromosomal localisation of mouse neurocan gene. Genomics 28: 405–410.

Nidogen

entactin

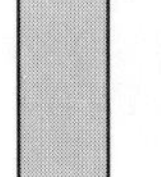

Entactin/nidogen is a sulphated glycoprotein that is an integral component of basement membranes. It associates specifically with both laminin (in a 1:1 molar ratio) and type IV collagen and is thought to play an important role in linking these two molecules together in the basement membrane.

Molecular structure

Entactin/nidogen is a single-chain polypeptide with a molecular weight of approximately 150 000. When isolated under denaturing conditions, the molecule resembles a dumb-bell when viewed by electron microscopy after rotary shadowing. However, recombinant molecules isolated from tissue culture medium under non-denaturing conditions appear to contain three rather than two globular domains separated by rod-like segments. The entactin/nidogen molecule contains a series of six-cysteine epidermal growth factor (EGF) repeats, broken up by repeats found in thyroglobulin and the low density lipoprotein (LDL) receptor [1,2].

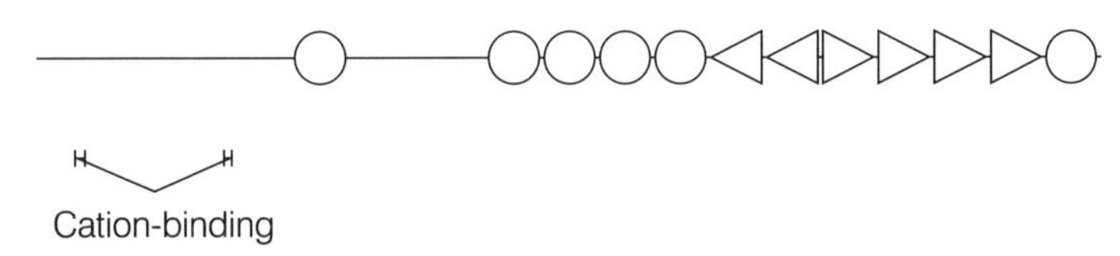

Isolation

Entactin/nidogen can be isolated from basement membranes in association with laminin by extraction with chelating agents. The entactin/nidogen–laminin complex can then be dissociated with buffers containing guanidine–HCl [3].

Accession number

P14543

Primary structure

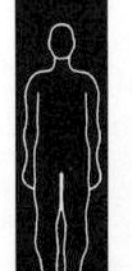

Ala	A	79	Cys	C	50	Asp	D	76	Glu	E	73
Phe	F	53	Gly	G	113	His	H	34	Ile	I	56
Lys	K	29	Leu	L	81	Met	M	8	Asn	N	45
Pro	P	87	Gln	Q	58	Arg	R	73	Ser	S	93
Thr	T	96	Val	V	86	Trp	W	13	Tyr	Y	44

Mol. wt (calc.) = 136 337 Residues = 1247

```
1     MLASSSRIRA  AWTRALLLPL  LLAGPVGCLS  RQELFPFGPG  QGDLELEDGD
51    DFVSPALELS  GALRFYDRSD  IDAVYVTTNG  IIATSEPPAK  ESHPGLFPPT
101   FGAVAPFLAD  LDTTDGLGKV  YYREDLSPSI  TQRAAECVHR  GFPEISFQPS
151   SAVVVTWESV  APYQGPSRDP  DQKGKRNTFQ  AVLASSDSSS  YAIFLYPEDG
201   LQFHTTFSKK  ENNQVPAVVA  FSQGSVGFLW  KSNGAYNIFA  NDRESIENLA
251   KSSNSGQQGV  WVFEIGSPAT  TNGVVPADVI  LGTEDGAEYD  DEDEDYDLAT
301   TRLGLEDVGT  TPFSYKALRR  GGADTYSVPS  VLSPRRAATE  RPLGPPTERT
```

```
351   RSFQLAVETF HQQHPQVIDV DEVEETGVVF SYNTDSRQTC ANNRHQCSVH
401   AECRDYATGF CCSCVAGYTG NGRQCVAEGS PQRVNGKVKG RIFVGSSQVP
451   IVFENTDLHS YVVMNHGRSY TAISTIPETV GYSLLPLAPV GGIIGWMFAV
501   EQDGFKNGFS ITGGEFTRQA EVTFVGHPGN LVIKQRFSGI DEHGHLTIDT
551   ELEGRVPQIP FGSSVHIEPY TELYHYSTSV ITSSSTREYT VTEPERDGAS
601   PSRIYTYQWR QTITFQECVH DDSRPALPST QQLSVDSVFV LYNQEEKILR
651   YAFSNSIGPV REGSPDALQN PCYIGTHGCD TNAACRPGPR TQFTCECSIG
701   FRGDGRTCYD IDECSEQPSV CGSHTICNNH PGTFRCECVE GYQFSDEGTC
751   VAVVDQRPIN YCETGLHNCD IPQRAQCIYT GGSSYTCSCL PGFSGDGQAC
801   QDVDECQPSR CHPDAFCYNT PGSFTCQCKP GYQGDGFRCV PGEVEKTRCQ
851   HEREHILGAA GATDPQRPIP PGLFVPECDA HGHYAPTQCH GSTGYCWCVD
901   RDGREVEGTR TRPGMTPPCL STVAPPIHQG PAVPTAVIPL PPGTHLLFAQ
951   TGKIERLPLE GNTMRKTEAK AFLHVPAKVI IGLAFDCVDK MVYWTDITEP
1001  SIGRASLHGG EPTTIIRQDL GSPEGIAVDH LGRNIFWTDS NLDRIEVAKL
1051  DGTQRRVLFE TDLVNPRGIV TDSVRGNLYW TDWNRDNPKI ETSYMDGTNR
1101  RILVQDDLGL PNGLHFDAFS SQLCWVDAGT NRAECLNPSQ PSRRKALEGL
1151  QYPFAVTSYG KNLYFTDWKM NSVVALDLAI SKETDAFQPH KQTRLYGITT
1201  ALSQCPQGHN YCSVNNGGCT HLCLATPGSR TCRCPDNTLG VDCIERK
```

Structural and functional sites

Signal peptide: 1–28
EGF (6C) repeats: 388–425, 670–711, 712–759, 760–803, 804–846, 1212–1247
Thyroglobulin repeats: 847–884, 885–921
LDL receptor repeats: 983–1028, 1029–1078, 1079–1128, 1129–1174
EF-hand-type divalent cation binding sites: 43–54, 278–289
Potential O-linked sulphation sites: 289, 296
Potential N-linked glycosylation site: 1137
Transglutaminase cross-linking site (to laminin): 756 (equivalent to Q726 in murine sequence which is a proven cross-linking site)
RGD cell adhesion site: 702–704 (not conserved in murine sequence)

Gene structure

The entactin/nidogen gene has been localized to human chromosome 1q43 [2].

References

[1] Nagayoshi, T. et al. (1989) Human nidogen: Complete amino acid sequence and structural domains deduced from cDNAs, and evidence for polymorphism of the gene. DNA 8: 581–594.

[2] Olsen, D.R. et al. (1989) Human nidogen: cDNA cloning, cellular expression, and mapping of the gene to chromosome 1q43. Am. J. Hum. Genet. 44: 876–855.

[3] Paulsson, M. et al. (1987) Laminin–nidogen complex. Extraction with chelating agents and structural characterisation. Eur. J. Biochem. 166: 467–478.

Osteocalcin

γ-carboxyglutamic acid-containing protein, bone Gla-protein

Osteocalcin is the most abundant non-collagenous protein in bone, synthesized by osteoblasts and odontoblasts. It is highly conserved among all vertebrate species studied and usually contains three γ-carboxyglutamic acid residues (the human protein may only contain two) which provide the molecule with its calcium-binding properties. Prior to mineralization, osteocalcin is found in only trace amounts, but its level increases markedly during periods of intense skeletal growth. Osteocalcin binds tightly to hydroxyapatite and may function in the assembly of mineralized bone by regulating apatite crystal growth. Its synthesis is vitamin K-dependent and is stimulated by 1,25-dihydroxyvitamin D3. Specific osteoblastic synthesis and presence in blood has led to its use as a diagnostic parameter of bone metabolism. Osteocalcin-deficient mice have a higher bone mass and bones of improved functional quality, which implies that osteocalcin is a negative regulator of bone formation.

Molecular structure

Osteocalcin is secreted as a precursor (approximate molecular weight 10 000) and processed to the active molecule of 46–50 residues stabilized by a single intrachain disulphide bond (approximate molecular weight 5800). The glutamic acid residues to be modified by vitamin K-dependent carboxylation are clustered in the centre of the molecule and are preceded by a putative γ-carboxylation recognition site. The amino acid sequence of this site shows identity with matrix Gla protein. A single hydroxyproline residue occurs in most species[1–6].

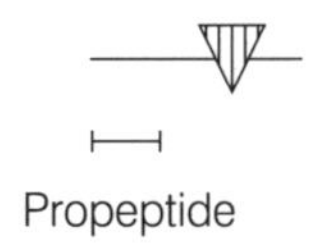

Isolation

Osteocalcin can be isolated from EDTA extracts of bone by gel filtration on Sephadex G-100 and DEAE–Sephadex ion exchange chromatography[1]. Protein degradation can be avoided by demineralizing in 20% formic acid, fractionating on Sephacryl S-200 and immunoabsorption[2].

Accession number

P02818

Primary structure

Ala	A	14	Cys	C	3	Asp	D	4	Glu	E	8
Phe	F	3	Gly	G	8	His	H	1	Ile	I	2
Lys	K	4	Leu	L	12	Met	M	1	Asn	N	1
Pro	P	9	Gln	Q	4	Arg	R	8	Ser	S	5
Thr	T	1	Val	V	6	Trp	W	1	Tyr	Y	5

Mol. wt (calc.) = 10 950 Residues = 100

```
1    MRALTLLALL ALAALCIAGQ AGAKPSGAES SKGAAFVSKQ EGSEVVKRPR
51   RYLYQWLGAP VPYPDPLEPR REVCELNPDC DELADHIGFQ EAYRRFYGPV
```

Structural and functional sites

Signal peptide: 1–23
Propeptide: 24–51
γ-Carboxyglutamic acid residues: 68 (9% of cases), 72, 75
Intra-chain disulphide bond: 74–80

Gene structure

The osteocalcin gene contains four exons. It has been mapped to the long arm of human chromosome 1 by Southern blot analysis [7]. The mouse genome contains an osteocalcin cluster of three genes arranged within 23 kb of genomic DNA [8,9].

References

1 Price, P.A. et al. (1976) Characterization of a γ-carboxyglutamic acid-containing protein from bone. Proc. Natl Acad. Sci. USA 73: 1447–1451.
2 Poser, J.W. et al. (1980) Isolation and sequence of the vitamin K-dependent protein from human bone. J. Biol. Chem. 255: 8685–8691.
3 Pan, L.C. and Price, P.A. (1985) The propeptide of rat bone γ-carboxyglutamic acid protein shares homology with other vitamin K-dependent protein precursors. Proc. Natl Acad. Sci. USA 82: 6109–6113.
4 Celeste, A.J. et al. (1986) Isolation of the human gene for bone Gla protein utilizing mouse and rat cDNA clones. EMBO J. 5: 1885–1890.
5 Price, P.A. (1987) Vitamin K-dependent bone proteins. In: Calcium Regulation and Bone Metabolism. Basic and Clinical Aspects, Vol. 9, ed. Cohn, D.V. et al., Elsevier, Amsterdam, pp. 419–425.
6 Kieffer, M.C. et al. (1990) The cDNA and derived amino acid sequences of human and bovine Gla protein. Nucleic Acids Res. 18: 1909.
7 Puchacz, E. et al. (1989) Chromosomal localization of the human osteocalcin gene. Endocrinology 124: 2648–2650.
8 Ducy, P. et al. (1996) Increased bone formation in osteocalcin-deficient mice. Nature 382: 448–452.
9 Desbois, C. et al. (1994) The mouse osteocalcin gene cluster contains three genes with two separate spatial and temporal patterns of expression. J. Biol. Chem. 269: 1183–1190.

Osteonectin

BM40, SPARC

Osteonectin was initially isolated as one of the main non-collagenous components of bone. However, a much broader functional role is indicated by its expression in a variety of tissues including endodermal, epidermal and soft connective tissues. The protein contains a calcium-binding region, but its function is unknown. Expression of osteonectin has been linked to bone formation, and to tissue differentiation and remodelling owing to an ability of some forms to bind collagens and hydroxyapatite. Osteonectin knockout mice develop normally but show delayed angiogenesis in skin wounds.

Molecular structure

Osteonectin is a single-chain polypeptide composed of four distinct domains. The NH_2-terminal domain (I) is encoded by exons 3 and 4, contains two glutamate-rich segments and can bind up to eight Ca^{2+} ions. Domain II is encoded by exons 5 and 6, shows some homology to ovomucoid and certain serine proteinase inhibitors, and is rich in intra-chain disulphide bonds. Domain III, encoded by exons 7 and 8, is predicted to be α-helical. Exon 9 encodes domain IV and contains a single EF-hand-type, helix-loop-helix cation-binding structure. The protein is glycosylated and some forms may be phosphorylated[1–7].

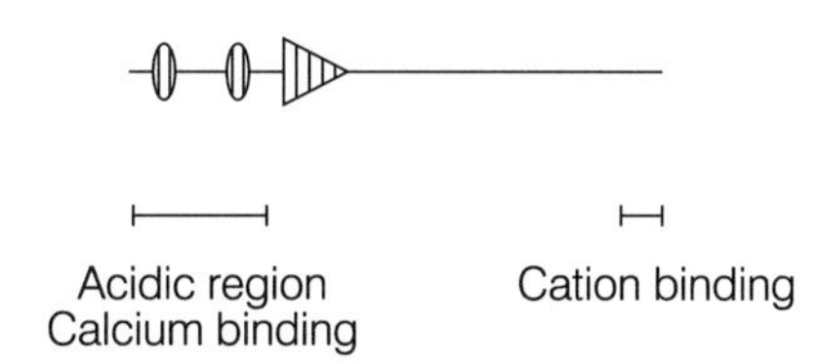

Isolation

Osteonectin has been isolated from both EDTA and guanidine–HCl/EDTA extracts of bone and can be purified by ion-exchange chromatography and gel permeation[1].

Accession number

P09486

Primary structure

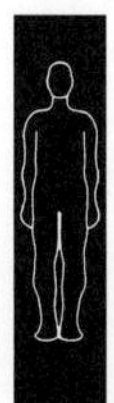

Ala	A	18	Cys	C	15	Asp	D	23	Glu	E	34
Phe	F	14	Gly	G	14	His	H	12	Ile	I	12
Lys	K	20	Leu	L	29	Met	M	5	Asn	N	14
Pro	P	18	Gln	Q	10	Arg	R	10	Ser	S	7
Thr	T	17	Val	V	20	Trp	W	4	Tyr	Y	7

Mol. wt (calc.) = 34 593 Residues = 303

```
1      MRAWIFFLLC  LAGRALAAPQ  QEALPDETEV  VEETVAEVTE  VSVGANPVQV
51     EVGEFDDGAE  ETEEEVVAEN  PCQNHHCKHG  KVCELDENNT  PMCVCQDPTS
101    CPAPIGEFEK  VCSNDNKTFD  SSCHFFATKC  TLEGTKKGHK  LHLDYIGPCK
151    YIPPCLDSEL  TEFPLRMRDW  LKNVLVTLYE  RDEDNNLLTE  KQKLRVKKIH
201    ENEKRLEAGD  HPVELLARDF  EKNYNMYIFP  VHWQFGQLDQ  HPIDGYLSHT
251    ELAPLRAPLI  PMEHCTTRFF  ETCDLDNDKY  IALDEWAGCF  GIKQKDIDKD
301    LVI
```

Structural and functional sites

Signal peptide: 1–17
Domain I: 23–70
Domain II: 71–155
Domain III: 185–221
Domain IV: 274–285
Ovomucoid/serine protease inhibitor repeat: 93–149
Potential N-linked glycosylation site: 116

Gene structure

The human gene is located on chromosome 5 at locus q31–q33 and contains ten exons. Although osteonectin is a single gene product, an extracellular matrix glycoprotein (SC1) of brain tissue contains a COOH-terminal segment that is highly homologous to the protein encoded by exons 5–9 of osteonectin [3–6].

References

[1] Romberg, R.W. et al. (1985) Isolation and characterization of native adult osteonectin. J. Biol. Chem. 260: 2728–2736.

[2] Lankat-Buttgereit, B. et al. (1988) Cloning and complete amino acid sequences of human and murine basement membrane protein BM-40 (SPARC-osteonectin). FEBS Lett. 236: 352–356.

[3] McVey, J.H. et al. (1988) Characterisation of the mouse SPARC/osteonectin gene. Intron/exon organisation and an unusual promoter region. J. Biol. Chem. 263: 11111–11116.

[4] Swaroop, A. et al. (1988) Molecular analysis of the cDNA for human SPARC/osteonectin/BM-40: Sequence, expression, and localization of the gene to chromosome 5q31–q33. Genomics 2: 37–47.

[5] Nomura, S. et al. (1989) Evidence for positive and negative regulatory elements in the 5′ flanking sequence of the mouse SPARC (osteonectin) gene. J. Biol. Chem. 264: 12201–12207.

[6] Young, M.F. et al. (1989) Osteonectin promoter. DNA sequence analysis and S1 endonuclease site potentially associated with transcriptional control in bone cells. J. Biol. Chem. 264: 450–456.

[7] Johnston, I.G. et al. (1990) Molecular cloning of SC1: A putative brain extracellular matrix glycoprotein showing partial similarity to osteonectin/BM40/SPARC. Neuron 2: 165–176.

Osteopontin

OPN, 2ar, secreted phosphoprotein (SPPI), bone sialoprotein I (BSP I)

Osteopontin was initially characterized as an abundant bone sialoprotein. Subsequently it was found in a number of tissues, including placenta, distal tubules of the kidney and the central nervous system. The molecule is expressed early in bone development, at high levels at sites of bone remodelling, and has been implicated in the process of osteogenesis. It binds to hydroxyapatite and appears to be associated with the attachment of osteoclasts, possibly via the integrin $\alpha V\beta 3$. Osteopontin expression is stimulated by 1,25-dihydroxyvitamin D3, growth factors and tumour-promoting factors.

Molecular structure

Osteopontin is a single-chain polypeptide with a molecular weight of approximately 32 600. It is glycosylated (five to six O-linked and one N-linked oligosaccharides in rat osteopontin), highly phosphorylated (12 phospho-Ser and one phospho-Thr in rat osteopontin) and sulphated. Its secondary structure predictions suggest an open, extended, flexible structure. The molecule contains a stretch of negatively charged amino acids, principally aspartic acid, and a cell-binding RGD sequence [1–8].

Isolation

Osteopontin is isolated from bone by extraction with 4 M guanidine–HCl, demineralization with 0.5 M EDTA, size fractionation, and hydroxyapatite and ion exchange chromatography [7].

Accession number

P10451

Primary structure

Ala	A	16	Cys	C	3	Asp	D	48	Glu	E	27
Phe	F	8	Gly	G	7	His	H	16	Ile	I	10
Lys	K	19	Leu	L	18	Met	M	5	Asn	N	12
Pro	P	15	Gln	Q	14	Arg	R	10	Ser	S	42
Thr	T	15	Val	V	19	Trp	W	2	Tyr	Y	8

Mol. wt (calc.) = 35 385 Residues = 314

```
1     MRIAVICFCL  LGITCAIPVK  QADSGSSEEK  QLYNKYPDAV  ATWLNPDPSQ
51    KQNLLAPQNA  VSSEETNDFK  QETLPSKSNE  SHDHMDDMDD  EDDDDHVDSQ
101   DSIDSNDSDD  VDDTDDSHQS  DESHHSDESD  ELVTDFPTDL  PATEVFTPVV
151   PTVDTYDGRG  DSVVYGLRSK  SKKFRRPDIQ  YPDATDEDIT  SHMESEELNG
201   AYKAIPVAQD  LNAPSDWDSR  GKDSYETSQL  DDQSAETHSH  KQSRLYKRKA
251   NDESNEHSDV  IDSQELSKVS  REFHSHEFHS  HEDMLVVDPK  SKEEDKHLKF
301   RISHELDSAS  SEVN
```

Structural and functional sites

Signal peptide: 1–16
Potential N-linked glycosylation sites: 79, 106
RGD cell adhesion site: 159–161

Gene structure

The osteopontin gene is located on human chromosome 4 [6].

References

[1] Oldberg, A. et al. (1986) Cloning and sequence analysis of rat bone sialoprotein (osteopontin) cDNA reveals an Arg-Gly-Asp cell binding sequence. Proc. Natl Acad. Sci. USA 83: 8819–8823.

[2] Fisher, L.W. et al. (1987) Purification and partial characterisation of small proteoglycans I and II, bone sialoproteins I and II and osteonectin from the mineral compartment of developing human bone. J. Biol. Chem. 262: 9702–9708.

[3] Prince, C.W. et al. (1987) Isolation, characterization and biosynthesis of a phosphorylated glycoprotein from rat bone. J. Biol. Chem. 262: 2900–2907.

[4] Yoon, K. et al. (1987) Tissue specificity and developmental expression of rat osteopontin. Biochem. Biophys. Res. Commun. 148: 1129–1136.

[5] Kiefer, M.C. et al. (1989) The cDNA and derived amino acid sequence for human osteopontin. Nucleic Acids Res. 17: 3306.

[6] Young, M.F. et al. (1990) cDNA cloning, mRNA distribution and heterogeneity, chromosomal location, and RFLP analysis of human osteopontin (OPN). Genomics 7: 491–502.

[7] Zhang, Q. et al. (1990) Characterisation of foetal porcine bone sialoproteins, secreted phosphoprotein I (SPPI, osteopontin), bone sialoprotein and a 23 kDa glycoprotein. J. Biol. Chem. 265: 7583–7589.

[8] Sodek, J. et al. (1992) Elucidating the functions of bone sialoprotein and osteopontin in bone formation. In: Chemistry and Biology of Mineralized Tissues, ed. Slavkin, H. and Price, P., Elsevier, Amsterdam, pp. 297–306.

Perlecan

basement membrane heparan sulphate proteoglycan

Perlecan is a large proteoglycan, and is a specific and integral component of all basement membranes. It endows basement membranes with fixed negative electrostatic charge and is responsible, in part, for the charge-selective ultrafiltration properties of this extracellular matrix. It interacts with other basement membrane components such as laminin and collagen type IV, and serves as an attachment substrate for cells.

Molecular structure

Perlecan consists of a large protein core of molecular weight 470 000 to which three heparan sulphate glycosaminoglycan side-chains are attached. The size of the side-chains varies depending upon the source of the proteoglycan. Perlecan appears to consist of a series of globular domains separated by thin segments by electron microscopy after rotary shadowing. The heparan sulphate side-chains are located near one end of the protein core. The perlecan core protein can be divided into five domains. Domain I is the NH_2-terminal globular domain that includes three putative heparan sulphate glycosaminoglycan attachment sites. Domain II contains four copies of internal repeats homologous to the ligand-binding domain of the low density lipoprotein (LDL) receptor. Domain II is separated from domain III by a short segment that shows homology to an immunoglobulin repeat. Domain III exhibits marked homology to the short arms of laminin chains. This domain includes four cysteine-rich subdomains containing eight cysteine epidermal growth factor (EGF) repeats (CR-1 to CR-4) separated by three apparently globular subdomains (G1–G3). These two types of subdomain exhibit homology to the laminin chain domains III/V and IV, respectively. Domain IV is composed of 21 consecutive immunoglobulin repeats similar to that separating domains II and III. The COOH-terminal domain V is similar to the COOH-terminal G domain of the laminin A chain. Domain V is composed of three homologous globular repeats (GR-1 to GR-3) separated by two double six-cysteine EGF repeats [1–3].

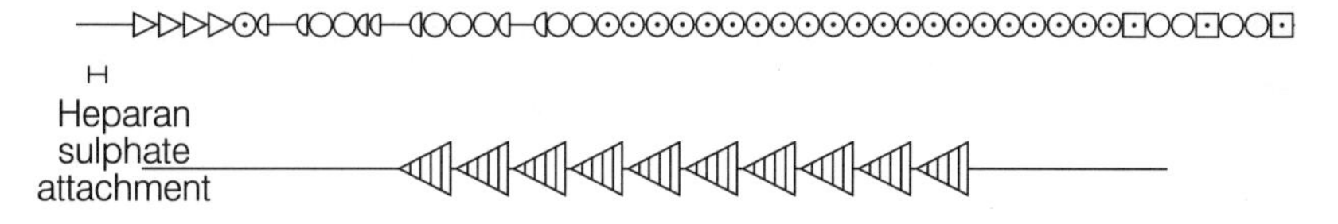

Isolation

Basement membrane heparan sulphate proteoglycan can be isolated after guanidine–HCl extraction by a combination of ion-exchange chromatography, size exclusion chromatography and caesium chloride density gradient centrifugation [1].

Accession number

X62515

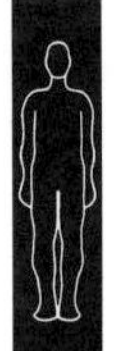

Primary structure

Ala	A	340	Cys	C	188	Asp	D	171	Glu	E	240
Phe	F	107	Gly	G	443	His	H	166	Ile	I	144
Lys	K	69	Leu	L	353	Met	M	42	Asn	N	106
Pro	P	349	Gln	Q	256	Arg	R	264	Ser	S	407
Thr	T	271	Val	V	326	Trp	W	41	Tyr	Y	110

Mol. wt (calc.) = 468 863   Residues = 4393

```
1     MGWRAPGALL LALLLHGRLL AVTHGLRAYD GLSLPEDIET VTASQMRWTH
51    SYLSDDEDML ADSISGDDLG SGDLGSGDFQ MVYFRALVNF TRSIEYSPQL
101   EDAGSREFRE VSEAVVDTLE SEYLKIPGDQ VVSVVFIKEL DGWVFVELDV
151   GSEGNADGAQ IQEMLLRVIS SGSVASYVTS PQGFQFRRLG TVPQFPRACT
201   EAEFACHSYN ECVALEYRCD RRPDCRDMSD ELNCEEPVLG ISPTFSLLVE
251   TTSLPPRPET TIMRQPPVTH APQPLLPGSV RPLPCGPQEA ACRNGHCIPR
301   DYLCDGQEDC EDGSDELDCG PPPPCEPNEF PCGNGHCALK LWRCDGDFDC
351   EDRTDEANCP TKRPEEVCGP TQFRCVSTNM CIPASFHCDE ESDCPDRSDE
401   FGCMPPQVVT PPRESIQASR GQTVTFTCVA IGVPAPFLIN WRLNWGHIPS
451   QPRVTVTSEG GRGTLIIRDV KESDQGAYTC EAMNARGMVF GIPDGVLELV
501   PQRAGPCPDG HFYLEHSAAC LPCFCFGITS VCQSTRRFRD QIRLRFDQPD
551   DFKGVNVTMP AQPGTPPLSS TQLQIDPSLH EFQLVDLSRR FLVHDSFWAL
601   PEQFLGNKVD SYGGSLRYNV RYELARGMLE PVQRPDVVLV GAGYRLLSRG
651   HTPTQPGALN QRQVQFSEEH WVHESGRPVQ RAELLQVLQS LEAVLIQTVY
701   NTKMASVGLS DIAMDTTVTH ATSHGRAHSV EECRCPIGYS GLSCESCDAH
751   FTRVPGGPYL GTCSGCSCNG HASSCDPVYG HCLNCQHNTE GPQCKKCKAG
801   FFGDAMKATA TSCRPCPCPY IDASRRFSDT CFLDTDGQAT CDACAPGYTG
851   RRCESCAPGY EGNPIQPGGK CRPVNQEIVR CDERGSMGTS GEACRCKNNV
901   VGRLCNECAD RSFHLSTRNP DGCLKCFCMG VSRHCTSSSW SRAQLHGASE
951   EPGHFSLTNA ASTHTTNEGI FSPTPGELGF SSFHRLLSGP YFWSLPSRFL
1001  GDKVTSYGGE LRFTVTQRSQ PGSTPLHGQP LVVLQGNNII LEHHVAQEPS
1051  PGQPSTFIVP FREQAWQRPD GQPATREHLL MALAGIDTLL IRASYAQQPA
1101  ESRVSGISMD VAVPEETGQD PALEVEQCSC PPGYRGPSCQ DCDTGYTRTP
1151  SGLYLGTCER CSCHGHSEAC EPETGACQGC QHHTEGPRCE QCQPGYYGDA
1201  QRGTPQDCQL CPCYGDPAAG QAAHTCFLDT DGHPTCDACS PGHSGRHCER
1251  CAPGYYGNPS QGQPCQRDSQ VPGPIGCNCD PQGSVSSQCD AAGQCQCKAQ
1301  VEGLTCSHCR PHHFHLSASN PDGCLPCFCM GITQQCASSA YTRHLISTHF
1351  APGDFQGFAL VNPQRNSRLT GEFTVEPVPE GAQLSFGNFA QLGHESFYWQ
1401  LPETYQGDKV AAYGGKLRYT LSYTAGPQGS PLSDPDVQIT GNNIMLVASQ
1451  PALQGPERRS YEIMFREEFW RRPDGQPATR EHLLMALADL DELLIRATFS
1501  SVPLVASISA VSLEVAQPGP SNRPRALEVE ECRCPPGYIG LSCQDCAPGY
1551  TRTGSGLYLG HCELCECNGH SDLCHPETGA CSQCQHNAAG EFCELCAPGY
1601  YGDATAGTPE DCQPCACPLT NPENMFSRTC ESLGAGGYRC TACEPGYTGQ
1651  YCEQCGPGYV GNPSVQGGQC LPETNQAPLV VEVHPARSIV PQGGSHSLRC
1701  QVSGRGPHYF YWSREDGRPV PSGTQQRHQG SELHFPSVQP SDAGVYICTC
1751  RNLHRSNTSR AELLVTEAPS KPITVTVEEQ RSQSVRPGAD VTFICTAKSK
1801  SPAYTLVWTR LHNGKLPTRA MDFNGILTIR NVQLSDAGTY VCTGSNMFAM
1851  DQGTATLHVQ ASGTLSAPVV SIHPPQLTVQ PGQLAEFRCS ATGSPTPTLE
1901  WTGGPGGQLP AKAQIHGGIL RLPAVEPTDQ AQYLCRAHSS AGQQVARAVL
1951  HVHGGGGPRV QVSPERTQVH AGRTVRLYCR AAGVPSATIT WRKEGGSLPP
2001  QARSERTDIA TLLIPAITTA DAGFYLCVAT SPAGTAQARI QVVVLSASDA
2051  SQPPVKIESS SPSVTEGQTL DLNCVVAGSA HAQVTWYRRG GSLPHHTQVH
2101  GSRLRLPQVS PADSGEYVCR VENGSGPKEA SITVSVLHGT HSGPSYTPVP
2151  GSTRPIRIEP SSSHVAEGQT LDLNCVVPGQ AHAQVTWHKR GGSLPARHQT
```

```
2201  HGSLLRLHQV  TPADSGEYVC  HVVGTSGPLE  ASVLVTIEAS  VIPGPIPPVR
2251  IESSSSTVAE  GQTLDLSCVV  AGQAHAQVTW  YKRGGSLPAR  HQVRGSRLYI
2301  FQASPADAGQ  YVCRASNGME  ASITVTVTGT  QGANLAYPAG  STQPIRIEPS
2351  SSQVAEGQTL  DLNCVVPGQS  HAQVTWHKRG  GSLPVRHQTH  GSLLRLYQAS
2401  PADSGEYVCR  VLGSSVPLEA  SVLVTIEPAG  SVPALGVTPT  VRIESSSSQV
2451  AEGQTLDLNC  LVAGQAHAQV  TWHKRGGSLP  ARHQVHGSRL  RLLQVTPADS
2501  GEYVCRVVGS  SGTQEASVLV  TIQQRLSGSH  SQGVAYPVRI  ESSSASLANG
2551  HTLDLNCLVA  SQAPHTITWY  KRGGSLPSRH  QIVGSRLRIP  QVTPADSGEY
2601  VCHVSNGAGS  RETSLIVTIQ  GSGSSHVPRV  SPPIRIESSS  PTVVEGQTLD
2651  LNCVVARQPQ  AIITWYKRGG  SLPSRHQTHG  SHLRLHQMSV  ADSGEYVCRA
2701  NNNIDALEAS  IVISVSPSAG  SPSAPGSSMP  IRIESSSSHV  AEGETLDLNC
2751  VVPGQAHAQV  TWHKRGGSLP  SYHQTRGSRL  RLHHVSPADS  GEYVCRVMGS
2801  SGPLEASVLV  TIEASGSSAV  HVPAPGGAPP  IRIEPSSSRV  AEGQTLDLKC
2851  VVPGQAHAQV  TWHKRGGNLP  ARHQVHGPLL  RLNQVSPADS  GEYSCQVTGS
2901  SGTLEASVLV  TIEPSSPGPI  PAPGLAQPIY  IEASSSHVTE  GQTLDLNCVV
2951  PGQAHAQVTW  YKRGGSLPAR  HQTHGSQLRL  HHVSPADSGE  YVCRAAGGPG
3001  PEQEASFTVT  VPPSEGSSYR  LRSPVISIDP  PSSTVQQGQD  ASFKCLIHDG
3051  AAPISLEWKT  RNQELEDNVH  ISPNGSIITI  VGTRPSNHGT  YRCVASNAYG
3101  VAQSVVNLSV  HGPPTVSVLP  EGPVWVKVGK  AVTLECVSAG  EPRSSARWTR
3151  ISSTPAKLEQ  RTYGLMDSHT  VLQISSAKPS  DAGTYVCLAQ  NALGTAQKQV
3201  EVIVDTGAMA  PGAPQVQAEE  AELTVEAGHT  ATLRCSATGS  PARTIHWSKL
3251  RSPLPWQHRL  EGDTLIIPRV  AQQDSGQYIC  NATSPAGHAE  ATIILHVESP
3301  PYATTVPEHA  SVQAGETVQL  QCLAHGTPPL  TFQWSRVGSS  LPGRATARNE
3351  LLHFERAAPE  DSGRYRCRVT  NKVGSAEAFA  QLLVQGPPGS  LPATSIPAGS
3401  TPTVQVTPQL  ETKSIGASVE  FHCAVPSDRG  TQLRWFKEGG  QLPPGHSVQD
3451  GVLRIQNLDQ  SCQGTYICQA  HGPWGKAQAS  AQLVIQALPS  VLINIRTSVQ
3501  TVVVGHAVEF  ECLALGDPKP  QVTWSKVGGH  LRPGIVQSGG  VVRIAHVELA
3551  DAGQYRCTAT  NAAGTTQSHV  LLLVQALPQI  SMPQEVRVPA  GSAAVFPCIA
3601  SGYPTPDISW  SKLDGSLPPD  SRLENNMLML  PSVQPQDAGT  YVCTATNRQG
3651  KVKAFAHLQV  PERVVPYFTQ  TPYSFLPLPT  IKDAYRKFEI  KITFRPDSAD
3701  GMLLYNGQKR  VPGSPTNLAN  RQPDFISFGL  VGGRPEFRFD  AGSGMATIRH
3751  PTPLALGHFH  TVTLLRSLTQ  GSLIVGDLAP  VNGTSQGKFQ  GLDLNEELYL
3801  GGYPDYGAIP  KAGLSSGFIG  CVRELRIQGE  EIVFHDLNLT  AHGISHCPTC
3851  RDRPCQNGGQ  CHDSESSSYV  CVCPAGFTGS  RCEHSQALHC  HPEACGPDAT
3901  CVNRPDGRGY  TCRCHLGRSG  LRCEEGVTVT  TPSLSGAGSY  LALPALTNTH
3951  HELRLDVEFK  PLAPDGVLLF  SGGKSGPVED  FVSLAMVGGH  LEFRYELGSG
4001  LAVLRTAEPL  ALGRWHRVSA  ERLNKDGSLR  VNGGRPVLRS  SPGKSQGLNL
4051  HTLLYLGGVE  PSVPLSPATN  MSAHFRGCVG  EVSVNGKRLD  LTYSFLGSQG
4101  IGQCYDSSPC  ERQPCQHGAT  CMPAGEYEFQ  CLCRDGIKGD  LCEHEENPCQ
4151  LREPCLHGGT  CQGTRCLCLP  GFSGPRCQQG  SGHGIAESDW  HLEGSGGNDA
4201  PGQYGAYFHD  DGFLAFPGHV  FSRSLPEVPE  TIELEVRTST  ASGLLLWQGV
4251  EVGEAGQGKD  FISLGLQDGH  LVFRYQLGSG  EARLVSEDPI  NDGEWHRVTA
4301  LREGRRGSIQ  VDGEELVSGR  SPGPNVAVNA  KGSIYIGGAP  DVATLTGGRF
4351  SSGITGCVKN  LVLHSARPGA  PPPQPLDLQH  RAQAGANTRP  CPS
```

Structural and functional sites

Signal peptide: 1–21
LDL receptor repeats: 194–234, 281–319, 320–359, 360–403
Immunoglobulin repeats: 404–506, 1679–1773, 1774–1867, 1868–1957, 1958–2052, 2053–2153, 2154–2246, 2247–2342, 2343–2438, 2439–2535, 2536–2631, 2632–2728, 2739–2828, 2829–2926, 2927–3023, 3024–3114, 3115–3213, 3214–3300, 3301–3401, 3402–3490, 3491–3576, 3577–3676

EGF (8C) repeats: 507–532 (partial), 733–765 (partial), 766–815, 816–873, 874–925 (partial), 926–935 (partial), 1127–1160 (partial),1161–1210, 1211–1276, 1277–1326, 1327–1336 (partial), 1532–1564 (partial), 1565–1614, 1615–1672
EGF (6C) repeats: 3850–3889, 3890–3930, 4110–4148, 4149–4184
G repeats: 3689–3849, 3952–4109, 4233–4393
Putative glycosaminoglycan attachment sites: 65, 71, 75
Potential N-linked glycosylation sites: 89, 556, 1757, 2123, 3074, 3107, 3281, 3782, 3838, 4070

Gene structure

The perlecan gene has been localized to human chromosome 1p35–36. The gene is composed of 94 exons and spans about 120 kb [2,4].

References

[1] Paulsson, M. et al. (1987) Structure of low density heparan sulphate proteoglycan isolated from a mouse tumor basement membrane. J. Mol. Biol. 197: 297–313.

[2] Kallunki, P. et al. (1991) Cloning of human heparan sulphate proteoglycan core protein, assignment of the gene (HSPG1) to 1p36.1–p35 and identification of a BamHI restriction fragment length polymorphism. Genomics 11: 389–396.

[3] Kallunki, P. and Tryggvason, K. (1992) Human basement membrane heparan sulphate proteoglycan core protein: A 467-kD protein containing multiple domains resembling elements of the low density lipoprotein receptor, laminin, neural cell adhesion molecules and epidermal growth factor. J. Cell Biol. 116: 559–571.

[4] Cohen, I.R. et al. (1993) Structural characterisation of the complete human perlecan gene and its promoter. Proc. Natl Acad. Sci. USA 90: 10404–10408.

Phosphophoryn

DMP2, DPP

Phosphophoryns are a family of aspartic acid, serine phosphate-rich proteins found as the major non-collagenous proteins of dentine matrix. There appears to be species-specific differences in phosphophoryn composition since in bovine phosphophoryn there is only one type of molecule while in the rat incisor version two or three types of molecule are found. These differences may be explained by age-dependent degradation *in situ*. Phosphophoryn has been shown to affect *in vitro* crystallization of hydroxyapatite markedly and, since the molecules are secreted directly at the mineral front, they are suggested to play a major role in dentinogenesis.

Molecular structure

In solution at neutral pH and physiological strength, the rat incisor phosphophoryn molecules are folded as loose globular structures, with diameters of about 20 nm. These molecules bind to both collagen monomers and collagen fibres. Each phosphophoryn molecule is a single polypeptide chain which is highly phosphorylated and rich in acidic amino acids. Aspartic acid and phosphoserine comprise over 80% of all residues in some molecules isolated. The carboxy-terminal domain contains 60 residue % serine and 31 residue % of aspartic acid with unique repetitive blocks of [DSS] and [SD] blocks. Conformational analysis shows that the phosphorylated form of the [DSS] is a non-planar, folded, modified trans-extended backbone which produces defined ridges of phosphates and carboxyl groups. Considerable discrepancies exist in the literature concerning molecular size, due mainly to high charge density and consequent anomalous biochemical behaviour [13].

Isolation

EDTA-soluble phosphophoryn can be purified from dentine by sequential calcium chloride precipitation, gel filtration in SDS buffer, anion-exchange chromatography and gel filtration in 4 M guanidine–HCl [7]. Human phosphophoryn has been purified from EDTA extracts of dentine by precipitation with calcium chloride, magnesium chloride and chromatography on DEAE–Sepharose CL-6B [9].

Accession number

U67916

Primary structure (mouse)

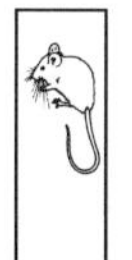

Ala	A	30	Cys	C	4	Asp	D	174	Glu	E	63
Phe	F	3	Gly	G	78	His	H	20	Ile	I	24
Lys	K	26	Leu	L	14	Met	M	3	Asn	N	48
Pro	P	21	Gln	Q	25	Arg	R	12	Ser	S	336
Thr	T	32	Val	V	20	Trp	W	2	Tyr	Y	5

Mol. wt (calc.) = 94 642 Residues = 940

```
1     MKMKIIIYIC  IWATAWAIPV  PQLVPLERDI  VENSVAVPLL  THPGTAAQNE
51    LSINSTTSNS  NDSPDGSEIG  EQVLSEDGYK  RDGNGSESIH  VGGKDFPTQP
101   ILVNEQGNTA  EEHNDIETYG  HDGVHARGEN  STANGIRSQV  GIVENAEEAE
151   SSVHGQAGQN  TKSGGASDVS  QNGDATLVQE  NEPPEASIKN  STNHEAGIHG
201   SGVATHETTP  QREGLGSENQ  GTEVTPSIGE  DAGLDDTDGS  PSGNGVEEDE
251   DTGSGDGEGA  EAGDGRESHD  GTKGQGGQSH  GGNTDHRGQS  SVSTEDDDSK
301   EQEGFPNGHN  GDNSSEENGV  EEGDSTQATQ  DKEKLSPKDT  RDAEGGIISQ
351   SEACPSGKSQ  DQGIETEGPN  KGNKSIITKE  SGKLSGSKDS  NGHQGVELDK
401   RNSPKQGESD  KPQGTAEKSA  AHSNLGHSRI  GSSSNSDGHD  SYEFDDESMQ
451   GDDPKSSDES  NGSDESDTNS  ESANESGSRG  DASYTSDESS  DDDNDSDSHA
501   GEDDSSDDSS  GDGDSDSNGD  GDSESEDKDE  SDSSDHDNSS  DSESKSDSSD
551   SSDDSSDSSD  SSDSSDSSDS  SDSSDSSDSS  DSSDSNSSSD  SSDSSGSSDS
601   SDSSDTCDSS  DSSDSSDSSD  SSDSSDSSDS  SDSSDSSDSS  DSSSSSDSSD
651   SSSCSDSSDS  SDSSDSSDSS  DSSDSSSSDS  SSSSNSSDSS  DSSDSSSSSD
701   SSDSSDSSDS  SDSSGSSDSS  DSSASSDSSS  SSDSSDSSSS  SDSSDSSDSS
751   DSSDSSESSD  SSNSSDSSDS  SDSSDSSDSS  DSSDSSDSSD  SSNSSDSSDS
801   SDSSDSSDSS  NSSDSSDSSD  SSDSSDSSDS  SDSSDSSDSS  DSSDSSDSSD
851   SSDSSDSSDS  SDSSDSSDSS  DSSDSSDSSN  SSDSSDSDSK  DSSSDSSDGD
901   SKSGNGNSDS  NSDSNSDSDS  DSEGSDSNHS  TSDDEIRENP
```

Structural and functional sites

Phosphophoryn represents amino acid residues 452–940 after translation of the mouse dentine sialoprotein gene.

Signal peptide: 1–17 [dentine sialoprotein]

Putative RGD cell adhesion site: 479–481 [28–30 after cleavage]

Gene structure

Evidence has been presented which localizes the gene to human chromosome 4. This gene codes for both dentine sialoprotein and phosphophoryn which are produced by proteolytic cleavage [8,11]. Data for other phosphophoryns are not available [13].

References

[1] DiMuzio, M.T. and Veis, A. (1978) Phosphoryns. Major noncollagenous proteins of rat incisor dentine. Calcified Tissue Res. 25: 169–178.

[2] Butler, W.T. et al. (1983) Multiple forms of rat dentin phosphoproteins. Arch. Biochem. Biophys. 225: 178–186.

[3] Nakamura, O. et al. (1985) Immunohistochemical studies with a monoclonal antibody on the distribution of phosphoryn in predentin and dentin. Calcified Tissue Int. 37: 491–500.

[4] Ibaraki, K. et al. (1991) An analysis of the biochemical and biosynthetic properties of dentin phosphoprotein. Matrix 11: 115–124.

[5] Sabsay, B. et al. (1991) Domain structure and sequence distribution in dentin phosphophoryn. Biochem. J. 276: 699–707.

[6] George, A. et al. (1993) Characterization of a novel dentin matrix acidic phosphoprotein. Implications for induction of biomineralization. J. Biol. Chem. 268: 12624–12630.

[7] Stetler-Stevenson, W.G. and Veis, A. (1983) Bovine dentin phosphoryn: Composition and molecular weight. Biochemistry 22: 4326–4335.

[8] Ritchie, H. and Wang, L.-H. (1997) A mammalian bicistronic transcript encoding two dentin-specific proteins. Biochem. Biophys. Res. Comm. 231: 425–428.
[9] Chang, S.R. et al. (1996) Characterisation and identification of a human dentin phosphoryn. Calc. Tiss. Int. 59: 149–153.
[10] MacDougall, M. et al. (1996) Assignment of DMP1 to human chromosome 4 band q21 by *in situ* hybridization. Cytogenet. Cell Genet. 74: 189.
[11] MacDougall, M. et al. (1997) Dentin phosphoprotein and dentin sialoprotein are cleavage products expressed from a single transcript coded by a gene on human chromosome 4. J. Biol. Chem. 272: 835–842.
[12] George, A. et al. (1996) The carboxyl-terminal domain of phosphophoryn contains unique extended triplet amino acid repeat sequences forming ordered carboxyl–phosphate interaction ridges that may be essential in the biomineralization process. J. Biol. Chem. 271: 32869–32873.
[13] George, A. et al. (1998) The phosphophoryn gene family: Identical domain structure at the carboxyl end. Eur. J. Oral Sci. (in press).

PRELP 55 kDa, prolargin

PRELP was originally isolated as a prominent component of bovine articular cartilage with a molecular weight of 58 kDa. The protein was shown to be present in many types of cartilage and in other tissues such as aorta, sclera, kidney, liver, skin and tendon [1,2]. In the human, protein and gene expression were high in juvenile and adult cartilage, but low in the fetus and neonate. Only adult lung in non-cartilaginous tissue contained appreciable message.

Molecular structure

PRELP has a calculated molecular mass of 41 646 Da. It contains ten well-conserved leucine-rich repeats and conserved cysteines at the N- and C-terminus. There are four potential *N*-glycosylation sites, but no consensus site for tyrosine sulphation. The protein shows most homology to fibromodulin (36%) and lumican (33%), in contrast to these it has a rather basic N-terminal extension, rich in arginine and proline [1–3].

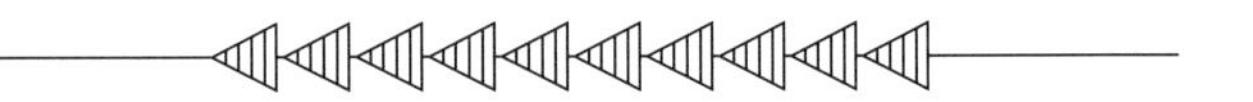

Isolation

PRELP has been isolated from bovine cartilage by 4 M guanidine hydrochloride (GuCl) extraction, density gradient centrifugation, size fractionation on Sephadex G-200, ion exchange chromatography on DEAE–cellulose and Superose 6 chromatography [1].

Accession number

P51888

Primary structure

Ala	A	11	Cys	C	7	Asp	D	17	Glu	E	15
Phe	F	16	Gly	G	14	His	H	11	Ile	I	23
Lys	K	14	Leu	L	59	Met	M	6	Asn	N	33
Pro	P	49	Gln	Q	14	Arg	R	30	Ser	S	23
Thr	T	11	Val	V	16	Trp	W	2	Tyr	Y	11

Mol. wt (calc.) = 43 810   Residues = 382

```
1     MRSPLCWLLP LLILASVAQG QPTRRPRPGT GPGRRPRPRP RPTPSFPQPD
51    EPAEPTDLPP PLPPGPPSIF PDCPRECYCP PDFPSALYCD SRNLRKVPVI
101   PPRIHYLYLQ NNFITELPVE SFQNATGLRW INLDNNRIRK IDQRVLEKLP
151   GLVFLYMEKN QLEEVPSALP RNLEQLRLSQ NHISRIPPGV FSKLENLLLL
201   DLQHNRLSDG VFKPDTFHGL KNLMQLNLAH NILRKMPPRV PTAIHQLYLD
251   SNKIETIPNG YFKSFPNLAF IRLNYNKLTD RGLPKNSFNI SNLLVLHLSH
301   NRISSVPAIN NRLEHLYLNN NSIEKINGTQ ICPNDLVAFH DFSSDLENVP
351   HLRYLRLDGN YLKPPIPLDL MMCFRLLQSV VI
```

Structural and functional sites

Signal peptide: 1–20
Leucine-rich repeats: 83–106, 107–130, 131–150, 151–175, 176–201, 202–222, 223–246, 247–271, 272–290, 291–317
Potential N-linked glycosylation sites: 124, 289, 320, 327

Gene structure

The human gene is located on chromosome 1q32 and encompasses at least 16 kb of genomic DNA. Two introns have been identified, one of 6.7 kb and the other of 2.6 kb. The protein is encoded by messages of 1.7, 4.6 and 6.7 kb. This message heterogeneity appears to reside in the length of the 3′-untranslated region [4].

References

1. Heinegard, D. et al. (1986) Two novel matrix proteins isolated from human cartilage show wide distributions among connective tissues. J. Biol. Chem. 261: 13866–13872.
2. Melching, L.I. and Roughley, P.J. (1990) A matrix protein of Mr 55 000 that accumulates in human cartilage with age. Biochim. Biophys. Acta 1036: 213–220.
3. Bengtsson, E. et al. (1995) The primary structure of a basic leucine-rich repeat protein, PRELP, found in connective tissues. J. Biol. Chem. 271: 9759–9763.
4. Grover, J. et al. (1996) The gene organisation, chromosome location, and expression of a 55-kDa matrix protein (PRELP) of human articular cartilage. Genomics 38: 109–117.

Procollagen C-proteinase

PCP, procollagen C-terminal peptidase, bone morphogenetic protein-1, BMP-1, mammalian tolloid (mTld), tolloid-like protein (mTll)

Procollagen C-proteinase specifically removes the C-propeptides of fibrillar procollagens (type I, II, III, V and XI procollagens)[1]. The most studied reaction is the removal of C-propeptides of type I procollagen and its role in collagen fibril assembly *in vitro*[2]. Removal of the C-propeptides catalyses the rate-limiting and committed step in the extracellular self-assembly of collagen into fibrils and larger fibres. The PCP is therefore critically important in the assembly of all connective tissues. PCP was thought to be a proteinase in its own right, with the enzyme cleaving only the fibrillar procollagens. However, genetic and biochemical data suggest that the molecules responsible for C-proteinase activity are those related to mammalian tolloid (mTld) and bone morphogenetic protein-1 (BMP-1)[3]. BMP-1 and mTld are encoded by alternative spliced transcripts which are differentially expressed in some tissues[4]. The recent discovery of a novel gene product (mammalian tolloid-like) with high sequence similarity to mammalian tolloid/BMP-1 raises the possibility that perhaps three, or more, proteins exhibit procollagen C-proteinase activity. Procollagen C-proteinase is a metalloproteinase which requires Zn^{2+} in its active site and Ca^{2+} for maximal activity. The enzyme is inhibited by metal chelators, such as EDTA and *o*-phanantholine, but not by serine proteinase inhibitors. The molecular weight of the mature enzyme was determined to be 97 kDa and 110 kDa by gel filtration and gel electrophoresis respectively. mTld and BMP-1 are synthesized as inactive proenzymes. The occurrence of potential furin-like proteinase cleavage sites at the junction of the prodomain and the metalloproteinase domain strongly suggests that the procollagen C-proteinase convertase is a furin-like serine proteinase.

Molecular structure

The larger molecule which exhibits C-proteinase activity is mTld[5] and is a modular protein consisting of a signal peptide, a prodomain (which renders the native molecule inactive), a metalloproteinase domain, five CUB domains (found in complement components C1r/C1s, the sea urchin protein, Uegf, and BMP-1) and two epidermal growth factor-like domains. The metalloproteinase domain is homologous to the metalloproteinase domain of astacin, a digestive proteinase from crayfish. BMP-1 is a spliced variant of mTld and also exhibits C-proteinase activity.

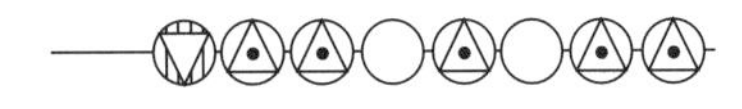

Pro-domain

Isolation

The mature, active enzyme (mTld and BMP-1) is purified from leg tendons of chick embryos[1] and from mouse fibroblast culture medium[6] by conventional liquid chromatography. The purification procedures have been compared and compiled[7]. Two forms of the enzyme are purified from chick tissue and these have been termed enzyme-A and enzyme-B. Both forms specifically cleave the C-propeptides of type I procollagen with equal efficiency. The two forms have molecular weights of 110 kDa and 95 kDa, respectively, and may represent different glycosylation variants of mTld. The mouse form of the mTld enzyme has a molecular weight of 125 kDa. The procollagen C-proteinase requires 5 mM Ca^{2+} for maximal activity *in vitro* and has an optimal pH of 7.4–8.0.

Accession number

P13497, L35279

Primary structure

Ala	A	49	Cys	C	41	Asp	D	67	Glu	E	65
Phe	F	53	Gly	G	87	His	H	28	Ile	I	36
Lys	K	61	Leu	L	60	Met	M	10	Asn	N	34
Pro	P	58	Gln	Q	37	Arg	R	72	Ser	S	81
Thr	T	52	Val	V	64	Trp	W	16	Tyr	Y	43

Mol. wt (calc.) = 114 598 Residues = 1014

```
1     MPGVARLPLL  LGLLLLPRPG  RPLDLADYTY  DLAEEDDSEP  LNYKDPCKAA
51    AFLGDIALDE  EDLRAFQVQQ  AVDLRRHTAR  KSSIKAAVPG  NTSTPSCQST
101   NGQPQRGACG  RWRGRSRSRR  AATSRPERVW  PDGVIPFVIG  GNFTGSQRAV
151   FRQAMRHWEK  HTCVTFLERT  DEDSYIVFTY  RPCGCCSYVG  RRGGGPQAIS
201   IGKNCDKFGI  VVHELGHVVG  FWHEHTRPDR  DRHVSIVREN  IQPGQEYNFL
251   KMEPQEVESL  GETYDFDSIM  HYARNTFSRG  IFLDTIVPKY  EVNGVKPPIG
301   QRTRLSKGDI  AQARKLYKCP  ACGETLQDST  GNFSSPEYPN  GYSAHMHCVW
351   RISVTPGEKI  ILNFTSLDLY  RSRLCWYDYV  EVRDGFWRKA  PLRGRFCGSK
401   LPEPIVSTDS  RLWVEFRSSS  NWVGKGFFAV  YEAICGGDVK  KDYGHIQSPN
451   YPDDYRPSKV  CIWRIQVSEG  FHVGLTFQSF  EIERHDSCAY  DYLEVRDGHS
501   ESSTLIGRYC  GYEKPDDIKS  TSSRLWLKFV  SDGSINKAGF  AVNFFKEVDE
551   CSRPNRGGCE  QRCLNTLGSY  KCSCDPGYEL  APDKRRCEAA  CGGFLTKLNG
601   SITSPGWPKE  YPPNKNCIWQ  LVAPTQYRIS  LQFDFFETEG  NDVCKYDFVE
651   VRSGLTADSK  LHGKFCGSEK  PEVITSQYNN  MRVEFKSDNT  VSKKGFKAHF
701   FSEKRPALQP  PRGRPHQLKF  RVQKRNRTPQ  DKDECSKDNG  GCQQDCVNTF
751   GSYECQCRSG  FVLHDNKHDC  KEAGCNHKVT  STSGTITSPN  WPDKYPSKKE
801   CTWAISSTPG  HRVKLTFMEM  DIESQPECAY  DHLEVFDGRD  AKAPVLGRFC
851   GSKKPEPVLA  TGSRMFLRFY  SDNSVQRKGF  QASHATECGG  QVRADVKTKD
901   LYSHAQFGDN  NYPGGVDCEW  VIVAEEGYGV  ELVFQTFEVE  EETDCGYDYM
951   ELFDGYDSTA  PSLGRYCGSG  PPEEVYSAGD  SVLVKFHSDD  TITKKGFHLR
1001  YTSTKFQDTL  HSRK
```

Structural and functional sites

Signal peptide: 1–22
Prodomain: 23–120
Catalytic domain: 121–321
CUB domain: 322–434, 435–546, 591–702, 775–887, 888–1003
EGF (6C) repeat: 547–590, 731–774
BMP-1 specific sequence: 703–730

Gene structure

The gene for mammalian tolloid is 46 kb, comprises 22 exons and is located on human chromosome 8 at position 8p21, between markers D8S298 and D8S5 on the genetic map[8]. A mammalian tolloid-like gene product (mTll) has been identified which has a domain structure identical to that of mTld. The protein has 76% identity with mTld for amino acid residues in all domains downstream of, and including, the metalloproteinase domain. In contrast, the N-terminal activation domains of the two proteins show little similarity[9].

References

[1] Hojima, Y. et al. (1985) Type I procollagen carboxyl-terminal proteinase from chick embryo tendons – purification and characterisation. J. Biol. Chem. 260: 15996–16003.

[2] Kadler, K.E. et al. (1987) Assembly of collagen fibrils *de novo* by enzymic cleavage of the type I pCcollagen by procollagen C-proteinase. Assay of critical concentration demonstrates that the process is an example of classical entropy-driven self assembly. J. Biol. Chem. 262: 15696–15701.

[3] Kessler, E. et al. (1996) Bone morphogenetic protein-1: The type I procollagen C-proteinase. Science 271: 360–362.

[4] Takahara, K. et al. (1994) Bone morphogenetic protein-1 and a mammalian tolloid homologue (mTld) are encoded by alternative spliced transcripts which are differentially expressed in some tissues. J. Biol. Chem. 269: 32572–32578.

[5] Li, S.-W. et al. (1996) The C-proteinase that processes procollagens to fibrillar collagens is identical to the protein previously identified as bone morphogenetic protein-1. Proc. Natl Acad. Sci. USA 93: 5127–5230.

[6] Kessler, E. and Adar, R. (1989) Type I procollagen C-proteinase from mouse fibroblasts. Purification and demonstration of a 55-kDa enhancer glycoprotein. Eur. J. Biochem. 186: 115–121.

[7] Kadler, K.E. and Watson, R.B. (1995) Methods Enzymol. 248: 771–781.

[8] Takahara, K. et al. (1995) Structural organisation and genetic localisation of the human bone morphogenetic protein-1/mammalian tolloid gene. Genomics 29: 9–15.

[9] Takahara, K. et al. (1996) Characterization of a novel gene product (mammalian tolloid-like) with high sequence similarity to mammalian tolloid/bone morphogenetic protein-1. Genomics 34: 157–165.

Procollagen I N-proteinase

procollagen type I/II N-proteinase, procollagen N-terminal proteinase, PNP

Procollagen N-proteinase (EC 3.4.24.14) cleaves the amino-propeptides of type I and type II procollagens during the conversion of procollagen to collagens [1]. Deficiencies in the activity of this enzyme result in fragility of skin in animals with dermatosparaxis and the equivalent disease in humans, the Ehlers–Danlos syndrome type VIIC. In these disorders, the extracellular matrix accumulates pN-collagen (a normal intermediate in the conversion of procollagen to collagen which contains the N-propeptides but not the C-propeptides) which is incorporated into collagen fibrils with an irregular cross-sectional shape [2]. The protein binds type XIV collagen [3]. The procollagen I N-proteinase requires that the type I procollagen and type II procollagen substrates are in a triple helical conformation (the enzyme will not cleave heat denatured procollagen molecules) [4]. Furthermore, the rate of cleavage of procollagen N-propeptides by the enzyme is highly sensitive to structural changes in the procollagen molecule which alter the phase (mutual alignment) of the three chains which comprise the triple helix of the procollagen molecule [5]. This requirement of the enzyme for a native triple helical procollagen substrate is used as an investigative tool to examine the triple helicity and structural integrity of procollagen molecules, particularly those from individuals with osteogenesis imperfecta [6,7] or the Ehlers–Danlos syndrome type VII [8] who have mutations in the genes for type I collagen. A separate enzyme cleaves the N-propeptides of type III procollagen [9]. There may still be further procollagen N-proteinases which specifically remove the N-propeptides of type V and XI procollagens.

Molecular structure

The protein comprises a single polypeptide chain of molecular weight 107 kDa. Sequence analysis shows that it is a modular protein that belongs to the M12 family of metalloproteinases, specifically to the clan MB of metallopeptidases. The enzyme therefore belongs to the same clan as the procollagen C-proteinase/BMP-1. PNP, however, is probably not part of the same sub-family as BMP-1 because the zinc-binding sequence in the catalytic subunit is different. PNP contains a potential integrin-binding sequence of RGD that is found in the disintegrins of the reprolysin sub-family. However, the distribution of cysteine residues around the RGD sequence in the protein is different from that found in disintegrins. At the N-terminus of the molecule are found two sequences of RTRR and RRRMRR which suggests that the enzyme is synthesized as an inactive proform which is converted to the mature enzyme by a furin-like serine proteinase. At the C-terminal domain of the PNP molecule are found patterns of cysteine residues which are homologous in terms of size and sequence to the six domains in properdin, suggesting a possible role in protein interactions. The properdin repeats are similar to those found in thrombospondin-1 [10].

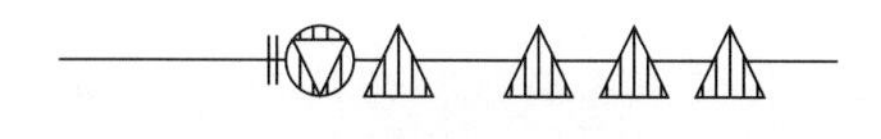

H

Potential transmembrane domain

Isolation

Procedures have been developed for the purification of the type I N-proteinase from chick leg tendons, and from bovine tendons and skin [3,4,11]. The procedures involve extraction of the protein into sodium chloride containing neutral buffers and subsequent liquid chromatography on columns of concanavalin A–Sepharose, heparin–Sepharose and Sephacryl S-300 and S-200 [12]. The enzyme is assayed by cleavage of ^{14}C-labelled type I procollagen and examining the reaction products by SDS–PAGE and autoradiography.

Accession number

X96389

Primary structure

Ala	A	75	Cys	C	55	Asp	D	79	Glu	E	73
Phe	F	30	Gly	G	101	His	H	47	Ile	I	36
Lys	K	50	Leu	L	101	Met	M	32	Asn	N	36
Pro	P	86	Gln	Q	43	Arg	R	88	Ser	S	84
Thr	T	59	Val	V	77	Trp	W	18	Tyr	Y	30

Mol. wt (calc.) = ? (7 residues are not assigned) Residues = 1207

```
1     MDPPAGAAGR LLCPALLLLL LLLPLPADAR LAAAAADPPG GPQGHGAAER
51    ILAVPVRTDA QGRLVSHVVS AATAPAGVRT RRAAPAQIPG LSGGSEEDPG
101   GRLFYNVTVF GRDLHLRLRP NARLVAPGAT VEWQGESGAT RVEPLLGTCL
151   YVGDVAGLAE SSSVALSNCD GLAGLIRMEE EEFFIEPLEK GLAAKEAEQG
201   RVHVVYHRPT TSRPPPLGGP QALDTGISAD FLDFLSRALG VLEERVNSSR
251   RRMRRHAABD DYNIEVLLGV DDSVVQFHGT EHVQKYLLTL MNIVNEIYHD
301   ESLGAHINVV LVRIILLVYG KSMSLIEIGN PSQSLENVCR WAYLQQKPDT
351   DHDEYHDHAI FLTRQDFGPS GMQGYAPVTG MCHPVRSCTL NHEDGFFSAF
401   VVAHETGHBL GMEHDGQGMR CGDDVRLGSI MAPLVQAASH RSHWSRCSQQ
451   ELSRYLHSYD CLRDDPVTHD WPALPQLPGL HYSMNEQSRF DFGLGYMMCT
501   AFRTFDPCKQ LWCSHPDNPY SCKTKKGPPL DGTMCAPGKH CFKGHCIWLT
551   PDILKRDGMW GAWSPFGSCS RTCGTGBKFR TRQCDMPHPA NGGRTCSGLA
601   YDFQLCNSQD CPDALADFRE EQCRQWDLYF EHGDAQHHWL PAEHRDAKER
651   CHLYCESKET GEVVSSKRNV HDGTRCSYKD AFSLCVRGDC RKVGCDGVIG
701   SSKQEDKCDV CGGDNSHCKV VKGTVSPSPK KLGYIKMFEI PAGARHLLIQ
751   EADTTSHHLA VKNLETGKFI LNEENDVDPN SKTFIAMGVE WEYRDEDGRE
801   TLQTMGPLHG TITVLVIPEG DARISLTYKY MIHVDSLMVD DMMVLEDDSV
851   GYEWALKKWS PCSKPCGGGS QFTKYGCRRR LDHKMVHRGV CDSBSKPKAI
901   RRTCMPQECQ PVWVTGEWEP SSRSGCGRTG MQVRSVRCVQ PLHNNTTRSB
951   HKKHCNDARP EGRRACNREL CPGRWRAGSW SQCSBTCGNG TQERPBLCRT
1001  ADDSFGVCRD ERPETARICR LGPCPRMTSD PSKKSYVVQW LSRPDPNSPV
1051  QETSSKGRCQ GDKSVFCRME VLSRYCSIPG YNKLCCKSCN PHDNLTDVDD
1101  RAEPPSGKHN DIEELMPTLS VPTLVMEVQP PPGIPLEVPL NTSSTNATED
1151  HPETNAVDVP YKIPGLEDEV QPPNLIPRRP SPYEKTRNQR IQELIDEMRK
1201  KEMLGKF
```

Structural and functional sites

Signal peptide: 1–36
Potential furin-like cleavage sites: 77–80, 248–253
Potential RGD cell adhesion site: 685–687
Potential transmembrane domain: 306–314
Catalytic domain: 315–551
Properdin repeat: 552–609, 847–907, 908–969, 970–1020
Potential N-linked glycosylation sites: 103, 244, 940, 941, 985, 1022, 1090, 1137, 1142

Gene structure

The gene for this protein is unknown.

References

[1] Lapiere, C.M. et al. (1971) Procollagen peptidase: An enzyme excising the coordination peptides of procollagen. Proc. Natl Acad. Sci. USA 68: 3054–3058.

[2] Holmes, D.F. et al. (1993) Ehlers Danlos syndrome type VIIB. Morphology of type I collagen fibrils is determined by the conformation of the N-propeptide. J. Biol. Chem. 268: 15758–15765.

[3] Colige, A. et al. (1995) Characterisation and partial amino acid sequencing of a 107 kDa procollagen-1 N-proteinase purified by affinity chromatography on immobilised type XIV collagen. J. Biol. Chem. 270: 16724–16730.

[4] Dombrowski, K.E. and Prockop, D.J. (1988) Cleavage of type I and type II procollagens by type I/II procollagen N-proteinase – correlation of kinetic constants with the predicted conformation of procollagen substrates. J. Biol. Chem. 263: 16545.

[5] Vogel, B.E. et al. (1988) A substitution of cysteine for glycine 748 of the α1 chain produces a 'kink' in the type I procollagen molecule. J. Biol. Chem. 263: 19249–19255.

[6] Lightfoot, S.J. et al. (1992) Type I procollagens containing substitutions of aspartate, arginine and cysteine for glycine in the proα1(I) chain are cleaved slowly by N-proteinase but only the cysteine substitution introduces a kink in the molecule. J. Biol. Chem. 267: 25521–25528.

[7] Lightfoot, S.J. et al. (1994) Substitution of serine for glycine 883 in the triple helix of the proα1(I) chain of type I procollagen in an individual with osteogenesis imperfecta type IV introduces a structural change in the triple helix that does not alter cleavage of the molecule by procollagen N-proteinase. J. Biol. Chem. 269: 30352–30357.

[8] Watson, R.B. et al. (1992) Ehlers Danlos syndrome type VIIB. Incomplete cleavage of the patient's abnormal type I procollagen by N-proteinase results in the formation of rough-bordered collagen fibrils characteristic of the disorder. J. Biol. Chem. 267: 9093–9100.

[9] Halila, R. and Peltonen, L. (1986) Purification of human procollagen type III N-proteinase from placenta and preparation of antiserum. Biochemistry 239: 47–52.

[10] Colige, A. et al. (1997) cDNA cloning and expression of bovine procollagen I N-proteinase: A new member of the super family of zinc-metalloproteinases with binding sites for cells and other matrix components. Proc. Natl Acad. Sci. USA 94: 2374–2379.
[11] Hojima, Y. et al. (1994) Characterisation of type I procollagen N-proteinase from fetal bovine tendon and skin – purification of the 500 kDa form of the enzyme from bovine tendon. J. Biol. Chem. 269: 11389–11390.
[12] Kadler, K.E. et al. (1995) Procollagen N-peptidases: procollagen N-proteinases. Methods Enzymol. 248: 756–771.

Tenascin-C

hexabrachion, myotendinous antigen, J1, cytotactin

Tenascin-C (TN-C) is a polymorphic, high molecular weight extracellular matrix glycoprotein. It is transiently expressed in many developing organs and reappears in the stroma of many tumours. This highly regulated expression suggests a possible function in cell-matrix adhesion and cell migration, and in modulation of growth and differentiation during morphogenesis. Current evidence suggests that tenascin exhibits an anti-adhesive activity. TN-C is conserved in all vertebrate species yet the knockout mouse developed normally and showed no apparent defect, although recent work suggests abnormal behaviour and neurotransmissions are shown [1–7].

Molecular structure

Tenascin is composed of six similar subunits joined at their NH_2-termini by disulphide bonds. The difference between the subunits is determined by alternative splicing. A striking feature of the sequence is its multiple repeats. Three different types of repeat have been identified: six-cysteine epidermal growth factor (EGF) repeats, fibronectin type III repeats, and a region homologous to the β and γ chains of fibrinogen. The junctional region is proposed to be formed by a triple coiled-coil of three subunits mediated by four heptad repeats of hydrophobic amino acids. The central globular domain is formed by a contribution from all six subunits and is the site at which subunits are fixed in place by disulphide cross-linking. The putative RGD cell adhesion sequence at position 876–878 is not conserved between species and is probably not functional [1–4].

Isolation

Tenascin can be isolated from conditioned medium after removal of fibronectin by gelatin–Sepharose affinity chromatography followed by hydroxyapatite chromatography. Final purification can be achieved by two sequential precipitations with 6% and 12.8% polyethylene glycol [8].

Accession number

A321160

Primary structure

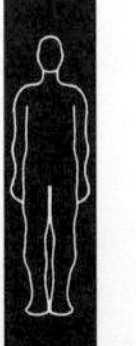

Ala	A	142	Cys	C	99	Asp	D	134	Glu	E	178
Phe	F	67	Gly	G	194	His	H	39	Ile	I	98
Lys	K	81	Leu	L	167	Met	M	23	Asn	N	94
Pro	P	112	Gln	Q	76	Arg	R	119	Ser	S	139
Thr	T	190	Val	V	154	Trp	W	28	Tyr	Y	65

Mol. wt (calc.) = 240 865 Residues = 2199

```
1     MGAMTQLLAG VFLAFLALAT EGGVLKKVIR HKRQSGVNAT LPEENQPVVF
51    NHVYNIKLPV GSQCSVDLES ASGEKDLAPP SEPSESFQEH TVDGENQIVF
101   THRINIPRRA CGCAAAPDVK ELLSRLEELE NLVSSLREQC TAGAGCCLQP
151   ATGRLDTRPF CSGRGNFSTE GCGCVCEPGW KGPNCSEPEC PGNCHLRGRC
201   IDGQCICDDG FTGEDCSQLA CPSDCNDQGK CVNGVCICFE GYAADCSREI
251   CPVPCSEEHG TCVDGLCVCH DGFAGDDCNK PLCLNNCYNR GRCVENECVC
301   DEGFTGEDCS ELICPNDCFD RGRCINGTCY CEEGFTGEDC GKPTCPHACH
351   TQGRCEEGQC VCDEGFAGVD CSEKRCPADC HNRGRCVDGR CECDDGFTGA
401   DCGELKCPNG CSGHGRCVNG QCVCDEGYTG EDCSQLRCPN DCHSRGRCVE
451   GKCVCEQGFK GYDCSDMSCP NDCHQHGRCV NGMCVCDDGY TGEDCRDRQC
501   PRDCSNRGLC VDGQCVCEDG FTGPDCAELS CPNDCHGQGR CVNGQCVCHE
551   GFMGKDCKEQ RCPSDCHGQG RCVDGQCICH EGFTGLDCGQ HSCPSDCNNL
601   GQCVSGRCIC NEGYSGEDCS EVSPPKDLVV TEVTEETVNL AWDNEMRVTE
651   YLVVYTPTHE GGLEMQFRVP GDQTSTIIRE LEPGVEYFIR VFAILENKKS
701   IPVSARVATY LPAPEGLKFK SIKETSVEVE WDPLDIAFET WEIIFRNMNK
751   EDEGEITKSL RRPETSYRQT GLAPGQEYEI SLHIVKNNTR GPGLKRVTTT
801   RLDAPSQIEV KDVTDTTALI TWFKPLAEID GIELTYGIKD VPGDRTTIDL
851   TEDENQYSIG NLKPDTEYEV SLISRRGDMS SNPAKETFTT GLDAPRNLRR
901   VSQTDNSITL EWRNGKAAID SYRIKYAPIS GGDHAEVDVP KSQQATTKTT
951   LTGLRPGTEY GIGVSAVKED KESNPATINA ATELDTPKDL QVSETAETSL
1001  TLLWKTPLAK FDRYRLNYSL PTGQWVGVQL PRNTTSYVLR GLEPGQEYNV
1051  LLTAEKGRHK SKPARVKAST EQAPELENLT VTEVGWDGLR LNWTAADQAY
1101  EHFIIQVQEA NKVEAARNLT VPGSLRAVDI PGLKAATPYT VSIYGVIQGY
1151  RTPVLSAEAS TGETPNLGEV VVAEVGWDAL KLNWTAPEGA YEYFFIQVQE
1201  ADTVEAAQNL TVPGGLRSTD LPGLKAATHY TITIRGVTQD FSTTPLSVEV
1251  LTEEVPDMGN LTVTEVSWDA LRLNWTTPDG TYDQFTIQVQ EADQVEEAHN
1301  LTVPGSLRSM EIPGLRAGTP YTVTLHGEVR GHSTRPLAVE VVTEDLPQLG
1351  DLAVSEVGWD GLRLNWTAAD NAYEHFVIQV QEVNKVEAAQ NLTLPGSLRA
1401  VDIPGLEAAT PYRVSIYGVI RGYRTPVLSA EASTAKEPEI GNLNVSDITP
1451  ESFNLSWMAT DGIFETFTIE IIDSNRLLET VEYNISGAER TAHISGLPPS
1501  TDFIVYLSGL APSIRTKTIS ATATTEALPL LENLTISDIN PYGFTVSWMA
1551  SENAFDSFLV TVVDSGKLLD PQEFTLSGTQ RKLELRGLIT GIGYEVMVSG
1601  FTQGHQTKPL RAEIVTEAEP EVDNLLVSDA TPDGFRLSWT ADEGVFDNFV
1651  LKIRDTKKQS EPLEITLLAP ERTRDITGLR EATEYEIELY GISKGRRSQT
1701  VSAIATTAMG SPKEVIFSDI TENSATVSWR APTAQVESFR ITYVPITGGT
1751  PSMVTVDGTK TQTRLVKLIP GVEYLVSIIA MKGFEESEPV SGSFTTALDG
1801  PSGLVTANIT DSEALARWQP AIATVDSYVI SYTGEKVPEI TRTVSGNTVE
1851  YALTDLEPAT EYTLRIFAEK GPQKSSTITA KFTTDLDSPR DLTATEVQSE
1901  TALLTWRPPR ASVTGYLLVY ESVDGTVKEV IVGPDTTSYS LADLSPSTHY
1951  TAKIQALNGP LRSNMIQTIF TTIGLLYPFP KDCSQAMLNG DTTSGLYTIY
2001  LNGDKAQALE VFCDMTSDGG GWIVFLRRKN GRENFYQNWK AYAAGFGDRR
2051  EEFWLGLDNL NKITAQGQYE LRVDLRDHGE TAFAVYDKFS VGDAKTRYKL
2101  KVEGYSGTAG DSMAYHNGRS FSTFDKDTDS AITNCALSTR GFWYRNCHRV
2151  NLMGRYGDNN HSQGVNWFHW KGHEHSIQFA EMKLRPSNFR NLEGRRKRA
```

Structural and functional sites

Signal peptide: 1–22

Heptad coiled-coil: 119–147 (4 repeats)

EGF (6C) repeats: 174–185 (partial), 186–216, 217–246, 247–278, 279–309, 310–340, 341–371, 372–402, 403–433, 434–464, 465–495, 496–526, 527–557, 558–588, 589–619

Fibronectin type III repeats: 622–710, 711–801, 802–891, 892–983, 984–1071, 1072–1162, 1163–1253, 1254–1344, 1345–1435, 1436–1526, 1527–1618, 1619–1708, 1708–1797, 1798–1885, 1886–1973

Alternatively spliced repeats: 983–1070, 1075–1252, 1253–1343, 1344–1434, 1435–1525, 1526–1616, 1617–1707

Potential N-linked glycosylation sites: 38, 166, 184, 326, 787, 1017, 1033, 1078, 1092, 1118, 1183, 1209, 1260, 1274, 1300, 1365, 1391, 1444, 1454, 1484, 1533, 1808, 2160

RGD cell adhesion site: 876–878 (not conserved in murine sequence)

Gene structure

The tenascin-C gene is located on human chromosome 9 at locus q32–34; however, Southern blot analyses indicate the presence of multiple related genes. The coding region of the tenascin gene spans approximately 80 kb and consists of 27 exons. The type III repeats are encoded by one or two exons. All alternatively spliced type III repeats are encoded by single exons. The fibrinogen-like domain is encoded by five exons and the EGF repeats by single exons [9,10].

References

1. Erickson, H.P. (1993) Tenascin-C, tenascin-R and tenascin-X: A family of talented proteins in search of functions. Current Opin. Cell Biol. 5: 869–876.
2. Gulcher, J.R. et al. (1989) Structure of the human hexabrachion (tenascin) gene. Proc. Natl Acad. Sci. USA 88: 9438–9442.
3. Chiquet-Ehrismann, R. et al. (1995) The complexity in regulating the tenascins. Bioessays 17: 873–878.
4. Chiquet-Ehrismann, R. (1991) Anti-adhesive molecules of the extracellular matrix. Curr. Opin. Cell Biol. 3: 800–804.
5. Saga, Y. et al. (1992) Mice develop normally without tenascin. Genes Dev. 6: 1821–1831.
6. Fukamauchi, F. et al. (1996) Abnormal behaviour and neurotransmissions of tenascin gene knockout mouse. Biochem. Biophys. Res. Commun. 221: 151–156.
7. Gherzi, R. et al. (1995) Human Tenascin gene: Structure of the 5′-region, identification and characterisation of the transcription regulatory sequences. J. Biol. Chem. 270: 3429–3434.
8. Saginata, M. et al. (1992) A simple procedure for tenascin purification. Eur. J. Biochem. 205: 545–549.
9. Gulcher, J.R. et al. (1989) An alternative spliced region of the human hexabrachion contains a novel repeat of potential *N*-glycosylation sites. Proc. Natl Acad. Sci. USA 86: 1588–1592.
10. Siri, A. et al. (1991) Human tenascin, primary structure, pre-mRNA splicing patterns and localisation of the epitope recognised by 2 monoclonal antibodies. Nucleic Acids Res. 19: 525–531.

Tenascin-R

janusin, J1–160/180, restrictin

Tenascin-R has been detected mainly in the central nervous system, and is localized around motor neurons and on motor axons in the spinal cord. It is associated with the surface of neurons, myelinating oligodendrocytes and type 2 astrocytes. In the chick brain it is detectable at embryonic days 6–16, but is barely detectable in the adult, suggesting a role in central nervous system development [1–3].

Molecular structure

Tenascin-R contains the same domains as the other members of the tenascin family, but is considerably shorter. It contains an N-terminal heptad repeat sequence and C-terminal fibrinogen domain similar to tenascin-C, but there are only 4.5 epidermal growth factor (EGF) repeats, and nine fibronectin type III repeats. Heptad repeats in the N-terminal junctional region are proposed to allow triple coiled-coil of three subunits, and the presence of an extra cysteine enables this trimer to form the characteristic hexabrachion structure. The putative RGD cell adhesion sequence at position 876–878 is not conserved between species and is probably not functional [4].

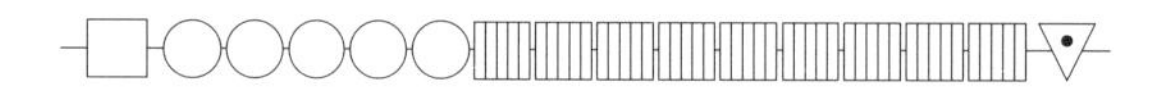

Isolation

Tenascin-R can be isolated from adult mouse brain. Brain tissue is homogenized in 10 mM EDTA, 10 mM EGTA, 1 M urea, pH 7.8 in the presence of proteinase inhibitors. After centrifugation, the supernatant was precipitated with 40% (w/v) ammonium sulphate. The precipitate was collected by centrifugation and dissolved in PBS, pH 7.2, and tenascin-R was purified by immunoaffinity chromatography [5].

Accession number

Z67996

Primary structure

Ala	A	78	Cys	C	41	Asp	D	72	Glu	E	107
Phe	F	39	Gly	G	102	His	H	25	Ile	I	66
Lys	K	33	Leu	L	101	Met	M	22	Asn	N	56
Pro	P	71	Gln	Q	58	Arg	R	71	Ser	S	125
Thr	T	112	Val	V	109	Trp	W	21	Tyr	Y	49

Mol. wt (calc.) = 149 620   Residues = 1358

```
1     MGADGETVVL  KNMLIGVNLI  LLGSMIKPSE  CQLEVTTERV  QRQSVEEEGG
51    IANYNTSSKE  QPVVFNHVYN  INVPLDNLCS  SGLEASAEQE  VSAEDETLAE
101   YMGQTSDHES  QVTFTHRINF  PKKACPCASS  AQVLQELLSR  IEMLEREVSV
151   LRDQCNANCC  QESAATEQLD  YIPHCSGHGN  FSFESCGCIC  NEGWFGKNCS
201   EPYCPLGCSS  RGVCVDGQCI  CDSEYSGDDC  SELRCPTDCS  SRGLCVDGEC
```

```
251  VCEEPYTGED CRELRCPGDC SGKGRCANGT CLCEEGYVGE DCGQRQCLNA
301  CSGRGQCEEG LCVCEEGYQG PDCSAVAPPE DLRVAGISDR SIELEWDGPM
351  AVTEYVISYQ PTALGGLQLQ QRVPGDWSGV TITELEPGLT YNISVYAVIS
401  NILSLPITAK VATHLSTPQG LQFKTITETT VEVQWEPFSF SFDGWEISFI
451  PKNNEGGVIA QVPSDVTSFN QTGLKPGEEY IVNVVALKEQ ARSPPTSASV
501  STVIDGPTQI LVRDVSDTVA FVEWIPPRAK VDFILLKYGL VGGEGGRTTF
551  RLQPPLSQYS VQALRPGSRY EVSVSAVRGT NESDSATTQF TTEIDAPKNL
601  RVGSRTATSL DLEWDNSEAE VQEYKVVYST LAGEQYHEVL VPRGIGPTTR
651  ATLTDLVPGT EYGVGISAVM NSQQSVPATM NARTELDSPR DLMVTASSET
701  SISLIWTKAS GPIDHYRITF TPSSGIASEV TVPKDRTSYT LTDLEPGAEY
751  IISVTAERGR QQSLESTVDA FTGFRPISHL HFSHVTSSSV NITWSDPSPP
801  ADRLILNYSP RDEEEEMMEV SLDATKRHAV LMGLQPATEY IVNLVAVHGT
851  VTSEPIVGSI TTGIDPPKDI TISNVTKDSV MVSWSPPVAS FDYYRVSYRP
901  TQVGRLDSSV VPNTVTEFTI TRLNPATEYE ISLNSVRGRE ESERICTLVH
951  TAMDNPVDLI ATNITPTEAL LQWKAPVGEV ENYVIVLTHF AVAGETILVD
1001 GVSEEFRLVD LLPSTHYTAT MYATNGPLTS GTISTNFSTL LDPPANLTAS
1051 EVTRQSALIS WQPPRAEIEN YVLTYKSTDG SRKELIVDAE DTWIRLEGLL
1101 ENTDYTVLLQ AAQDTTWSSI TSTAFTTGGR VFPHPQDCAQ HLMNGDTLSG
1151 VYPIFLNGEL SQKLQVYCDM TTDGGGWIVF QRRQNGQTDF FRKWADYRVG
1201 FGNVEDEFWL GLDNIHRITS QGRYELRVDM RDGQEAAFAS YDRFSVEDSR
1251 NLYKLRIGSY NGTAGDSLSY HQGRPFSTED RDNDVAVTNC AMSYKGAWWY
1301 KNCHRTNLNG KYGESRHSQG INWYHWKGHE FSIPFVEMKM RPYNHRLMAG
1351 RKRQSLQF
```

Structural and functional sites

Signal peptide: 1–23
EGF (6C) repeats: 188–199, 200–230, 231–261, 262–292, 293–323
Fibronectin type III repeats: 324–413, 414–502, 503–592, 593–684, 685–772, 773–862, 863–951, 952–1039, 1040–1127
Alternatively spliced repeats: 773–862
Potential N-linked glycosylation sites: 55, 179, 198, 278, 391, 471, 580, 791, 807, 874, 963, 1037, 1046, 1261.

Gene structure

The tenascin-R gene is located on human chromosome 1 at locus q23–24. The semi-dominant mouse mutation loop-tail (Lp), which has been proposed as a model for a subset of human neural tube defects, also maps to the equivalent area of chromosome 1 [4]. The gene spans 85 kb and consists of 21 exons, the position of introns is precisely conserved in tenascin-C and tenascin-X [6].

References

[1] Chiquet-Ehrismann, R. (1995) Tenascins, a growing family of extracellular matrix proteins. Experientia 51: 9–10.

[2] Rathjen, F.G. et al. (1991) Restrictin: A chick neural extracellular matrix protein involved in cell attachment co-purifies with the cell recognition molecule F11. Development 113: 151–164.

[3] Erickson, H.P. (1994) Evolution of the tenascin family – implications for the function of the C-terminal fibrinogen-like domain. Persp. Dev. Neurobiol. 2: 9–19.

[4] Carnemolla, B. et al. (1996) Human tenascin-R: Complete primary structure, pre-mRNA alternative splicing and gene localisation on chromosome 1q23–q24. J. Biol. Chem. 271: 8157–8160.
[5] Pesheva, P. et al. (1989) JI-160 and JI-180 are oligodendrocyte-secreted nonpermissive substrates for cell adhesion. J. Cell Biol. 109: 1765–1778.
[6] Leprini, A. et al. (1996) The human tenascin-R gene. J. Biol. Chem. 271: 31251–31254.

Tenascin-X

Tenascin-X is a member of the tenascin modular protein family, and differs from tenascin-C and -R in the number of modules, in being poorly glycosylated and in not containing an RGD sequence. It is expressed in nearly all tissues and is especially widely expressed in developing fetal tissues. In the mouse, antibodies to TN-X are most prominent in skeletal and heart muscle. No function has yet been ascribed to this molecule, although its high expression in muscle may indicate an important role in muscle development and functioning. Tenascin-X deficiency is associated with Ehlers Danlos syndrome.

Molecular structure

Tenascin-X monomer contains four types of domains. At the N-terminal there are three heptad repeats capable of forming α-helices that allow triple stranded coiled coils to form. There then follows 18.5 epidermal growth factor (EGF)-like repeats and at least 29 fibronectin type III repeats. The C-terminus comprises a fibrinogen-like domain. The purified mouse subunit has an apparent molecular weight of approximately 500 000 Da. Tenascin-X is missing a cysteine residue at the N-terminal end, and while three chains can associate, it appears unlikely that two trimers can connect as found in tenascin-C and -R which exist as six-armed molecules. It is likely that TN-X can form pairs or trimers, but not higher order polymers [1–6].

Isolation

Mouse tenascin-X has been isolated from conditioned medium of confluent Ren-Ca cells. The 50% ammonium sulphate pellet was passed through gelatin-agarose and the flow-through absorbed on to a Tn-X antibody affinity column [6].

Accession number

A40701, A33725, C42175

Primary structure

Ala	A	167	Cys	C	122	Asp	D	211	Glu	E	257
Phe	F	98	Gly	G	418	His	H	77	Ile	I	49
Lys	K	95	Leu	L	284	Met	M	40	Asn	N	52
Pro	P	329	Gln	Q	150	Arg	R	248	Ser	S	260
Thr	T	257	Val	V	305	Trp	W	46	Tyr	Y	101

Mol. wt (calc.) = 385 619 Residues = 3566

```
1     MMPAQYALTS  SLVLLVLLST  ARAGPFSSRS  NVTLPAPRPP  PQPGGHTVGA
51    GVGSPSSQLY  EHTVEGGEKQ  VVFTHRINLP  PSTGCGCPPG  TEPPVLASEV
101   QALRVRLEIL  EELVKGLKEQ  CTGGCCPASA  QAGTGQTDVR  TLCSLHGVFD
151   LSRCTCSCEP  GWGGPTCSDP  TDAEIPPSSP  PSASGSCPDD  CNDQGRCVRG
201   RCVCFPGYTG  PSCGWPSCPG  DCQGRGRCVQ  GVCVCRAGFS  GPDCSQRSCP
```

251	RGCSQRGRCE	GGRCVCDPGY	TGDDCGMRSC	PRGCSQRGRC	ENGRCVCNPG
301	YTGEDCGVRS	CPRGCSQRGR	CKDGRCVCDP	GYTGEDCGTR	SCPWDCGEGG
351	RCVDGRCVCW	PGYTGEDCST	RTCPRDCRGR	GRCEDGECIC	DTGYSGDDCG
401	VRSCPGDCNQ	RGRCEDGRCV	CWPGYTGTDC	GSRACPRDCR	GRGRCENGVC
451	VCNAGYSGED	CGVRSCPGDC	RGRGRCESGR	CMCWPGYTGR	DCGTRACPGD
501	CRGRGRCVDG	RCVCNPGFTG	EDCGSRRCPG	DCRGHGLCED	GVCVCDAGYS
551	GEDCSTRSCP	GGCRGRGQCL	DGRCVCEDGY	SGEDCGVRQC	PNDCSQHGVC
601	QDGVCICWEG	YVSEDCSIRT	CPSNCHGRGR	CEEGRCLCDP	GYTGPTCATR
651	MCPADCRGRG	RCVQGVCLCH	VGYGGEDCGQ	EEPPASACPG	GCGPRELCRA
701	GQCVCVEGFR	GPDCAIQTCP	GDCRGRGECH	DGSCVCKDGY	AGEDCGEVID
751	GPQDLRVVSV	TPTTLELGWL	RPQAEVDRFV	VSYVSADNQR	VRLEVPPETD
801	GTLLTDLMPG	VEYVVTVTAE	RGRAVSYPAS	VRANTDPGRK	YKFVLYGFVG
851	KKRHGPLVAE	AKILPQSDPS	PGTPPRLGNL	WVTDPTPDSL	HLSWTVPEGQ
901	FDTFMVQYRD	RDGRPQVVPV	EGPERSFVVS	SLDPDHKYRF	TLFGIANKKR
951	YGPLTADGTT	APERKEEPPR	PEFLEQPLLG	ELTVTGVTPD	SLRLSWTVAQ
1001	GPFDSFMVQY	KDAQGQPQAV	PVAGDENEVT	VPGLDPDRKY	KMNLYGLRGR
1051	QRVGPESVVA	KTDPQPHRTG	TEAPESPEEP	LLGELTVTGS	SPDSLSLFWT
1101	VPQGSFDSFT	VQYKDRDGRP	RAVRVGGKES	EVTVGGLEPG	HKYKMHLYGL
1151	HEGQRVGPVS	AVGLTALEPR	LGELTVTDIT	PDSVGLSWTV	PEGEFDSFVV
1201	QYKDRDGQPQ	VVPVAADQRE	VTIPDLEPST	KYKFLLFGIQ	DGKRRSPVSV
1251	EAKTAILSWT	VPEASSTQKV	VQFKDKDGPQ	VVPVEGHERS	VTVTPLDAGR
1301	KYRFLLYGLL	GKKRHGPLTA	DGTSGEGHSL	QVTTVTQNSV	GLSWTVPEGQ
1351	FDSFVVQYKD	RDGQPQVVPV	EGSLREVSVP	GLDPAHRYKL	LLYGLHHGKR
1401	VGPISAVAIT	APEPHLGELT	VEEATSHTLH	LSWMVTEGEF	DSFEIQYTDR
1451	DGQLQMVRIG	GDRNDITLSG	LESDHRYLVT	LYGFSDGKHV	GPVHVEALTE
1501	PPTATPEPPI	KPRLGELTVT	DATPDSLSLS	WTVPEGQFDH	FLVQYRNGDG
1551	QPKAVRGPGH	EEGVTISGLE	PDHKYKMNLY	GFHGGQRMGP	VSVVGVTAAE
1601	EETPSPTEPS	MEAPEPAEEP	LLGELTVTGS	SPDSLSLSWT	VPQGRFDSFT
1651	VQYKDRDGRP	QVVRVGGPEE	ESPDAPLAKL	RLGQMTVRDI	TSDSLSLSWT
1701	VPEGQFDHFL	VQFKNGDGQP	KALRVPGHED	GAEEETPSPT	EPSMEAPEPP
1751	EEPLLGELTV	TGSSPDSLSL	SWTVPQGRFD	SFTVQYKDRD	GRPQVVRVGG
1801	EESEVTVGGL	EPGRKYLMHL	YGLHEGRRVG	PVSAVGVTEP	GTEAPEPPEE
1851	PLLGELTVTG	SSPDSLSLSW	TVPQGRFDSF	TVQYKDRDGR	PQAVRVGGQE
1901	SKVTVRGLEP	GRKYKMHLYG	LHEGRRLGPV	SAVGVTEDEA	ETTQAVPTMT
1951	PEPPIKPRLG	ELTMTDATPD	SLSLSWTVPE	GQFDHFLVQY	RNGDGQPKAA
2001	RVPGHEDGVT	ISGLEPDHKY	KMNLYGFHGG	QRVGPISVIG	VTEEETPSPT
2051	ELSTEAPEPP	EEPLLGELTV	TGSSPDSLSL	SWTIPQGHFD	SFTVQYKDRD
2101	GRPQVMRVRG	EERQVTVGGL	EPGRKYKMHL	YGLHEGRRVG	PVSTVGVTED
2151	EAETTQAVPT	TTPEPPNKPS	LGELTVTDAT	PDSLSLSWMV	RGQFDHFLVQ
2201	YRNGDGQPKV	VRWPGHEDGV	TISGLEPDHK	YKMNLYGFHG	GQRVGPISVI
2251	GVTAAEEETP	APTEPSTEAP	EPPEEPLLGE	LTVTGSSPDS	LSLSWTIPQG
2301	RFDSFTVQYK	DRDGRPQVVR	VRGEESEVTV	GGLEPGRKYK	MHLYGLHEGQ
2351	RVGPVSAVGV	TAPEDEAETT	QAVPTTTPEP	PNKPRLGELT	VTDATPDSLS
2401	LSWMVPEGQF	DHFLVQYRNG	DGQPKVVRWP	GHEDGVTISG	LEPDHKYKMN
2451	LYGFHGGQRV	GPVSAIGVSE	EETPSPTEPS	TEAPEAPEEP	LLGELTVTGS
2501	SPDSLSLSWT	VPQGRFDSFT	VQYKDRDGQP	QVVRVRGEES	EVTVGGLEPG
2551	RKYKMHLYGL	HEGQRVGPVS	TVGITAPLPT	PLPVEPRLGE	LAVEAVTSDS
2601	VGLSWTVAQG	PFDSFLVQYR	DAQGQPQAVP	VSGDLRAVAV	SGLVPARKYK
2651	FLLFGLQNGK	RHGPVPVEAR	TAPDTKPSPR	LGELTVTDAT	PDSVGLSWTV
2701	PEGEFDSFVV	QYKDKDGRLQ	VVPVAANQRE	VTVQALEPSR	KYRFLLYGLS
2751	GRKRLGPISA	DSTTAPLEKE	LPPHLGELTV	AEETSSSLRL	SWTVAQGPFD
2801	SFVVQYRDTD	GQPRAVPVAA	DQRTVTVEDL	EPGKKYKFLL	YGLLGGKRLG
2851	PVSALGMTAP	EEDTPAPELA	PEAPEPPEEP	RLGVLTVTDT	TPDSMRLSWS
2901	VAQGPFDSFV	VQYEDTNGQP	QALLVDGDQS	KILISGLEPS	TPYRFLLYGL

```
2951 HEGKRLGPLS AEGTTGLAPA GQTSEESRPR LSQLSVTDVT TSSLRLNWEA
3001 PPGAFDSFLL RFGVPSPSTL EPHPRPLLQR ELMVPGTRHS AVLRDLRSGT
3051 LYSLTLYGLR GPHKADSIQG TARTLSPVLE SPRDLQFSEI RETSAKVNWM
3101 PPPSRADSFK VSYQLADGGE PQSVQVDGQA RTQKLQGLIP GARYEVTVVS
3151 VRGFEESEPL TGFLTTVPDG PTQLRALNLT EGFAVLHWKP PQNPVDTYDI
3201 QVTAPGAPPL QAETPGSAVD YPLHDLVLHT NYTATVRGLR GPNLTSPASI
3251 TFTTGLEAPR DLEAKEVTPR TALLTWTEPP VRPAGYLLSF HTPGGQNQIL
3301 LPGGITSHQL LGLFPSTSYN ARLQAMWGQS LLPPVSTSFT TGGLRIPFPR
3351 DCGEEMQNGA GASRTSTIFL NGNRERPLNV FCDMETDGGG WLVFQRRMDG
3401 QTDFWRDWED YAHGFGNISG EFWLGNEALH SLTQAGDYSI RVDLRAGDEA
3451 VFAQYDSFHV DSAAEYYRLH LEGYHGTAGD SMSYHSGSVF SARDRDPNSL
3501 LISCAVSYRG AWWYRNCHYA NLNGLYGSTV DHQGVSWYHW KGFEFSVPFT
3551 EMKLRPRNFR SPAGGG
```

Structural and functional sites

Signal peptide: 1–23

EGF (6C) repeats: 166–212, 213–243, 244–274, 275–305, 306–336, 337–367, 368–391, 399–429, 430–460, 461–491, 492–523, 524–553, 554–584, 585–615, 616–646, 647–677, 678–713, 714–743

Fibronectin type III repeats: 748–828, 829–856, 873–953, 975–1055, 1078–1158, 1167–1247, 1248–1317, 1323–1403, 1412–1492, 1510–1590, 1618–1676, 1678–1749, 1751–1831, 1849–1929, 1955–2035, 2061–2141, 2167–2246, 2274–2354, 2382–2462, 2488–2568, 2584–2664, 2677–2757, 2771–2851, 2878–2958, 2977–3067, 3078–3159, 3167–3247, 3255–3334, 3349–3557

Potential N-linked glycosylation sites: 31, 3178, 3231, 3243, 3417

Gene structure

The human tenascin-X gene spans about 100 kb, overlapping the Creb-rp and P450c21B genes at its 5′- and 3′-ends, respectively, and is in the human major histocompatibility locus on chromosome 6. This region is duplicated in the human genome, and the XA gene became truncated by non-homologous recombination that led to duplication; this pseudogene does not appear to encode protein. Transcription of the human Tn-X gene can be initiated from three different promoters. The coding region spans approximately 65 kb and consists of 39 exons, encoding a 12–16 kb mRNA in muscle and adrenal. The type III repeats are encoded by one or two exons. The fibrinogen-like domain is encoded by five exons and the EGF repeats by single exons. There is no evidence of alternative splicing of the fibronectin type III repeats[2,7–9].

References

[1] Burch, G.H. et al. (1997) Tenascin-X deficiency is associated with Ehlers Danlos syndrome. Nature Genetics 17: 104–108.

[2] Bristow, J. et al. (1993) Tenascin-X: A novel extracellular matrix protein encoded by the human XB gene overlapping P450c21B. J. Cell Biol. 122: 265–278.

[3] Erickson, H.P. (1993) Tenascin-C, tenascin-R and tenascin-X: A family of talented proteins in search of functions. Curr. Opin. Cell Biol. 5: 869–876.

[4] Geffrotin, C. (1995) Distinct tissue distribution in pigs of tenascin-X and tenascin-C transcripts. Eur. J. Biochem. 231: 83–92.
[5] Chiquet-Ehrismann, R. et al. (1994) The tenascin gene family. Perspect. Dev. Neurobiol. 2: 3–7.
[6] Matsumoto, K. et al. (1994) The distribution of tenascin-X is distinct and often reciprocal to that of tenascin-C. J. Cell Biol. 125: 483–493.
[7] Morel, Y. et al. (1989) Transcript encoded on the opposite strand of the human steroid 21-hydroxylase/complement component of C4 gene locus. Proc. Natl Acad. Sci. 86: 6582–6586.
[8] Matsumoto, K. et al. (1992) Extracellular matrix protein tenascin-like gene found in human MHC class III region. Immunogenetics 36: 400–403.
[9] Speek, M. et al. (1996) Alternate promoters and alternate splicing of human tenascin-X, a gene with 5′ and 3′ ends buried in other genes. Hum. Molec. Genet. 5: 1749–1758.

Tenascin-Y

The tenascins are polymorphic, high molecular weight extracellular matrix glycoproteins, which to date contain three members in vertebrates TN-C, TN-R and TN-X. In the chicken TN-C and TN-R are present, but TN-Y appears to replace TN-X. TN-Y is predominantly expressed in embryonic and adult chicken heart and skeletal muscle.

Molecular structure

Tenascin-Y is composed of a polypeptide of 1914 amino acids. The signal peptide is followed by a short proline-rich region, 3.5 heptad repeats and a cysteine-rich region followed by an epidermal growth factor (EGF)-like region. A fibronectin type III domains is separated from the other type III repeats by a unique serine/proline-rich motif. A short proline-rich peptide leads to the carboxy-terminal fibrinogen-like globular portion. Only monomeric TN-Y could be purified from antibody columns of muscle extracts, but oligomeric forms from muscle fibroblast conditioned medium[1–4].

Isolation

Tenascin-Y can be isolated from washed adult chicken hearts (20 mM Tris, pH 7.5) by extraction in 25 mM Tris pH 8.0, 400 mM NaCl and 10 mM EDTA, centrifugation, concentration and precipitation at 45% ammonium sulphate. The pellet was dissolved in TBS, applied to a gelatin agarose column and the flow through passed over an anti-TN-Y affinity column, and eluted with 8 M LiCl, pH 11.0[4].

Accession number

X99062

Primary structure (chicken)

Ala	A	130	Cys	C	19	Asp	D	80	Glu	E	111
Phe	F	57	Gly	G	193	His	H	42	Ile	I	26
Lys	K	41	Leu	L	174	Met	M	15	Asn	N	33
Pro	P	217	Gln	Q	99	Arg	R	124	Ser	S	178
Thr	T	121	Val	V	159	Trp	W	40	Tyr	Y	55

Mol. wt (calc.) = 206 582 Residues = 1914

```
1    MTPGLPLLLL ALALRRAQGN AGASPGPPPV PPELSCGAEV LEAVLGRLRA
51   LEGEVRALQG QCGDSLGPQA GTGSTAARRL CDTPGAGGCC GCRLGESGDG
101  LPCPAPRCPL DCSDQGRCRA GRCHCFEGFT GPFCATPVCP PGRGGPHCTL
151  EIPSVTLRLA ARNQTSFRVT WPRPAKPVDG YEVAVIPMDE PAALTTHELP
201  GSAVTFEVTG LTPGQAFEIF IQAQREQHLG APGTLRVRTL LAQSLPNHGG
251  PRGTPTFLAS PVAPASPSAR QSPASPWSPV SLGSLGSPAS LRSPVSPESS
301  AAPASLESPW SPASPRSPES PASPVSPRSP TSPWSPASPQ SPLSPASPIS
351  PVLPNVPSLH ELGVKLSSYN GSLLQRLESH LRATNFPLRG NQTVPAVARA
```

401	ILSYLLRRSP	ASLRYQFLRH	LQQNPHPKPQ	VLPGAAGEAL	VDLDGLRGHA
451	ETVVIRYRLL	EEPEGDEGEV	RVPGDTTVAR	VPGLVPGATY	RVEVHGVVRG
501	RVSKSYTSLV	TAGLGDTSEP	PPEWENLYDM	EVTEPQGAMA	KAAPSEEEPP
551	QRPRLGMLTV	SHVTPSSIQL	EWSVLEGTFD	SFTVQYRDAQ	GQPQALAVDG
601	GSRTVTVPGL	SPSRHYKFNL	YGVWGRKRIG	FISTDAVTDS	AKPKEELPSQ
651	PRLGELTASH	VSPDSVQLEW	SVPEGSFDSF	TVQYKDAQGQ	PQVVPVDGGL
701	RTVTVPGLSP	SRRYKFNLYG	VWGRKRLGPM	STDAVTAAEE	TEEEPPSQPR
751	LGELTASHVS	PNSVQLEWSI	PEGSFDSFTV	QYIDVQGQPQ	ELHLDSGSRT
801	VTVSGLLPSH	PYKFNLYGVW	GQRRLGPIST	DTITAAAPAQ	KEPPSQPRLG
851	ELTASHVGPD	SVQLEWSVPE	GSFDSFTVQY	KDAQGQPQVV	PVDGGLRTVT
901	VPGLSPSRRY	KFNLYGVWGR	KRLGPMSTDA	VTAPAQKEPP	SQPLLGELTA
951	SHVGPDSVQL	EWSVPEGSFD	SFTVQYKDAQ	GQPQVVPVDG	GLRTVTVPGL
1001	SPSRRYKFNL	YGVWGRKRLG	PMSTDAVTAA	APAQEEPPSP	PRLGELTASH
1051	VGPDSVQLEW	SVPEGSFDSF	TVQYKDAQGQ	PQVVPVDGGL	RTVTVPGLSP
1101	SRRYKFNLYG	VWGRKRLGPM	STDAVTAAEE	TEEEPPSQPR	LGELTASHVS
1151	PNSVQLEWSI	PEGSFDSFTV	QYIDVQGQPQ	ELHLDSGSRT	VTVSGLLPSH
1201	PYKFNLYGVW	GQTRLGPIST	DTITAAAPAQ	EEPPSPPRLG	ELTASHVGPD
1251	SVQLEWSVPE	GSFDSFTVQY	KDAQGQPQVV	PVDGGLRTVT	VPGLSLSRRY
1301	KFNLYGVWGR	KRLGPMSTDA	VTAPAQKEPP	SQPLLGELTA	SHVGPDSVQL
1351	EWSVPEGSFD	SFTVQYKDAQ	GQPQVVPVDG	GLRTVTVPGL	SPSRRYKFNL
1401	YGVWGRKRLG	PMSTDAVTAA	EETEEEPPSQ	PRLGELTASH	VSPNSVQLEW
1451	SIPEGSFDSF	TVQYIDVQGQ	PQELHLDSGS	RTVTVSGLLP	SHPYKFNLYG
1501	VWGQRRLGPI	STDTITAAAP	AQKEPPSQPR	LGVLTASHVG	PDSVQLEWSV
1551	PEGSFDSFTV	QYKDARGQPQ	VVPVDGGLRT	VTVPGLSPSR	RYKFNLYGVW
1601	GRKRLGPMST	DAVTGAPGTL	WVGTLWPRSA	HLHWAPPHVP	PEGYNLIYGP
1651	PGGPVKTLQL	PPEATSKELW	GLEPSGRYRV	QLWGRGLEPL	ETTFDTPPLP
1701	HPHPRDCAEE	QLNGPGPSRE	VLIFLGGDRQ	RPLHVFCDME	SNGGGWLVFQ
1751	RRMDGGTDFW	RGWEEYVHGF	GNVSGEFWLG	NAALHTLTAS	GPTELRVDLR
1801	TPSDSAFARY	RDFAVSGPED	NFRLHLGAYS	GTAGDALSYH	AGSPFSTRDH
1851	DPRGRPRPCA	VAYTGAWWYR	NCHYANLNGR	YGVPYDHQGI	NWYPWKGFEY
1901	SIPFTEMKLR	PQRD			

Structural and functional sites

Signal peptide: 1–19
EGF (6C) repeats: 90–150
Fibronectin type III repeats: 151–239, 513–638, 639–736, 737–834, 835–932, 933–1028, 1029–1126, 1127–1224, 1225–1322, 1323–1418, 1419–1516, 1517–1614, 1615–1696
Potential N-linked glycosylation sites: 163, 370, 391, 1772

Gene structure

Northern blot showed two major bands at 6.5 and 9.5 kb in both embryonic and adult tissues[4].

References

[1] Chiquet-Ehrismann, R. et al. (1994) The tenascin gene family. Perspect. Dev. Neurobiol. 2: 3–7.

[2] Chiquet-Ehrismann, R. (1995) Tenascins, a growing family of extracellular matrix proteins. Experientia 51: 853–862.

[3] Chiquet-Ehrismann, R. (1990) What distinguishes tenascin from fibronectin? FASEB J. 4: 2598–2604.
[4] Hagios, C. et al. (1996) Tenascin-Y: A protein of novel domain structure is secreted by differentiated fibroblasts of muscle connective tissue. J. Cell Biol. 134: 1499–1512.

Thrombospondin-1

Thrombospondin-1 (TSP1) is the most abundant protein component of platelet α granules and is rapidly secreted upon platelet activation at sites of injury and thrombosis. Thrombospondin-1 is synthesized and secreted by a variety of cell types, including fibroblasts and smooth muscle cells, and has been implicated in the regulation of cell migration and proliferation during development, wound healing, angiogenesis and tumorigenesis. Its adhesive interactions with cells are complicated and are mediated by a variety of receptors including integrins, CD36, proteoglycans and sulphatides. In addition, thrombospondin binds fibrinogen, fibronectin, laminin and collagens. TSP1 knockout mice show delayed organization and prolonged neovascularization of skin wounds.

Molecular structure

Thrombospondin-1 is a trimer made up of three identical subunits joined by disulphide bonds. In platelets, the subunits have a molecular weight of approximately 180 000. In line with other extracellular adhesive glycoproteins, the subunits are made up of discrete structural domains. There are three repeating motifs; the three type 1 motifs are homologous to the complement components C6–C9 and properdin, and contain the major cell-binding site, VTCG, which probably interacts with the counter-receptor CD36. There are three type 2 or six-cysteine epidermal growth factor (EGF) repeats, and eight cysteine-containing type 3 repeats, which include either one or two copies of an EF-hand-type loop and which probably constitute the major calcium-binding site in the molecule. In addition to the three repeating motifs, there are several distinct segments to the molecule. These include a heparin-binding domain which participates in thrombospondin-mediated disruption of focal contacts and regulation of proliferation, a pair of cysteine residues that cross-link the three subunits, a 70 amino acid segment that is homologous to the N-propeptide of type I collagen, and a COOH-terminal domain that interacts with platelets and other cells. Preliminary reports suggest that alternative splicing can produce two variant chains of molecular weight 140 000 and 50 000 [1–5].

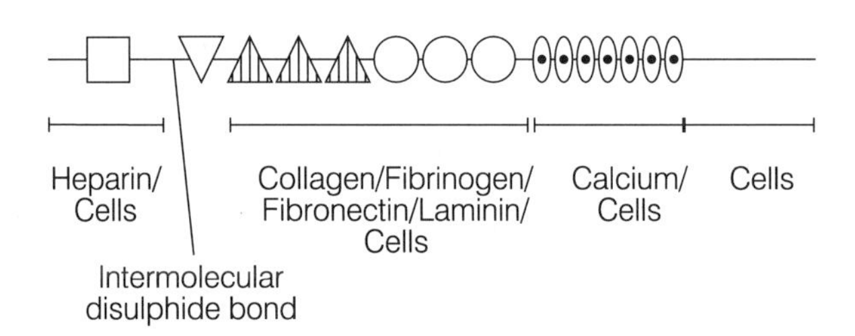

Isolation

Thrombospondin can be purified from the supernatant of activated platelets by passage through a gelatin–Sepharose column and retention on a heparin–Sepharose column. Thrombospondin can be further purified by size exclusion on Bio-Gel A-0.5 m [6].

Accession number

P07996

Primary structure

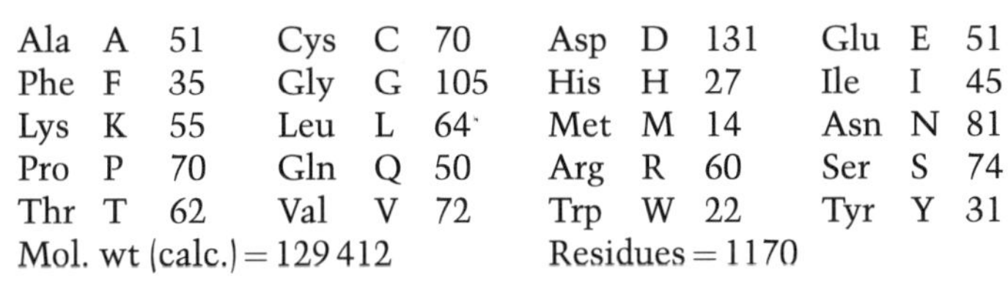

Ala	A	51	Cys	C	70	Asp	D	131	Glu	E	51
Phe	F	35	Gly	G	105	His	H	27	Ile	I	45
Lys	K	55	Leu	L	64	Met	M	14	Asn	N	81
Pro	P	70	Gln	Q	50	Arg	R	60	Ser	S	74
Thr	T	62	Val	V	72	Trp	W	22	Tyr	Y	31

Mol. wt (calc.) = 129 412 Residues = 1170

```
1     MGLAWGLGVL FLMHVCGTNR IPESGGDNSV FDIFELTGAA RKGSGRRLVK
51    GPDPSSPAFR IEDANLIPPV PDDKFQDLVD AVRTEKGFLL LASLRQMKKT
101   RGTLLALERK DHSGQVFSVV SNGKAGTLDL SLTVQGKQHV VSVEEALLAT
151   GQWKSITLFV QEDRAQLYID CEKMENAELD VPIQSVFTRD LASIARLRIA
201   KGGVNDNFQG VLQNVRFVFG TTPEDILRNK GCSSSTSVLL TLDNNVVNGS
251   SPAIRTNYIG HKTKDLQAIC GISCDELSSM VLELRGLRTI VTTLQDSIRK
301   VTEENKELAN ELRRPPLCYH NGVQYRNNEE WTVDSCTECH CQNSVTICKK
351   VSCPIMPCSN ATVPDGECCP RCWPSDSADD GWSPWSEWTS CSTSCGNGIQ
401   QRGRSCDSLN NRCEGSSVQT RTCHIQECDK RFKQDGGWSH WSPWSSCSVT
451   CGDGVITRIR LCNSPSPQMN GKPCEGEARE TKACKKDACP INGGWGPWSP
501   WDICSVTCGG GVQKRSRLCN NPTPQFGGKD CVGDVTENQI CNKQDCPIDG
551   CLSNPCFAGV KCTSYPDGSW KCGACPPGYS GNGIQCTDVD ECKEVPDACF
601   NHNGEHRCEN TDPGYNCLPC PPRFTGSQPF GQGVEHATAN KQVCKPRNPC
651   TDGTHDCNKN AKCNYLGHYS DPMYRCECKP GYAGNGIICG EDTDLDGWPN
701   ENLVCVANAT YHCKKDNCPN LPNSGQEDYD KDGIGDACDD DDDNDKIPDD
751   RDNCPFHYNP AQYDYDRDDV GDRCDNCPYN HNPDQADTDN NGEGDACAAD
801   IDGDGILNER DNCQYVYNVD QRDTDMDGVG DQCDNCPLEH NPDQLDSDSD
851   RIGDTCDNNQ DIDEDGHQNN LDNCPYVPNA NQADHDKDGK GDACDHDDDN
901   DGIPDDKDNC RLVPNPDQKD SDGDGRGDAC KDDFDHDSVP DIDDICPENV
951   DISETDFRRF QMIPLDPKGT SQNDPNWVVR HQGKELVQTV NCDPGLAVGY
1001  DEFNAVDFSG TFFINTERDD DYAGFVFGYQ SSSRFYVVMW KQVTQSYWDT
1051  NPTRAQGYSG LSVKVVNSTT GPGEHLRNAL WHTGNTPGQV RTLWHDPRHI
1101  GWKDFTAYRW RLSHRPKTGF IRVVMYEGKK IMADSGPIYD KTYAGGRLGL
1151  FVFSQEMVFF SDLKYECRDP
```

Structural and functional sites

Signal peptide: 1–18
Procollagen homology region: 303–372
Properdin repeats: 379–434, 435–491, 492–548
EGF (6C) repeats: 549–587, 588–645, 646–689
Type 3 repeats: 724–758, 759–781, 782–817, 818–840, 841–878, 879–914, 915–950
Potential N-linked glycosylation sites: 248, 360, 708, 1067
Heparin-binding sites: 41–47, 99–102
RGD cell adhesion site: 926–928
VTCG cell adhesion sites: 449–452, 506–509
Platelet adhesion sites: 1034–1063, 1084–1138
Alternatively spliced domain: 19–292

Gene structure

Thrombospondin-1 is located on human chromosome 15q15 [7].

References

1 Lawler, J. and Hynes, R.O. (1986) The structure of human thrombospondin, an adhesive glycoprotein with multiple calcium binding sites and homologies with several different proteins. J. Cell Biol. 103: 1635–1648.

2 Paul, L. et al. (1989) Identification of an alternatively spliced product of the human thrombospondin gene. J. Cell Biol. 109: 200a.

3 Bornstein, P. (1992) Thrombospondins: Structure and regulation of expression. FASEB J. 6: 3290–3299.

4 Kosfeld, M.D. and Frazier, W.A. (1992) Identification of active peptide sequences in the carboxyl-terminal cell binding domain of human thrombospondin-1. J. Biol. Chem. 267: 16230–16236.

5 Adams, J.C. and Lawler, J. (1993) The thrombospondin family. Current Biol. 3: 188–190.

6 Santoro, S.A. and Frazier, W.D. (1987) Isolation and characterisation of thrombospondin. Methods Enzymol. 144: 438–446.

7 Wolf, F.W. et al. (1990) Structure and chromosomal localisation of the human thrombospondin gene. Genomics 6: 685–691.

Thrombospondin-2

corticotrophin-induced secreted protein (CISP)

Thrombospondins (TSPs) are extracellular glycoproteins associated with a variety of cellular processes including growth and embryogenesis. TSP2 shows different expression during embryogenesis to TSP1, it was not expressed in human umbilical vein endothelial cells and, in those cells where both TSP1 and 2 were secreted, TSP1 was always more abundant. TSP2 shows conservation of some of the functionally critical residues present in TSP1 and shows many of the same cell and ligand binding characteristics. TSP2 is an active inhibitor of angiogenesis *in vitro* and *in vivo* [1].

Molecular structure

Conservation of the two cysteines involved in inter-chain disulphide linkage and those in the procollagen domain suggests that TSP2 trimerization and oligomerization are similar to TSP1. Isolation of TSP2 from bovine adrenocortical cell cultures showed it ran as a 600 kDa protein under non-reducing conditions and as a 195–200 kDa protein when reduced. The trimeric structure was also observed by rotary shadowing. Trimer formation for TSP2 has been shown for recombinantly produced molecules. It is unclear at present if heterotrimer containing chains of both TSP1 and TSP2 can exist. The three repeating motifs found in TSP1 are found in TSP2. The three type 1 motifs are homologous to the complement components C6–C9 and properdin, and contain the major cell-binding site, VTCG, which probably interacts with the counter receptor CD36. There are three type 2 or six-cysteine epidermal growth factor (EGF) repeats, and eight cysteine-containing type 3 repeats which include either one or two copies of an EF-hand-type loop, which probably constitutes the major calcium-binding site in the molecule [2–6].

Isolation

Thrombospondin-2 has been purified from adrenocortical cell cultures using heparin–Sepharose and Mono Q chromatography [7].

Accession number

P35442

Primary structure

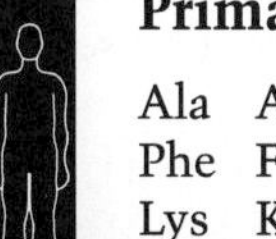

Ala	A	51	Cys	C	68	Asp	D	120	Glu	E	59
Phe	F	40	Gly	G	106	His	H	31	Ile	I	41
Lys	K	44	Leu	L	67	Met	M	12	Asn	N	68
Pro	P	69	Gln	Q	59	Arg	R	57	Ser	S	73
Thr	T	69	Val	V	80	Trp	W	26	Tyr	Y	32

Mol. wt (calc.) = 129 955 Residues = 1172

```
1     MVWRLVLLAL WVWPSTQAGH QDKDTTFDLF SISNINRKTI GAKQFRGPDP
51    GVPAYRFVRF DYIPPVNADD LSKITKIMRQ KEGFFLTAQL KQDGKSRGTL
101   LALEGPGLSQ RQFEIVSNGP ADTLDLTYWI DGTRHVVSLE DVGLADSQWK
151   NVTVQVAGET YSLHVGCDLI GPVALDEPFY EHLQAEKSRM YVAKGSARES
201   HFRGLLQNVH LVFENSVEDI LSKKGCQQGQ GAEINAISEN TETLRLGPHV
251   TTEYVGPSSE RRPEVCERSC EELGNMVQEL SGLHVLVNQL SENLKRVSND
301   NQFLWELIGG PPKTRNMSAC WQDGRFFAEN ETWVVDSCTT CTCKKFKTIC
351   HQITCPPATC ASPSFVEGEC CPSCLHSVDG EEGWSPWAEW TQCSVTCGSG
401   TQQRGRSCDV TSNTCLGPSI QTRACSLSKC DTRIRQDGGW SHWSPWSSCS
451   VTCGVGNITR IRLCNSPVPQ MGGKNCKGSG RETKACQGAP CPIDGRWSPW
501   SPWSACTVTC AGGIRERTRV CNSPEPQYGG KACVGDVQER QMCNKRSCPV
551   DGCLSNPCFP GAQCSSFPDG SWSCGFCPVG FLGNGTHCED LDECALVPDI
601   CFSTSKVPRC VNTQPGFHCL PCPPRYRGNQ PVGVGLEAAK TEKQVCEPEN
651   PCKDKTHNCH KHAECIYLGH FSDPMYKCEC QTGYAGDGLI CGEDSDLDGW
701   PNLNLVCATN ATYHCIKDNC PHLPNSGQED FDKDGIGDAC DDDDDNDGVT
751   DEKDNCQLLF NPRQADYDKD EVGDRCDNCP YVHNPAQIDT DNNGEGDACS
801   VDIDGDDVFN ERDNCPYVYN TDQRDTDGDG VGDHCDNCPL VHNPDQTDVD
851   NDLVGDQCDN NEDIDDDGHQ NNQDNCPYIS NANQADHDRD GQGDACDPDD
901   DNDGVPDDRD NCRLVFNPDQ EDLDGDGRGD ICKDDFDNDN IPDIDDVCPE
951   NNAISETDFR NFQMVPLDPK GTTQIDPNWV IRHQGKELVQ TANSDPGIAV
1001  GFDEFGSVDF SGTFYVNTDR DDDYAGFVFG YQSSSRFYVV MWKQVTQTYW
1051  EDQPTRAYGY SGVSLKVVNS TTGTGEHLRN ALWHTGNTPG QVRTLWHDPR
1101  NIGWKDYTAY RWHLTHRPKT GYIRVLVHEG KQVMADSGPI YDQTYAGGRL
1151  GLFVFSQEMV YFSDLKYECR DI
```

Structural and functional sites

Signal peptide: 1–18
Potential N-linked glycosylation sites: 151, 316, 330, 457, 584, 710, 1069
Putative RGD cell attachment site: 928–930
Type I repeats: 381–432, 437–493, 494–550
EGF (6C) repeats: 549–589, 590–647, 648–692
Type III repeats: 725–760, 761–783, 784–819, 820–842, 843–880, 881–916, 917–952

Gene structure

Thrombospondin-2 is located on human chromosome 6q27. Northern analysis shows hybridization to a 7.5 kb message [5].

References

[1] Volpert, O.V. et al. (1995) Inhibition of angiogenesis by thrombospondin-2. Biochem. Biophys. Res. Commun. 217: 326–332

[2] Bornstein, P. et al. (1991) A second thrombospondin gene in the mouse is similar in organisation to thrombospondin 1 but does not respond to serum. Proc. Natl Acad. Sci. USA 88: 8636–8640.

[3] Bornstein, P. et al. (1991) A second expressed thrombospondin gene (Thbs2) exists in the mouse genome. J. Biol. Chem. 266: 12821–12824.

[4] Laherty, C.D. et al. (1992) Characterisation of mouse thrombospondin 2 sequence and expression during cell growth and development. J. Biol. Chem. 267: 3274–3281.

[5] LaBell, T.L. et al. (1992) Thrombospondin II: Partial cDNA sequence, chromosome location, and expression of a second member of the thrombospondin gene family in humans. Genomics 12: 421–429.
[6] Pellerin, S. et al. (1993) The molecular structure of corticotrophin-induced secreted protein, a novel member of the thrombospondin family. J. Biol. Chem. 268: 18810–18817.
[7] Pellerin, S. et al. (1993) Corticotrophin-induced secreted protein and ACTH-induced protein secreted by adrenocortical cells is structurally related to thrombospondin. J. Biol. Chem. 268: 4304–4310.

Thrombospondin-3

In the developing mouse thrombospondin-3 (TSP3) expression differs from that of TSP1 and 2, and is limited largely to the hippocampus of the brain, lung and cartilage. Although it is similar in sequence to TSP1 and 2 at the carboxyl-terminal, it is distinctly different at the amino-terminal. While its function is not known, it would be expected that some functions would overlap with TSP1 and 2, but that there would be unique functions.

Molecular structure

The cDNA encodes a polypeptide of 956 amino acids that is highly acidic. Homology to TSP1 and 2 is found in the seven type 3 repeats and four type 2 (or six-cysteine EGF) repeats. The amino-terminus is unique in not containing either the procollagen or the type I repeats found in both TSP1 and 2. Native TSP3 migrates as an oligomer in the absence of reducing agents and, as a monomer on reduction, mutation of cysteines 245 and 248 there was no oligomerization. Rotary shadowing showed that, unlike TSP1 and 2, which are trimeric molecules, TSP3 exists as a pentamer. The presence of two cysteines separated by two amino acids may be a determinant in the fact that TSP3, 4 and 5 all form pentameric structures [1–4].

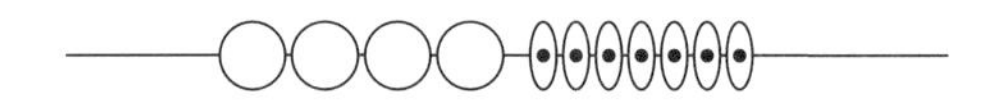

Isolation

Recombinant thrombospondin-3 has been purified from a serum-free conditioned medium of transfected 293T cells by heparin–Sepharose affinity chromatography [3].

Accession number

P49746

Primary structure

Ala	A	52	Cys	C	48	Asp	D	103	Glu	E	41
Phe	F	29	Gly	G	93	His	H	23	Ile	I	31
Lys	K	25	Leu	L	76	Met	M	15	Asn	N	60
Pro	P	59	Gln	Q	56	Arg	R	45	Ser	S	50
Thr	T	49	Val	V	69	Trp	W	10	Tyr	Y	22

Mol. wt (calc.) = 104 201 Residues = 956

```
1     METQELRGAL  ALLLLCFFTS  ASQDLQVIDL  LTVGESRQMV  AVAEKIRTAL
51    LTAGDIYLLS  TFRLPPKQGG  VLFGLYSRQD  NTRWLEASVV  GKINKVLVRY
101   QREDGKVHAV  NLQQAGLADG  RTHTVLLRLR  GPSRPSPALH  LYVDCKLGDQ
151   HAGLPALAPI  PPAEVDGLEI  RTGQKAYLRM  QGFVESMKII  LGGSMARVGA
201   LSECPFQGDE  SIHSAVTNAL  HSILGEQTKA  LVTQLTLFNQ  ILVELRDDIR
251   DQVKEMSLIR  NTIMECQVCG  FHEQRSHCSP  NPCFRGVDCM  EVYEYPGYRC
301   GPCPPGLQGN  GTHCSDINEC  AHADPCFPGS  SCINTMPGFH  CEACPRGYKG
351   TQVSGVGIDY  ARASKQVCND  IDECNDGNNG  GCDPNSICTN  TVGSFKCGPC
```

```
401  RLGFLGNQSQ GCLPARTCHS PAHSPCHIHA HCLFERNGAV SCQCNVGWAG
451  NGNVCGTDTD IDGYPDQALP CMDNNKHCKQ DNCLLTPNSG QEDADNDGVG
501  DQCDDDADGD GIKNVEDNCR LFPNKDQQNS DTDSFGDACD NCPNVPNNDQ
551  KDTDGNGEGD ACDNDVDGDG IPNGLDNCPK VPNPLQTDRD EDGVGDACDS
601  CPEMSNPTQT DADSDLVGDV CDTNEDSDGD GHQDTKDNCP QLPNSSQLDS
651  DNDGLGDECD GDDDNDGIPD YVPPGPDNCR LVPNPNQKDS DGNGVGDVCE
701  DDFDNDAVVD PLDVCPESAE VTLTDFRAYQ TVVLDPEGDA QIDPNWVVLN
751  QGMEIVQTMN SDPGLAVGYT AFNGVDFEGT FHVNTVTDDD YAGFLFSYQD
801  SGRFYVVMWK QTEQTYWQAT PFRAVAQPGL QLKAVTSVSG PGEHLRNALW
851  HTGHTPDQVR LLWTDPRNVG WRDKTSYRWQ LLHRPQVGYI RVKLYEGPQL
901  VADSGVIIDT SMRGGRLGVF CFSQENIIWS NLQYRCNDTV PEDFEPFRRQ
951  LLQGRV
```

Structural and functional sites

Signal peptide: 1–21
Potential N-linked glycosylation sites: 310, 407, 644, 937
EGF (6C) repeats: 274–315, 316–369, 370–413, 414–456
Type III repeats: 488–523, 524–546, 547–582, 583–605, 606–643, 644–683, 684–719

Gene structure

Thrombospondin-3 is located on human chromosome 1q21–24. The gene is more than 12 kb with the 3.1 kb of coding region divided into 23 exons [2].

References

[1] Vos, H.L. et al. (1992) Thrombospondin 3 (Thbs3), a new member of the thrombospondin gene family. J. Biol. Chem. 267: 12192–12196.
[2] Adolph, K.W. et al. (1995) Structure and organisation of the human thrombospondin 3 gene (THBS3). Genomics 27: 329–336.
[3] Qabar, A. et al. (1995) Thrombospondin 3 is a pentameric molecule held together by interchain disulfide linkage involving two cysteine residues. J. Biol. Chem. 270: 12725–12729.
[4] Iruela-Arispe, M.L. et al. (1993) Differential expression of thrombospondin 1, 2 and 3 during murine development. Dev. Dyn. 197: 40–56.

Thrombospondin-4

Thrombospondin-4 (TSP4) is expressed in high levels in heart and skeletal muscles. It has been shown to associate with cell surfaces, be incorporated in the extracellular matrix, is a preferred substrate for neurons and promotes neurite outgrowth. It binds to both heparin and calcium, and associates with certain synapse-rich structures. It may be involved in local signalling in the developing and adult nervous system [1,2].

Molecular structure

The cDNA encodes a polypeptide of 961 amino acids, with a calculated molecular weight of 103 326. It contains four type 2 (or six-cysteine EGF) repeats and seven type 3 repeats, but lacks the region with homology with procollagen and the type I repeats. Human thrombospondin contains an RGD sequence in the third type 3 repeat. Recombinant protein shows a band of about 550 kDa with a reduced subunit of about 140. Rotary shadowing of both recombinant protein and that isolated from tendon suggest TSP4 exists as a pentamer [1–5].

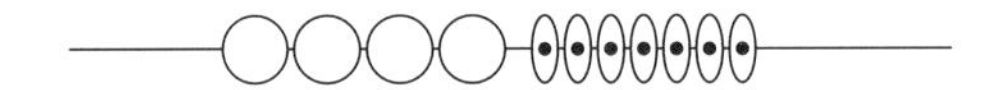

Isolation

Recombinant thrombospondin-4 has been purified from culture supernatant by heparin–Sepharose and anti-thrombospondin-4 antibody–Affi-gel affinity chromatography [3].

Accession number

P35443

Primary structure

Ala	A	48	Cys	C	47	Asp	D	110	Glu	E	38
Phe	F	39	Gly	G	83	His	H	12	Ile	I	38
Lys	K	35	Leu	L	73	Met	M	9	Asn	N	58
Pro	P	57	Gln	Q	59	Arg	R	50	Ser	S	56
Thr	T	49	Val	V	67	Trp	W	10	Tyr	Y	23

Mol. wt (calc.) = 105 802 Residues = 961

```
1    MLAPRGAAVL LLHLVLQRWL AAGAQATPQV FDLLPSSSQR LNPGALLPVL
51   TDPALNDLYV ISTFKLQTKS SATIFGLYSS TDNSKYFEFT VMGRLSKAIL
101  RYLKNDGKVH LVVFNNLQLA DGRRHRILLR LSNLQRGAGS LELYLDCIQV
151  DSVHNLPRAF AGPSQKPETI ELRTFQRKPQ DFLEELKLVV RGSLFQVASL
201  QDCFLQQSEP LAATGTGDFN RQFLGQMTQL NQLLGEVKDL LRQQVKETSF
251  LRNTIAECQA CGPLKFQSPT PSTVVAPAPP APPTRPPRRC DSNPCFRGVQ
301  CTDSRDGFQC GPCPEGYTGN GITCIDVDEC KYHPCYPGVH CINLSPGFRC
351  DACPVGFTGP MVQGVGISFA KSNKQVCTDI DECRNGACVP NSICVNTLGS
401  YRCGPCKPGY TGDQIRGCKV ERNCRNPELN PCSVNAQCIE ERQGDVTCVC
451  GVGWAGDGYI CGKDVDIDSY PDEELPCSAR NCKKDNCKYV PNSGQEDADR
```

```
501  DGIGDACDED ADGDGILNEQ DNCVLIHNVD QRNSDKDIFG DACDNCLSVL
551  NNDQKDTDGD GRGDACDDDM DGDGIKNILD NCPKFPNRDQ RDKDGDGVGD
601  ACDSCPDVSN PNQSDVDNDL VGDSCDTNQD SDGDGHQDST DNCPTVINSA
651  QLDTDKDGIG DECDDDDDND GIPDLVPPGP DNCRLVPNPA QEDSNSDGVG
701  DICESDFDQD QVIDRIDVCP ENAEVTLTDF RAYQTVGLDP EGDAQIDPNW
751  VVLNQGMEIV QTMNSDPGLA VGYTAFNGVD FEGTFHVNTQ TDDDYAGFIF
801  GYQDSSSFYV VMWKQTEQTY WQATPFRAVA EPGIQLKAVK SKTGPGEHLR
851  NSLWHTGDTS DQVRLLWKDS RNVGWKDKVS YRWFLQHRPQ VGYIRVRFYE
901  GSELVADSGV TIDTTMRGGR LGVFCFSQEN IIWSNLKYRC NDTIPEDFQE
951  FQTQNFDRFD N
```

Structural and functional sites

Signal peptide: 1–21
Potential N-linked glycosylation sites: 612, 941
Putative RGD cell adhesion site: 562–564
EGF (6C) repeats: 286–325, 326–378, 379–419, 420–462
Type III repeats: 492–527, 528–550, 551–586, 587–609, 610–647, 648–687, 688–723

Gene structure

Thrombospondin-4 encodes a 3.4 kb message [2].

References

[1] Lawler, J. et al. (1993) Identification and characterisation of thrombospondin-4, a new member of the thrombospondin gene family. J. Cell Biol. 120: 1059–1067.

[2] Arber, S. and Caroni, P. (1995) Thrombospondin-4, an extracellular matrix protein expressed in the developing and adult nervous system promotes neurite outgrowth. J. Cell Biol. 131: 1083–1094.

[3] Lawler, J. et al. (1995) Characterisation of human thrombospondin-4. J. Biol. Chem. 270: 2809–2814.

[4] Lawler, J. et al. (1993) Evolution of the thrombospondin gene family. J. Mol. Evol. 36: 509–516.

[5] Hauser, N. et al. (1995) Tendon extracellular matrix contains pentameric thrombospondin-4 (TSP-4). FEBS Lett. 368: 307–310.

Thrombospondin-5

cartilage oligomeric matrix protein (COMP), high molecular weight cartilage matrix glycoprotein

Thrombospondin-5 (TSP5) is found in cartilage and appears at specific times, distinct from type II collagen, during chondrogenesis. Anchorage in the cartilage matrix appears to be dependent on divalent cations. It is preferentially localized in the territorial matrix surrounding chondrocytes. TSP5 is found in all cartilages and in the vitreous of the eye. Mutations in the gene cause pseudochondroplasia and multiple epiphyseal dysplasia [1,2].

Molecular structure

TSP5 consists of five disulphide-bonded subunits (each of approximate molecular weight 100 000) which give a molecular mass of 524 kDa by sedimentation equilibrium centrifugation. Each subunit contains a peripheral globular domain, a flexible strand of epidermal growth factor (EGF) and type III repeats, and an assembly domain where the five arms meet. There are four six-cysteine EGF repeats and seven type 3 repeats [3–7].

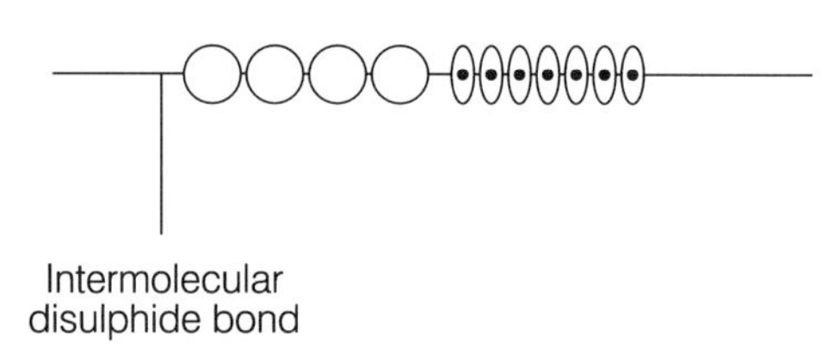

Isolation

TSP5 can be isolated from 4 M guanidine–HCl extracts of cartilage by CsCl-density gradient centrifugation, Sephadex G-200 chromatography, DEAE–cellulose chromatography, and can be purified by passing through gelatin–Sepharose and heparin–Sepharose columns [5]. It has also been extracted from adult human cartilage in EDTA-containing buffers, DEAE–Sepharose and Sepharose CL 4-B chromatography, and heparin affinity chromatography [7].

Accession number

P49747

Primary structure

Ala	A	45	Cys	C	46	Asp	D	99	Glu	E	33
Phe	F	26	Gly	G	73	His	H	14	Ile	I	19
Lys	K	20	Leu	L	37	Met	M	8	Asn	N	43
Pro	P	48	Gln	Q	51	Arg	R	46	Ser	S	36
Thr	T	35	Val	V	56	Trp	W	11	Tyr	Y	11

Mol. wt (calc.) = 82 832   Residues = 757

```
1      MVPDTACVLL  LTLAALGASG  QGQSPLGSDL  GPQMLRELQE  TNAALQDVRD
51     WLRQQVREIT  FLKNTVMECD  ACGMQQSVRT  GLPSVRPLLH  CAPGFCFPGV
101    ACIQTESGGR  CGPCPAGFTG  NGSHCTDVNE  CNAHPCFPRV  RCINTSPGFR
151    CEACPPGYSG  PTHQGVGLAF  AKANKQVCTD  INECETGQHN  CVPNSVCINT
```

```
201  RGSFQCGPCQ PGFVGDQASG CQRGAQRFCP DGSPSECHEH ADCVLERDGS
251  RSCVCRVGWA GNGILCGRDT DLDGFPDEKL RCPEPQCRKD NCVTVPNSGQ
301  EDVDRDGIGD ACDPDADGDG VPNEKDNCPL VRNPDQRNTD EDKWGDACDN
351  CRSQKNDDQK DTDQDGRGDA CDDDIDGDRI RNQADNCPRV PNSDQKDSDG
401  DGIGDACDNC PQKSNPDQAD VDHDFVGDAC DSDQDQDGDG HQDSRDNCPT
451  VPNSAQEDSD HDGQGDACDD DDDNDGVPDS RDNCRLVPNP GQEDADRDGV
501  GDVCQDDFDA DKVVDKIDVC PENAEVTLTD FRAFQTVVLD PEGDAQIDPN
551  WVVLNQGREI VQTMNSDPGL AVGYTAFNGV DFEGTFHVNT VTDDDYAGFI
601  FGYQDSSSFY VVMWKQMEQT YWQANPFRAV AEPGIQLKAV KSSTGPGEQL
651  RNALWHTGDT ESQVRLLWKD PRNVGWKDKK SYRWFLQHRP QVGYIRVRFY
701  EGPELVADSN VVLDTTMRGG RLGVFCFSQE NIIWANLRYR CNDTIPEDYE
751  THQLRQA
```

Structural and functional sites

Signal peptide: 1–20
EGF (6C) repeats: 87–126, 127–179, 180–222, 225–267
TSP type III repeats: 297–332, 333–355, 356–391, 392–414, 415–452, 453–488, 489–524
Potential N-linked glycosylation sites: 121, 742
Putative RGD cell adhesion site: 367–369

Gene structure

TSP5 is located on human chromosome 19 in band p13.1.

References

[1] Hecht, J.T. et al. (1995) Mutations in exon 17B of cartilage oligomeric matrix protein (COMP) cause pseudochondroplasia. Nature Genet. 10: 325–329.

[2] Briggs, M.D. et al. (1995) Pseudochondroplasia and multiple epiphyseal dysplasia due to mutations in the cartilage oligomeric matrix protein gene. Nature Genet. 10: 330–336.

[3] Morgelin, M. et al. (1992) Electron microscopy of native cartilage oligomeric matrix protein purified from the swarm rat chondrosarcoma reveals a five-armed structure. J. Biol. Chem. 267: 6137–6141.

[4] Oldberg, A. et al. (1992) COMP (cartilage oligomeric matrix protein) is structurally related to the thrombospondins. J. Biol. Chem. 267: 22346–22350.

[5] Hedbom, E. et al. (1992) Cartilage matrix proteins. An acidic oligomeric protein (COMP) detected only in cartilage. J. Biol. Chem. 267: 6132–6136.

[6] Newton, G. et al. (1994) Characterisation of human and mouse cartilage oligomeric matrix protein. Genomics 24: 435–439.

[7] DiCesare, P.E. et al. (1994) Cartilage oligomeric matrix protein and thrombospondin 1. Eur. J. Biochem. 223: 927–937.

TIMP1 tissue inhibitor of metalloproteinase-1, erythroid potentiating activity, EPA

TIMPs (tissue inhibitor of metalloproteinases) regulate the activation and proteolytic activity of the matrix metalloproteinases (MMPs) and therefore play a fundamental role in controlling extracellular matrix degradation. TIMPs have roles in important biological processes such as development, erythroid-potentiation, cell-growth regulation, cancer cell invasion and metastasis, and degenerative diseases such as arthritis. The TIMPs comprise four members which show extensive sequence homology, particularly in the position and relative spacing of the 12 cysteine residues which underpin the forming of the characteristic six-loop structure of the molecules.

Molecular structure

TIMP1 is a secreted 184 amino acid glycoprotein having a molecular mass of 28 500 Da [1] contains two sites for N-linked glycosylation, with a variable sialic acid substitution which is not necessary for inhibitory function [2,3]. The molecule comprises six loops held in place by six disulphide bonds. The locations of the 12 cysteine residues involved in forming this structure are highly conserved in the four TIMPs identified to date.

Isolation

TIMP1 can be purified from the conditioned medium of fibroblasts stimulated with phorbol myristate acetate or cytokines such as interleukin-1. TIMP1 has been isolated from the culture medium of fetal lung and other fibroblasts [4,5].

Accession number

P01033

Primary structure

Ala	A	15	Cys	C	12	Asp	D	5	Glu	E	10
Phe	F	11	Gly	G	13	His	H	6	Ile	I	9
Lys	K	8	Leu	L	24	Met	M	4	Asn	N	4
Pro	P	13	Gln	Q	12	Arg	R	11	Ser	S	15
Thr	T	16	Val	V	9	Trp	W	4	Tyr	Y	6

Mol. wt (calc.) = 23 171 Residues = 207

```
1     MAPFEPLASG  ILLLLWLIAP  SRACTCVPPH  PQTAFCNSDL  VIRAKFVGTP
51    EVNQTTLYQR  YEIKMTKMYK  GFQALGDAAD  IRFVYTPAME  SVCGYFHRSH
101   NRSEEFLIAG  KLQDGLLHIT  TCSFVAPWNS  LSLAQRRGFT  KTYTVGCEEC
151   TVFPCLSIPC  KLQSGTHCLW  TDQLLQGSEK  GFQSRHLACL  PREPGLCTWQ
201   SLRSQIA
```

Structural and functional sites

Signal peptide: 1–23
Potential N-linked glycosylation sites: 53, 101

Gene structure

The human gene is located at Xp11.3–p11.23[6]. Further information can be obtained from GeneCard by pointing a web browser at http://bioinfo.weizmann.ac.il/cgi-bin/lvrebhan/carddisp?TIMP1.

References

1. Docherty, A.J.P. et al. (1985) Sequence of human tissue inhibitor of metalloproteinases and its identity to erythroid-potentiating activity. Nature 318: 66–69.
2. Kishi, J.-L. and Hayakawa, T.J. (1984) Purification and characterization of bovine dental pulp collagenase inhibitor. J. Biochem. 96: 395–404.
3. Tolley S.P. et al. (1993) Crystallization and preliminary X-ray analysis of nonglycosylated tissue inhibitor of metalloproteinases-1, N30QN78Q TIMP-1. Proteins Struct. Funct. Genet. 17: 435–7.
4. Cawston, T.E. et al. (1981) Purification of rabbit bone inhibitor of collagenase. Biochem. J. 195: 159–165.
5. Murphy, G. and Willenbrock, F. (1995) Tissue inhibitors of matrix metalloproteinases. Methods Enzymol. 248: 496–510.
6. Spurr et al. (1987) Chromosomal assignment of the gene encoding the human tissue inhibitor of metalloproteinases to Xp11.1–p11.4. Ann. Hum. Genet. 51: 189–194.

TIMP2 tissue inhibitor of metalloproteinase-2

TIMPs (tissue inhibitor of metalloproteinases) regulate the activation and proteolytic activity of the matrix metalloproteinases (MMPs) and therefore play a fundamental role in controlling extracellular matrix degradation. TIMPs have roles in important biological processes such as development, erythroid-potentiation, cell-growth regulation, cancer cell invasion and metastasis, and degenerative diseases such as arthritis. The TIMPs comprise four members which show extensive sequence homology, particularly in the position and relative spacing of the 12 cysteine residues which underpin the forming of the characteristic six-loop structure of the molecules.

Molecular structure

Like other TIMPs, TIMP2 is comprised of a six-loop structure held together by six disulphide bonds. The position and relative spacing of the 12 cysteine residues involved in forming the loop structures are highly conserved between members of the TIMP family of proteins.

Isolation

A detailed procedure for the purification of TIMP2 and how the procedure differs from that of TIMP1 has been described[1]. In brief, TIMP2 is purified from the conditioned media of tumour cells and fibroblasts, usually complexed with progelatinase A (MMP2), by gelatin–agarose affinity chromatography, concanavalin A–Sepharose chromatography, reverse-phase HPLC and by gel filtration chromatography.

Accession number

P16035

Primary structure

Ala	A	20	Cys	C	12	Asp	D	14	Glu	E	12
Phe	F	7	Gly	G	15	His	H	4	Ile	I	17
Lys	K	18	Leu	L	16	Met	M	6	Asn	N	6
Pro	P	13	Gln	Q	8	Arg	R	9	Ser	S	13
Thr	T	9	Val	V	10	Trp	W	4	Tyr	Y	7

Mol. wt (calc.) = 24 399 Residues = 220

```
1     MGAAARTLRL  ALGLLLLATL  LRPADACSCS  PVHPQQAFCN  ADVVIRAKAV
51    SEKEVDSGND  IYGNPIKRIQ  YEIKQIKMFK  GPEKDIEFIY  TAPSSAVCGV
101   SLDVGGKKEY  LIAGKAEGDG  KMHITLCDFI  VPWDTLSTTQ  KKSLNHRYQM
151   GCECKITRCP  MIPCYISSPD  ECLWMDWVTE  KNINGHQAKF  FACIKRSDGS
201   CAWYRGAAPP  KQEFLDIEDP
```

Structural and functional sites

Signal peptide: 1–26

Gene structure

The human gene is encoded by five exons spanning 83 kb of genomic DNA [2] located at 17q25 [3]. Further information can be obtained by pointing a web browser at the GeneCard for TIMP2 (http://bioinfo.weizmann.ac.il/cgi-bin/lvrebhan/carddisp?TIMP2).

References

1 Murphy, G. and Willenbrock, F. (1995) Tissue inhibitors of matrix metalloproteinases. Methods Enzymol. 248: 496–510.

2 Hammani, K. et al. (1996) Structure and characterization of the human tissue inhibitor of metalloproteinases-2 gene. J. Biol. Chem. 271: 25498–25505.

3 De Clerck et al. (1992) The gene for tissue inhibitor of metalloproteinases-2 is localized on human chromosome arm 17q25. Genomics 14: 782–784.

TIMPs (tissue inhibitor of metalloproteinases) regulate the activation and proteolytic activity of the matrix metalloproteinases (MMPs) and therefore play a fundamental role in controlling extracellular matrix degradation. TIMPs have roles in important biological processes such as development, erythroid potentiation, cell-growth regulation, cancer cell invasion and metastasis, and degenerative diseases such as arthritis. TIMP3 is located in colorectal adenocarcinoma, retina and choroid. It is an inhibitor of angiogenesis and is the major metalloproteinase inhibitor in the decidualizing murine uterus. TIMP3 inhibits MMP1, stromelysin-1, and gelatinases A and B. TIMP3 and TIMP1 inhibition are quantitatively similar, implying that all TIMPs are equally efficient in MMP inhibition [1]. TIMP3 is found exclusively in the extracellular matrix of a large number of cultured human cells but not in the conditioned media [2]. It has an apparent molecular mass of 24–25 kDa, as determined by its migration on protease–substrate gels, is intermediate between TIMP1 (28.5 kDa) and TIMP2 (21 kDa). The protein has a potential site for *N*-glycosylation that is apparently not used, as determined by treatment with *N*-glycosidase-F.

Molecular structure

The TIMPs comprise four members which show extensive sequence homology, particularly in the position and relative spacing of the 12 cysteine residues which underpin the forming of the characteristic six-loop structure of the molecules. Like other TIMP molecules, TIMP3 comprises six loops held in place by six disulphide bonds.

Isolation

TIMP3 can be partially purified from the extracellular matrix deposited by cultured cells including FHs 173We cells (normal human whole embryo), CEF (chicken embryo fibroblasts), neuroblastoma cells (SK-N-SH), cervix epitheloid carcinoma (HeLa), colon adenocarcinoma (Caco-2), ileocecal adenocarcinoma (HCT-8) and fibrosarcoma (SW 684) as described [2].

Accession number

P35625

Primary structure

Ala	A	8	Cys	C	12	Asp	D	8	Glu	E	7
Phe	F	6	Gly	G	19	His	H	5	Ile	I	9
Lys	K	17	Leu	L	20	Met	M	6	Asn	N	8
Pro	P	10	Gln	Q	8	Arg	R	8	Ser	S	14
Thr	T	12	Val	V	11	Trp	W	7	Tyr	Y	16

Mol. wt (calc.) = 24 145 Residues = 211

```
1     MTPWLGLIVL  LGSWSLGDWG  AEACTCSPSH  PQDAFCNSDI  VIRAKVVGKK
51    LVKEGPFGTL  VYTIKQMKMY  RGFTKMPHVQ  YIHTEASESL  CGLKLEVNKY
101   QYLLTGRVYD  GKMYTGLCNF  VERWDQLTLS  QRKGLNYRYH  LGCNCKIKSC
151   YYLPCFVTSK  NECLWTDMLS  NFGYPGYQSK  HYACIRQKGG  YCSWYRGWAP
201   PDKSIINATD  P
```

Structural and functional sites

Signal peptide: 1–23

Gene structure

TIMP3 is encoded by five exons extending over approximately 55 kb of genomic DNA located at 22q12.1–q13.2[3]. Mutations in the gene for TIMP3 have been shown to be the cause of Sorsby fundus dystrophy. Further information can be obtained by pointing a web browser at the GeneCard for TIMP3 (http://bioinfo.weizmann.ac.il/cgi-bin/lvrebhan/carddisp?TIMP3).

References

[1] Apte, S.S. et al. (1995) The gene structure of tissue inhibitor of metalloproteinases (TIMP)-3 and its inhibitory activities define the distinct TIMP gene family. J. Biol. Chem. 270: 14313–14318.

[2] Kishnani, N.S. et al. (1995) Identification and characterization of human tissue inhibitor of metalloproteinase-3 and detection of three additional metalloproteinase inhibitor activities in extracellular matrix. Matrix Biol. 14: 479–488.

[3] Apte, S.S. et al. (1994) Cloning of the cDNA encoding human tissue inhibitor of metalloproteinases-3 (TIMP3) and mapping of the TIMP3 gene to chromosome 22. Genomics 19: 86–90.

TIMPs (tissue inhibitor of metalloproteinases) regulate the activation and proteolytic activity of the matrix metalloproteinases (MMPs) and therefore play a fundamental role in controlling extracellular matrix degradation. TIMPs have roles in important biological processes such as development, erythroid potentiation, cell-growth regulation, cancer cell invasion and metastasis, and degenerative diseases such as arthritis. The TIMPs comprise four members which show extensive sequence homology, particularly in the position and relative spacing of the 12 cysteine residues which underpin the forming of the characteristic six-loop structure of the molecules. The mRNA for TIMP4 is expressed in relatively high amounts in adult heart, with very low levels expressed in a variety of other tissues including kidney, placenta, colon and testes. No TIMP4 mRNA was detected in liver, brain, lung, thymus and spleen [1].

Molecular structure

TIMP4 shares 37% sequence identity with TIMP1, and 51% identity with TIMP2 and TIMP3. The protein has a deduced isoelectric point of 7.34.

Isolation

No information is available for the isolation and purification of TIMP4 (see entries of TIMP1, 2 and 3 for the purification of other TIMPs).

Accession number

U76456

Primary structure

Ala	A	13	Cys	C	12	Asp	D	9	Glu	E	12
Phe	F	5	Gly	G	11	His	H	10	Ile	I	11
Lys	K	14	Leu	L	26	Met	M	4	Asn	N	5
Pro	P	16	Gln	Q	12	Arg	R	9	Ser	S	15
Thr	T	9	Val	V	15	Trp	W	5	Tyr	Y	11

Mol. wt (calc.) = 25 503 Residues = 224

```
1     MPGSPRPAPS  WVLLLRLLAL  LRPPGLGEAC  SCAPAHPQQH  ICHSALVIRA
51    KISSEKVVPA  SADPADTEKM  LRYEIKQIKM  FKGFEKVKDV  QYIYTPFDSS
101   LCGVKLEANS  QKQYLLTGQV  LSDGKVFIHL  CNYIEPWEDL  SLVQRESLNH
151   HYHLNCGCQI  TTCYTVPCTI  SAPNECLWTD  WLLERKLYGY  QAQHYVCMKH
201   VDGTCSWYRG  HLPLRKEFVD  IVQP
```

Structural and functional sites

Signal peptide: 1–29

Gene structure

No information is available.

Reference

[1] Greene, J. et al. (1996) Molecular cloning and characterisation of human tissue inhibitor of metalloproteinase 4. J. Biol. Chem. 271: 30375–30380.

Versican

Versican is a large chondroitin sulphate proteoglycan secreted by fibroblasts. Versican contains domains that are highly homologous to aggrecan in having a hyaluronan-binding domain at its NH_2-terminal end and two epidermal growth factor (EGF) repeats, a lectin repeat and a complement control protein repeat at its COOH-terminal end. Versican may play a role in intracellular signalling, cell recognition and connecting extracellular matrix components and cell surface glycoproteins.

Molecular structure

The versican core protein is highly negatively charged and has a calculated pI of 4.2. The NH_2-terminal domain is similar to the three-loop structure of link protein and to the link-like sequences of aggrecan. On the COOH-side of the hyaluronan-binding domain is a 200 amino acid domain containing two cysteines and a cluster of glutamic acid residues that may be important in the interaction of versican with hydroxyapatite in bone. The COOH-terminal domain of versican contains two EGF repeats in tandem and sequences homologous to the complement control protein repeats of murine factor H and human C4 binding protein [1–4].

Isolation

Versican is present in the culture medium of human IMR-90 lung fibroblasts. It has been purified from this source by ammonium sulphate precipitation and DEAE–cellulose chromatography [3].

Accession number

P13611

Primary structure

Ala	A	146	Cys	C	35	Asp	D	141	Glu	E	258
Phe	F	95	Gly	G	133	His	H	57	Ile	I	114
Lys	K	96	Leu	L	139	Met	M	37	Asn	N	73
Pro	P	155	Gln	Q	94	Arg	R	79	Ser	S	253
Thr	T	259	Val	V	159	Trp	W	24	Tyr	Y	62

Mol. wt (calc.) = 264 759 Residues = 2409

```
1     MFINIKSILW  MCSTLIVTHA  LHKVKVGKSP  PVRGSLSGKV  SLPCHFSTMP
51    TLPPSYNTSE  FLRIKWSKIE  VDKNGKDLKE  TTVLVAQNGN  IKIGQDYKGR
101   VSVPTHPEAV  GDASLTVVKL  LASDAGLYRC  DVMYGIEDTQ  DTVSLTVDGV
151   VFHYRAATSR  YTLNFEAAQK  ACLDVGAVIA  TPEQLFAAYE  DGFEQCDAGW
201   LADQTVRYPI  RAPRVGCYGD  KMGKAGVRTY  GFRSPQETYD  VYCYVDHLDG
251   DVFHLTVPSK  FTFEEAAKEC  ENQDARLATV  GELQAAWRNG  FDQCDYGWLS
301   DASVRHPVTV  ARAQCGGGLL  GVRTLYRFEN  QTGFPPPDSR  FDAYCFKRRM
351   SDLSVIGHPI  DSESKEDEPC  SEETDPVHDL  MAEILPEFPD  IIEIDLYHSE
```

```
401   ENEEEEEECA NATDVTTTPS VQYINGKHLV TTVPKDPEAA EARRGQFESV
451   APSQNFSDSS ESDTHPFVIA KTELSTAVQP NESTETTESL EVTWKPETYP
501   ETSEHFSGGE PDVFPTVPFH EEFESGTAKK GAESVTERDT EVGHQAHEHT
551   EPVSLFPEES SGEIAIDQES QKIAFARATE VTFGEEVEKS TSVTYTPTIV
601   PSSASAYVSE EEAVTLIGNP WPDDLLSTKE SWVEATPRQV VELSGSSSIP
651   ITEGSGEAEE DEDTMFTMVT DLSQRNTTDT LITLDTSRII TESFFEVPAT
701   TIYPVSEQPS AKVVPTKFVS ETDTSEWISS TTVEEKKRKE EEGTTGTAST
751   FEVYSSTQRS DQLILPFELE SPNVATSSDS GTRKSFMSLT TPTQSEREMT
801   DSTPVFTETN TLENLGAQTT EHSSIHQPGV QEGLTTLPRS PASVFMEQGS
851   GEAAADPETT TVSSFSLNVE YAIQAEKEVA GTLSPHVETT FSTEPTGLVL
901   STVMDRVVAE NITQTSREIV ISERLGEPNY GAEIRGFSTG FPLEEDFSGD
951   FREYSTVSHP IAKEETVMME GSGDAAFRDT QTSPSTVPTS VHISHISDSE
1001  GPSSTMVSTS AFPWEEFTSS AEGSGEQLVT VSSSVVPVLP SAVQKFSGTA
1051  SSIIDEGLGE VGTVNEIDRR STILPTAEVE GTKAPVEKEE VKVSGTVSTN
1101  FPQTIEPAKL WSRQEVNPVR QEIESETTSE EQIQEEKSFE SPQNSPATEQ
1151  TIFDSQTFTE TELKTTDYSV LTTKKTYSDD KEMKEEDTSL VNMSTPDPDA
1201  NGLESYTTLP EATEKSHFFL ATALVTESIP AEHVVTDSPI KKEESTKHFP
1251  KGMRPTIQES DTELLFSGLG SGEEVLPTLP TESVNFTEVE QINNTLYPHT
1301  SQVESTSSDK IEDFNRMENV AKEVGPLVSQ TDIFEGSGSV TSTTLIEILS
1351  DTGAEGPTVA PLPFSTDIGH PQNQTVRWAE EIQTSRPQTI TEQDSNKNSS
1401  TAEINETTTS STDFLARAYG FEMAKEFVTS APKPSDLYYE PSGEGSGEVD
1451  IVDSFHTSAT TQATRQESST TFVSDGSLEK HPEVPSAKAV TADGFPTVSV
1501  MLPLHSEQNK SSPDPTSTLS NTVSYERSTD GSFQDRFREF EDSTLKPNRK
1551  KPTENIIIDL DKEDKDLILT ITESTILEIL PELTSDKNTI IDIDHTKPVY
1601  EDILGMQTDI DTEVPSEPHD SNDESNDDST QVQEIYEAAV NLSLTEETFE
1651  GSADVLASYT QATHDESMTY EDRSQLDHMG FHFTTGIPAP STETELDVLL
1701  PTATSLPIPR KSATVIPEIE GIKAEAKALD DMFESSTLSD GQAIADQSEI
1751  IPTLGQFERT QEEYEDKKHA GPSFQPEFSS GAEEALVDHT PYLSIATTHL
1801  MDQSVTEVPD VMEGSNPPYY TDTTLAVSTF AKLSSQTPSS PLTIYSGSEA
1851  SGHTEIPQPS ALPGIDVGSS VMSPQDSFKE IHVNIEATFK PSSEEYLHIT
1901  EPPSLSPDTK LEPSEDDGKP ELLEEMEASP TELIAVEGTE ILQDFQNKTD
1951  GQVSGEAIKM FPTIKTPEAG TVITTADEIE LEGATQWPHS TSASATYGVE
2001  AGVVPWLSPQ TSERPTLSSS PEINPETQAA LIRGQDSTIA ASEQQVAARI
2051  LDSNDQATVN PVEFNTEVAT PPFSLLETSN ETDFLIGINE ESVEGTAIYL
2101  PGPDRCKMNP CLNGGTCYPT ETSYVCTCVP GYSGDQCELD FDECHSNPCR
2151  NGATCVDGFN TFRCLCLPSY VGALCEQDTE TCDYGWHKFQ GQCYKYFAHR
2201  RTWDAAEREC RLQGAHLTSI LSHEEQMFVN RVGHDYQWIG LNDKMFEHDF
2251  RWTDGSTLQY ENWRPNQPDS FFSAGEDCVV IIWHENGQWN DVPCNYHLTY
2301  TCKKGTVACG QPPVVENAKT FGKMKPRYEI NSLIRYHCKD GFIQRHLPTI
2351  RCLGNGRWAI PKITCMNPSA YQRTYSMKYF KNSSSAKDNS INTSKHDHRW
2401  SRRWQESRR
```

Structural and functional sites

Putative signal sequence: 1–20
Hyaluronic acid binding domain: 48–346
  Immunoglobulin repeat: 48–148
  Link protein repeats: 149–244, 250–346
Glutamate-rich region: 400–408
Potential N-linked glycosylation sites: 57, 330, 411, 455, 481, 676, 911, 1192, 1285, 1293, 1373, 1398, 1405, 1509, 1641, 1947, 2080, 2382, 2392
EGF (6C) repeats: 2103–2140, 2141–2178
Lectin repeat: 2179–2305
CCP repeat: 2306–2366

Gene structure

The human versican gene is found on the long arm of chromosome 5 at locus 5q12–5q14[5].

References

[1] Mole, J.E. et al. (1984) Complete primary structure for the zymogen of human complement factor B. J. Biol. Chem. 259: 3407–3412.

[2] Kristensen, T. and Tack, B.F. (1986) Murine protein-H is comprised of 20 repeating units, 61 amino acids in length. Proc. Natl Acad. Sci. USA 83: 3963–3967.

[3] Krusius, T. et al. (1987) A fibroblast chondroitin sulfate proteoglycan core protein contains lectin-like and growth factor-like sequences. J. Biol. Chem. 262: 13120–13125.

[4] Zimmermann, D.R. and Ruoslahti, E. (1989) Multiple domains of the large fibroblast proteoglycan, versican. EMBO J. 8: 2975–2981.

[5] Iozzo, R.V. et al. (1992) Mapping of the versican proteoglycan gene (CSPG2) to the long arm of human chromosome 5 (5q12–5q14). Genomics 14: 845–851.

Vitronectin

serum spreading factor, S-protein

Vitronectin is a cell adhesion and spreading factor found in plasma and the extracellular matrix. The molecule binds to proteins at the terminal stages of both complement and coagulation pathways and inhibits cytolysis. It also interacts with cells through a number of integrin receptors containing the αV subunit, principally $\alpha V\beta 3$. Vitronectin participates in a variety of protective events including haemostasis, phagocytosis, tissue repair and immune function. It is synthesized predominantly in the liver, but platelets, macrophages and smooth muscle cells can produce a similar molecule. Immunofluorescence localization suggests that vitronectin is deposited in a fibrillar form in a number of connective tissues.

Molecular structure

In plasma, vitronectin exists in two forms, a single chain (molecular weight approximately 75 000) and an endogenously clipped, two-chain form held together by disulphide bonds (molecular weights approximately 65 000 and 10 000). Vitronectin is an asymmetrically shaped molecule with a large content of predicted β-sheet structure. Conformational transitions may lead to activation of binding sites. The NH_2-terminal somatomedin B domain is an independently folded structural module which precedes the RGD-containing cell-binding domain and a highly acidic domain homologous to hemopexin. Somatomedin B is a growth hormone-dependent serum factor with proteinase-inhibiting activity. In the two domains that are homologous to hemopexin, there are only six typical repeats as opposed to eight in hemopexin itself. The two domains are connected by a flexible hinge facilitating ligand-induced conformational changes. A polycationic cluster is located at the COOH-terminal end of the molecule and constitutes the major heparin-binding domain [1–5].

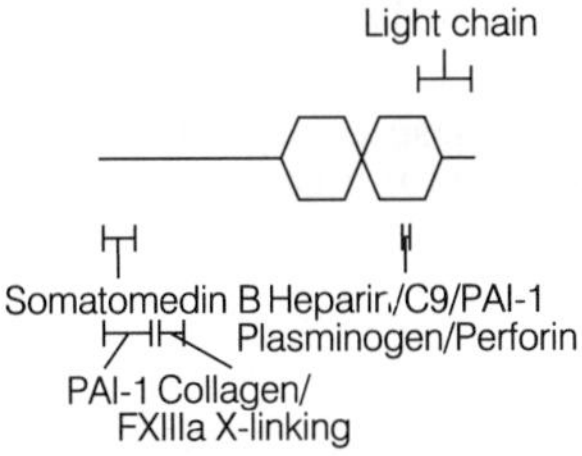

Light chain

Somatomedin B Heparin/C9/PAI-1

Plasminogen/Perforin

PAI-1 Collagen/

FXIIIa X-linking

Isolation

Vitronectin is readily adsorbed on to a variety of surfaces and can be extracted most conveniently on a column of glass beads or by affinity chromatography with anti-vitronectin antibody or heparin–agarose. Major difficulties encountered relate to its sensitivity towards proteases and denaturation, and its tendency to form disulphide-linked oligomers [6]. The molecules isolated by each procedure may exist in different conformational states.

Accession number

P04004, P01141

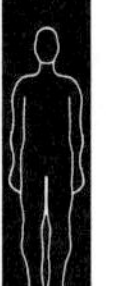

Primary structure

Ala	A	34	Cys	C	14	Asp	D	32	Glu	E	34
Phe	F	23	Gly	G	37	His	H	9	Ile	I	14
Lys	K	20	Leu	L	32	Met	M	7	Asn	N	15
Pro	P	37	Gln	Q	26	Arg	R	36	Ser	S	34
Thr	T	19	Val	V	22	Trp	W	10	Tyr	Y	23

Mol. wt (calc.) = 54 245 Residues = 478

```
1     MAPLRPLLIL ALLAWVALAD QESCKGRCTE GFNVDKKCQC DELCSYYQSC
51    CTDYTAECKP QVTRGDVFTM PEDEYTVYDD GEEKNNATVH EQVGGPSLTS
101   DLQAQSKGNP EQTPVLKPEE EAPAPEVGAS KPEGIDSRPE TLHPGRPQPP
151   AEEELCSGKP FDAFTDLKNG SLFAFRGQYC YELDEKAVRP GYPKLIRDVW
201   GIEGPIDAAF TRINCQGKTY LFKGSQYWRF EDGVLDPDYP RNISDGFDGI
251   PDNVDAALAL PAHSYSGRER VYFFKGKQYW EYQFQHQPSQ EECEGSSLSA
301   VFEHFAMMQR DSWEDIFELL FWGRTSAGTR QPQFISRDWH GVPGQVDAAM
351   AGRIYISGMA PRPSLAKKQR FRHRNRKGYR SQRGHSRGRN QNSRRPSRAT
401   WLSLFSSEES NLGANNYDDY RMDWLVPATC EPIQSVFFFS GDKYYRVNLR
451   TRRVDTVDPP YPRSIAQYWL GCPAPGHL
```

Structural and functional sites

Signal peptide: 1–19
Somatomedin B domain: 20–63
Hemopexin repeats: 150–287, 288–478
RGD cell adhesion site: 64–66
Potential N-linked glycosylation sites: 86, 169, 242
Protease cleavage site: 398–399
Intra-chain disulphide bond: 293–430
Phosphorylation site for cAMP-dependent protein kinase: 397
Sulphation sites: 75, 78
Factor XIIIa transglutaminase-catalysed cross-linking site: 112
Heparin-binding domain: 362–395

Gene structure

The human vitronectin gene is 4.5–5 kb long, comprises eight exons and generates a 1.7 kb mRNA transcript. It is located in the centromeric region of human chromosome 17q. There is no evidence of alternative splicing [7].

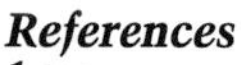

References

[1] Hayman, E.G. et al. (1985) Vitronectin – a major cell attachment-promoting protein in foetal bovine serum. Exp. Cell Res. 160: 245–258.

[2] Suzuki, S. et al. (1985) Complete amino acid sequence of human vitronectin deduced from cDNA. Similarity of cell attachment sites in vitronectin and fibronectin. J. Biol. Chem. 259: 15307–15314.

[3] Jenne, D. and Stanley, K.K. (1987) Nucleotide sequence and organisation of the human S-protein gene: Repeating motifs in the 'pexin' family and a model for their evolution. Biochemistry 26: 6735–6742.

[4] Preissner, K.T. (1991) Structure and biological role of vitronectin. Annu. Rev. Cell Biol. 7: 275–310.
[5] Felding-Habermann and Cheresh, D.A. (1993) Vitronectin and its receptors. Curr. Opin. Cell Biol. 5: 864–868.
[6] Yatohgo, T. et al. (1988) Novel purification of vitronectin from human plasma by heparin affinity chromatography. Cell Struct. Funct. 13: 281–292.
[7] Fink, T.M. et al. (1992) The human vitronectin (complement S-protein) gene maps to the centromeric region of 17q. Hum. Genet. 88: 569–572.

von Willebrand factor

von Willebrand factor has an important function in the maintenance of haemostasis by promoting platelet–vessel wall interactions at sites of vascular injury. The molecule is a multimeric plasma glycoprotein synthesized by endothelial cells and megakaryotes; it serves as both a carrier for factor VIII and as a mediator of initial platelet adhesion to the subendothelium. Multimerization provides a greater density of platelet-binding sites. von Willebrand factor interacts with two known platelet receptors, the membrane glycoprotein complexes Ib/IX and IIbIIIa (integrin $\alpha IIb\beta 3$), as well as with collagen and heparin. IIbIIIa recognition involves a single RGD motif. Ib/IX recognition is induced by the viper venom protein botrocetin.

Molecular structure

In plasma, von Willebrand factor circulates as multimers ranging in size from dimers of about 500 kDa to multimers of 20 000 kDa, at a concentration of 5–10 μg/ml. The molecule is synthesized as a pre-propolypeptide which is cleaved and modified. Its cysteine residues (which account for 8.3% of the total amino acids) are all involved in inter- and intra-chain disulphide bonds and are clustered at the NH_2- and COOH-termini of the molecule. After translocation, the pro-von Willebrand factor forms dimers which then multimerize in the Golgi as a result of disulphide bonding. The propeptide is cleaved from most, but not all, of the subunits producing the range of multimers that are seen in the circulation. The majority of the protein is made up of four types of repeating module (A–D). The A-type repeats are found in a number of other proteins, including cartilage matrix protein and collagen type VI. The C-type repeats share some similarity with the N-propeptide of procollagen type I (also found in thrombospondin)[1–8].

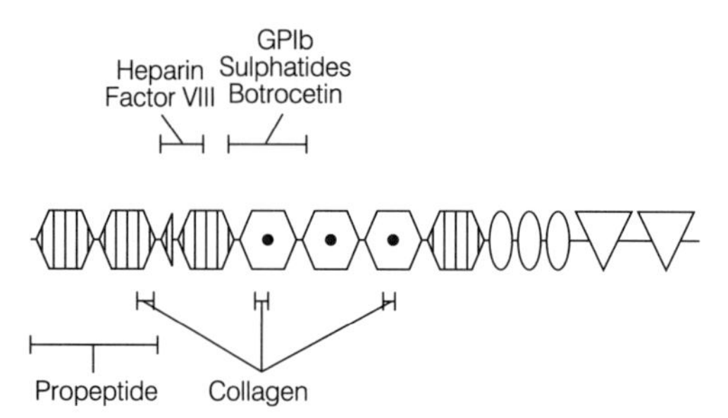

Isolation

von Willebrand factor and factor VIII can be isolated from plasma by cryoprecipitation and agarose gel filtration. It is possible to separate factor VIII from von Willebrand factor by gel filtration in high ionic strength buffers[4].

Accession number

P04275

Primary structure

Ala	A	154	Cys	C	233	Asp	D	157	Glu	E	182
Phe	F	88	Gly	G	206	His	H	68	Ile	I	95
Lys	K	108	Leu	L	228	Met	M	56	Asn	N	98
Pro	P	176	Gln	Q	132	Arg	R	144	Ser	S	208
Thr	T	151	Val	V	225	Trp	W	26	Tyr	Y	78

Mol. wt (calc.) = 308 913 Residues = 2813

```
1     MIPARFAGVL LALALILPGT LCAEGTRGRS STARCSLFGS DFVNTFDGSM
51    YSFAGYCSYL LAGGCQKRSF SIIGDFQNGK RVSLSVYLGE FFDIHLFVNG
101   TVTQGDQRVS MPYASKGLYL ETEAGYYKLS GEAYGFVARI DGSGNFQVLL
151   SDRYFNKTCG LCGNFNIFAE DDFMTQEGTL TSDPYDFANS WALSSGEQWC
201   ERASPPSSSC NISSGEMQKG LWEQCQLLKS TSVFARCHPL VDPEPFVALC
251   EKTLCECAGG LECACPALLE YARTCAQEGM VLYGWTDHSA CSPVCPAGME
301   YRQCVSPCAR TCQSLHINEM CQERCVDGCS CPEGQLLDEG LCVESTECPC
351   VHSGKRYPPG TSLSRDCNTC ICRNSQWICS NEECPGECLV TGQSHFKSFD
401   NRYFTFSGIC QYLLARDCQD HSFSIVIETV QCADDRDAVC TRSVTVRLPG
451   LHNSLVKLKH GAGVAMDGQD VQLPLLKGDL RIQRTVTASV RLSYGEDLQM
501   DWDGRGRLLV KLSPVYAGKT CGLCGNYNGN QGDDFLTPSG LAEPRVEDFG
551   NAWKLHGDCQ DLQKQHSDPC ALNPRMTRFS EEACAVLTSP TFEACHRAVS
601   PLPYLRNCRY DVCSCSDGRE CLCGALASYA AACAGRGVRV AWREPGRCEL
651   NCPKGQVYLQ CGTPCNLTCR SLSYPDEECN EACLEGCFCP PGLYMDERGD
701   CVPKAQCPCY YDGEIFQPED IFSDHHTMCY CEDGFMHCTM SGVPGSLLPD
751   AVLSSPLSHR SKRSLSCRPP MVKLVCPADN LRAEGLECTK TCQNYDLECM
801   SMGCVSGCLC PPGMVRHENR CVALERCPCF HQGKEYAPGE TVKIGCNTCV
851   CRDRKWNCTD HVCDATCSTI GMAHYLTFDG LKYLFPGECQ YVLVQDYCGS
901   NPGTFRILVG NKGCSHPSVK CKKRVTILVE GGEIELFDGE VNVKRPMKDE
951   THFEVVESGR YIILLLGKAL SVVWDRHLSI SVVLKQTYQE KVCGLCGNFD
1001  GIQNNDLTSS NLQVEEDPVD FGNSWKVSSQ CADTRKVPLD SSPATCHNNI
1051  MKQTMVDSSC RILTSDVFQD CNKLVDPEPY LDVCIYDTCS CESIGDCACF
1101  CDTIAAYAHV CAQHGKVVTW RTATLCPQSC EERNLRENGY ECEWRYNSCA
1151  PACQVTCQHP EPLACPVQCV EGCHAHCPPG KILDELLQTC VDPEDCPVCE
1201  VAGRRFASGK KVTLNPSDPE HCQICHCDVV NLTCEACQEP GGLVVPPTDA
1251  PVSPTTLYVE DISEPPLHDF YCSRLLDLVF LLDGSSRLSE AEFEVLKAFV
1301  VDMMERLRIS QKWVRVAVVE YHDGSHAYIG LKDRKRPSEL RRIASQVKYA
1351  GSQVASTSEV LKYTLFQIFS KIDRPEASRI ALLLMASQEP QRMSRNFVRY
1401  VQGLKKKKVI VIPVGIGPHA NLKQIRLIEK QAPENKAFVL SSVDELEQQR
1451  DEIVSYLCDL APEAPPPTLP PDMAQVTVGP GLLGVSTLGP KRNSMVLDVA
1501  FVLEGSDKIG EADFNRSKEF MEEVIQRMDV GQDSIHVTVL QYSYMVTVEY
1551  PFSEAQSKGD ILQRVREIRY QGGNRTNTGL ALRYLSDHSF LVSQGDREQA
1601  PNLVYMVTGN PASDEIKRLP GDIQVVPIGV GPNANVQELE RIGWPNAPIL
1651  IQDFETLPRE APDLVLQRCC SGEGLQIPTL SPAPDCSQPL DVILLLDGSS
1701  SFPASYFDEM KSFAKAFISK ANIGPRLTQV SVLQYGSITT IDVPWNVVPE
1751  KAHLLSLVDV MQREGGPSQI GDALGFAVRY LTSEMHGARP GASKAVVILV
1801  TDVSVDSVDA AADAARSNRV TVFPIGIGDR YDAAQLRILA GPAGDSNVVK
1851  LQRIEDLPTM VTLGNSFLHK LCSGFVRICM DEDGNEKRPG DVWTLPDQCH
1901  TVTCQPDGQT LLKSHRVNCD RGLRPSCPNS QSPVKVEETC GCRWTCPCVC
1951  TGSSTRHIVT FDGQNFKLTG SCSYVLFQNK EQDLEVILHN GACSPGARQG
2001  CMKSIEVKHS ALSVELHSDM EVTVNGRLVS VPYVGGNMEV NVYGAIMHEV
2051  RFNHLGHIFT FTPQNNEFQL QLSPKTFASK TYGLCGICDE NGANDFMLRD
2101  GTVTTDWKTL VQEWTVQRPG QTCQPILEEQ CLVPDSSHCQ VLLLPLFAEC
2151  HKVLAPATFY AICQQDSSHQ EQVCEVIASY AHLCRTNGVC VDWRTPDFCA
```

```
2201 MSCPPSLVYN HCEHGCPRHC DGNVSSCGDH PSEGCFCPPD KVMLEGSCVP
2251 EEACTQCIGE DGVQHQFLEA WVPDHQPCQI CTCLSGRKVN CTTQPCPTAK
2301 APTCGLCEVA RLRQNADQCC PEYECVCDPV SCDLPPVPHC ERGLQPTLTN
2351 PGECRPNFTC ACRKEECKRV SPPSCPPHRL PTLRKTQCCD EYECACNCVN
2401 STVSCPLGYL ASTATNDCGC TTTTCLPDKV CVHRSTIYPV GQFWEEGCDV
2451 CTCTDMEDAV MGLRVAQCSQ KPCEDSCRSG FTYVLHEGEC CGRCLPSACE
2501 VVTGSPRGDS QSSWKSVGSQ WASPENPCLI NECVRVKEEV FIQQRNVSCP
2551 QLEVPVCPSG FQLSCKTSAC CPSCRCERME ACMLNGTVIG PGKTVMIDVC
2601 TTCRCMVQVG VISGFKLECR KTTCNPCPLG YKEENNTGEC CGRCLPTACT
2651 IQLRGGQIMT LKRDETLQDG CDTHFCKVNE RGEYFWEKRV TGCPPFDEHK
2701 CLAEGGKIMK IPGTCCDTCE EPECNDITAR LQYVKVGSCK SEVEVDIHYC
2751 QGKCASKAMY SIDINDVQDQ CSCCSPTRTE PMQVALHCTN GSVVYHEVLN
2801 AMECKCSPRK CSK
```

Structural and functional sites

Signal peptide: 1–22
Propeptide: 23–763
A-type repeats: 1260–1479, 1480–1672, 1673–1874
B-type repeats: 2296–2330, 2340–2365, 2375–2399
C-type repeats: 2400–2515, 2544–2662
D-type repeats: 23–295, 363–652, 657–741 (partial), 842–1130, 1934–2203
Potential N-linked glycosylation sites: 857, 1147, 1231, 1248, 1255, 1256, 1263, 1468, 1477, 1486, 1487, 1515, 1574, 1679, 2223, 2290, 2298, 2357, 2400, 2546, 2585, 2790
Sulphation site: 652
RGD cell adhesion site: 2507–2509
GPIb binding sites: 1237–1251, 1457–1471

Gene structure

The von Willebrand factor gene spans 178 kb on the short arm (p) of human chromosome 12, contains 52 exons and produces an 8.7 kb mRNA. A partial unprocessed von Willebrand factor pseudogene is located on chromosome 22 (locus q11–13); this spans 21–29 kb and corresponds to exons 23–34 of the authentic von Willebrand factor gene [6].

References

1. Bonthron, D. et al. (1986) Nucleotide sequence of pre-pro-von Willebrand factor cDNA. Nucleic Acids Res. 14: 7125–7127.
2. Verweij, C.L. et al. (1986) Full length von Willebrand factor (vWF) cDNA encodes a highly repetitive protein considerably larger than the mature vWF subunit. EMBO J. 5: 1839–1847.
3. Zimmerman, T.S. and Meyer, D. (1987) Structure and function of factor VIII and von Willebrand factor. In: Haemostasis and Thrombosis, 2nd edition, ed. Bloom, A.L. and Thomas, D.P., Churchill Livingstone, pp. 131–147.
4. Mohri, H. et al. (1988) Structure of the von Willebrand factor domain interacting with glycoprotein Ib. J. Biol. Chem. 263: 17901–17904.
5. Mancuso, D.J. et al. (1989) Structure of the gene for human von Willebrand factor. J. Biol. Chem. 264: 19514–19527.

[6] Sadler, J.E. (1991) Von Willebrand factor. J. Biol. Chem. 266: 22777–22780.
[7] Ruggeri, Z.M. and Zimmerman, T.S. (1987) Von Willebrand factor and von Willebrand disease. Blood 70: 895–904.
[8] Ginsburg, D. et al. (1992) Fine mapping of monoclonal antibody epitopes on human von Willebrand factor using a recombinant peptide. Thromb. Haemost. 67: 166–171.

Index